Introduction to Linear Elasticity

Phillip L. Gould • Yuan Feng

Introduction to Linear Elasticity

Fourth Edition

 Springer

Phillip L. Gould
Department of Mechanical Engineering
and Materials Science
Washington University
St. Louis, MO, USA

Yuan Feng
School of Biomedical Engineering
Shanghai Jiao Tong University
Shanghai, China

ISBN 978-3-030-08878-1   ISBN 978-3-319-73885-7   (eBook)
https://doi.org/10.1007/978-3-319-73885-7

Printed on acid-free paper

This Springer imprint is published by the registered company Springer International Publishing AG part of
Springer Nature.
The registered company address is: Gewerbestrasse 11, 6330 Cham, Switzerland

Dedication from P.L. Gould
To my children
Elizabeth Sue Gould (of blessed memory)
Nathan Charles Gould
Rebecca Blair Carlisle
Joshua Robert Gould

Dedication from Yuan Feng
To my family
mother Huiling Li
father Zijin Feng
wife Long Huang
and my son Samuel Baisheng Feng

Preface

The fourth edition of this book is inspired by the enduring importance of classical elasticity in the emerging fields of solid mechanics, especially biomechanics and materials engineering, as well as in the traditional fields of structural, mechanical, and aerospace engineering.

To repeat a quote from the preface of earlier editions, "Elasticity is one of the crowning achievements of Western culture" by our late colleague Professor George Zahalak. It is this sentence, eloquently expressing my admiration for the creative efforts of the premier physicists, mathematicians, and mechanicians of the nineteenth and twentieth centuries, that led us to try and popularize the basis of solid mechanics in this introductory text. Although most of the names mentioned in the course of the development of the theory in the early chapters, such as Hooke, Euler, Lagrange, St. Venant, Poisson, Boussinesq, Lamé, and Rayleigh, date to the nineteenth century and before, at least two Nobel Prize–winning physicists in the twentieth century made contributions to elasticity. They are Albert Einstein with the summation convention and Max Born who wrote his doctoral thesis on a topic in elasticity.

The book is intended to provide a thorough grounding in tensor-based theory of elasticity, which is rigorous in treatment but limited in scope. While the traditional audience for elasticity is graduate students in the fields mentioned above, advanced undergraduates have increasingly become interested in the subject. This is reflected by the emergence of a basic course in elasticity as a replacement for the traditional Advanced Strength of Materials in some curriculums. The authors are indebted to many historical and some contemporary contributors to this subject, and have endeavored to select, organize, and present the material in a concise manner.

As before, the text is strongly influenced by a new generation of students who sought out modern applications of elasticity to emerging fields. Eric Clayton, Kevin Kilgallon, Kevin Derendorf, Rebecca Veto, Chris Ward, and Michael Van Dusseldorp have made significant contributions. Also the careful collating of the text by Linda Buckingham is appreciated.

The larger blocks of new materials in the third edition of this text dealt with functionally graded materials and viscoelasticity. These materials were drawn from the works of Professors B.V. Sakkar of the University of Florida and David Roylance of MIT, respectively, and their active assistance is appreciated. Many corrections to the second edition were submitted by Professor Charles Bert of the University of Oklahoma and by our late colleague Professor S. Sridharan who also class-tested the book over several years. To broaden the scope of the book to introduce some topics not traditionally covered in an introductory text but of importance to contemporary researchers, a brief discussion of nonlinear elasticity is presented along with the aforementioned chapter on viscoelasticity. Also the chapter on plates was expanded to include an introduction to thin shell theory based on earlier works of the senior author.

Computational tools have been playing an important role in the research and applications of elasticity theories, both analytically and numerically. Therefore, in this edition, we introduce the usage of MATLAB in solving classical elasticity problems. The basics of using MATLAB are introduced in the first chapter with common commands summarized in appendices. To illustrate the application of the computational tools, sample problems are solved and presented as the last section of each chapter. We hope that this addition of computational materials will enhance the understanding of the elasticity theories and provide an opportunity to learn numerical methods and tools via solving classical elasticity problems.

We continue to believe that calibration with classical analytical solutions is essential to establish confidence, to gain efficiency, and to quantify errors in numerically based results and trust that the students who undertake the study of this subject will find it stimulating and rewarding.

St. Louis, MO, USA   Phillip L. Gould
Shanghai, China   Yuan Feng

Contents

List of Figures

Chapter 1
Introduction and Mathematical Preliminaries

Abstract The *theory of elasticity* comprises a consistent set of equations which uniquely describe the state of stress, strain, and displacement at each point within an elastic deformable body. Engineering approaches are often based on a strength-of-materials formulation with its various specialized derivatives such as the theories of rods, beams, plates, and shells. The distinguishing feature between the various alternative approaches and the theory of elasticity is the *pointwise* description, as opposed to sectional description embodied in elasticity. The basic elements of the theory are *equilibrium* equations relating the stresses, *kinematic* equations relating the strains and displacements, *constitutive* equations relating the stresses and strains, *boundary conditions* relating to the physical domain, and *uniqueness* constraints relating to the applicability of the solution. In this chapter, the mathematical preliminaries for the presentation, such as vector algebra, integral theorems, indicial notation, and Cartesian tensors, are introduced.

1.1 Scope

The *theory of elasticity* comprises a consistent set of equations which uniquely describe the state of stress, strain, and displacement at each point within an elastic deformable body. Solutions of these equations fall into the realm of *applied mathematics*, while applications of such solutions are of *engineering* interest. When elasticity is selected as the basis for an engineering solution, a rigor is accepted that distinguishes this approach from the alternatives, which are mainly based on the strength-of-materials with its various specialized derivatives such as the theories of rods, beams, plates, and shells. The distinguishing feature between the various alternative approaches and the theory of elasticity is the *pointwise* description embodied in elasticity, without resort to expedients such as Navier's hypothesis of plane sections remaining plane.

The theory of elasticity contains *equilibrium* equations relating the stresses, *kinematic* equations relating the strains and displacements, *constitutive* equations relating the stresses and strains, *boundary conditions* relating to the physical domain, and *uniqueness* constraints relating to the applicability of the solution. Origination of

© Springer International Publishing AG, part of Springer Nature 2018
P. L. Gould, Y. Feng, *Introduction to Linear Elasticity*,
https://doi.org/10.1007/978-3-319-73885-7_1

the theory of elasticity is attributed to Claude Louis Marie Henri Navier, Siméon Denis Poisson, and George Green in the first half of the nineteenth century [1].

In subsequent chapters, each component of the theory will be developed in full from fundamental principles of physics and mathematics. Some limited applications will then be presented to illustrate the potency of the theory as well as its limitations.

1.2   Vector Algebra

A *vector* is a directed line segment in the physical sense. Referred to the unit basis vectors ($\mathbf{e}_x$, $\mathbf{e}_y$, $\mathbf{e}_z$) in the Cartesian coordinate system (x, y, z), an arbitrary vector $\mathbf{A}$ may be written in component form as

$$\mathbf{A} = A_x\mathbf{e}_x + A_y\mathbf{e}_y + A_z\mathbf{e}_z. \tag{1.1}$$

Alternately, the Cartesian system could be numerically designated as (x_1, x_2, x_3), whereupon

$$\mathbf{A} = A_1\mathbf{e}_1 + A_2\mathbf{e}_2 + A_3\mathbf{e}_3. \tag{1.2}$$

The latter form is common in elasticity. An example is vector $\mathbf{r}$ in Fig. 1.1, where the unit vectors $\mathbf{e}_1$, $\mathbf{e}_2$, and $\mathbf{e}_3$ are identified [2]. Beyond the physical representation, it is often sufficient to deal with the components alone as ordered triples,

$$\mathbf{A} = (A_1, A_2, A_3). \tag{1.3}$$

The length or magnitude of $\mathbf{A}$ is given by

$$|\mathbf{A}| = \sqrt{A_1^2 + A_2^2 + A_3^2}. \tag{1.4}$$

Vector equality, addition, and subtraction are trivial, while vector multiplication has two forms. The *inner*, *dot*, or *scalar* product is

$$\begin{aligned} \mathbf{C} &= \mathbf{A} \cdot \mathbf{B} \\ &= A_1B_1 + A_2B_2 + A_3B_3 \\ &= |\mathbf{A}||\mathbf{B}| \cos \theta_{AB}. \end{aligned} \tag{1.5}$$

Also, there is the *outer*, *cross*, or *vector* product

$$\begin{aligned} \mathbf{C} &= \mathbf{A} \times \mathbf{B} \\ &= (A_2B_3 - A_3B_2)\mathbf{e}_1 + (A_3B_1 - A_1B_3)\mathbf{e}_2 + (A_1B_2 - A_2B_1)\mathbf{e}_3, \end{aligned} \tag{1.6a}$$

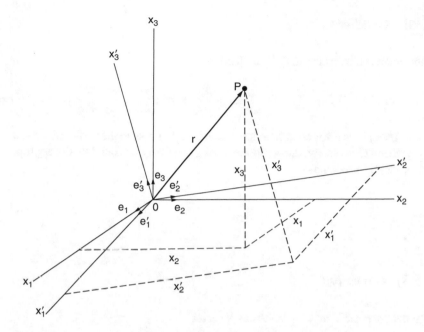

Fig. 1.1 Cartesian coordinate systems (After Tauchert [2]). Reproduced by permission

which is conveniently evaluated as a determinant

$$C = A \times B = \begin{vmatrix} \mathbf{e}_1 & \mathbf{e}_2 & \mathbf{e}_3 \\ A_1 & A_2 & A_3 \\ B_1 & B_2 & B_3 \end{vmatrix}; \tag{1.6b}$$

C is perpendicular to the plane containing **A** and **B**.

1.3   Scalar and Vector Fields

1.3.1   Definitions

A *scalar* quantity expressed as a function of the Cartesian coordinates such as

$$f(x_1, x_2, x_3) = \text{constant} \tag{1.7}$$

is known as a *scalar field*. An example is the temperature at a point. A *vector* quantity similarly expressed, such as $\mathbf{A}(x_1, x_2, x_3)$, is called a *vector field*. An example is the velocity of a particle. We are especially concerned with changes or derivatives of these fields.

1.3.2 Gradient

The *gradient* of a scalar field f is defined as

$$\mathbf{grad}\, f = \nabla f = \frac{\partial f}{\partial x_1}\mathbf{e}_1 + \frac{\partial f}{\partial x_2}\mathbf{e}_2 + \frac{\partial f}{\partial x_3}\mathbf{e}_3 = \left(\frac{\partial f}{\partial x_1}, \frac{\partial f}{\partial x_2}, \frac{\partial f}{\partial x_3}\right), \qquad (1.8)$$

where $\mathbf{grad}\, f$ is a *vector* point function which is orthogonal to the surface f = constant, everywhere. Conversely, the components of $\mathbf{grad}\, f$ may be found by the appropriate dot product, for example,

$$\frac{\partial f}{\partial x_1} = \mathbf{e}_1 \cdot \nabla f. \qquad (1.9)$$

1.3.3 Operators

The del operator ∇ may be treated as a vector

$$\nabla(\ ) = \frac{\partial(\ )}{\partial x_1}\mathbf{e}_1 + \frac{\partial(\ )}{\partial x_2}\mathbf{e}_2 + \frac{\partial(\ )}{\partial x_3}\mathbf{e}_3 \qquad (1.10a)$$

while the higher-order operators are written as

$$\nabla^2(\ ) = \nabla \cdot \nabla(\ ) = \frac{\partial^2}{\partial x_1^2}(\ ) + \frac{\partial^2}{\partial x_2^2}(\ ) + \frac{\partial^2}{\partial x_3^2}(\ ) \qquad (1.10b)$$

and

$$\nabla^4(\ ) = \nabla^2[\nabla^2(\ )]$$
$$= \frac{\partial^4}{\partial x_1^4}(\ ) + \frac{\partial^4}{\partial x_2^4}(\ ) + \frac{\partial^4}{\partial x_3^4}(\ ) \qquad (1.10c)$$
$$+ 2\frac{\partial^4}{\partial x_1^2 \partial x_2^2}(\ ) + 2\frac{\partial^4}{\partial x_1^2 \partial x_3^2}(\ ) + 2\frac{\partial^4}{\partial x_2^2 \partial x_3^2}(\ ).$$

Both $\nabla^2(\ )$ and $\nabla^4(\ )$ are scalars and are used frequently in the following chapters. These operators are general because the operation performed is independent of any particular coordinate system or *invariant*. However, the forms given in (1.10a), (1.10b), and (1.10c) are for Cartesian coordinates; for curvilinear coordinates, such as cylindrical coordinates, the operators must be appropriately transformed. This is developed in Sect. 7.4.

1.3.4   Divergence

The *divergence* of a vector field **A**, (1.2), is defined as

$$\operatorname{div} A = \boldsymbol{\nabla} \cdot A$$
$$= \frac{\partial A_1}{\partial x_1} + \frac{\partial A_2}{\partial x_2} + \frac{\partial A_3}{\partial x_3}, \tag{1.11}$$

which is a scalar conveniently written as ΔA.

1.3.5   Curl

Since there are two forms of vector multiplication, it is natural to expect another derivative form of **A**. The *curl* of **A** is defined as

$$\operatorname{curl} A = \boldsymbol{\nabla} \times A$$
$$= \begin{vmatrix} \mathbf{e}_1 & \mathbf{e}_2 & \mathbf{e}_3 \\ \dfrac{\partial (\ )}{\partial x_1} & \dfrac{\partial (\ )}{\partial x_2} & \dfrac{\partial (\ )}{\partial x_3} \\ A_1 & A_2 & A_3 \end{vmatrix} \tag{1.12}$$

in determinant form.

1.3.6   Integral Theorems

Two integral theorems relating vector fields are particularly useful for transforming between contour, area, and volume integrals: *Green's theorem* and the *divergence theorem*.

First, considering two functions $P(x, y)$ and $Q(x, y)$, which are continuous and have continuous first partial derivatives (C^1 continuous) in a two-dimensional domain D, Green's theorem states that

$$\oint_C (P\mathrm{d}x + Q\mathrm{d}y) = \int_A \left(\frac{\partial Q}{\partial x} - \frac{\partial P}{\partial y} \right) \mathrm{d}x\mathrm{d}y, \tag{1.13}$$

where A is a closed region within D bounded by C.

Then, considering a continuously differentiable vector point function **G** in D, the divergence theorem states that

$$\int_V \nabla \cdot \mathbf{G}\, dV = \int_A \mathbf{n} \cdot \mathbf{G}\, dA, \tag{1.14}$$

where V is the volume bounded by the oriented surface A and $\mathbf{n}$ is the positive normal to A.

1.4   Indicial Notation

One of the conveniences of modern treatments of the theory of elasticity is the use of shorthand notation to facilitate the mathematical manipulation of lengthy equations.

Referring to the ordered triple representation for $\mathbf{A}$ in (1.3), the three Cartesian components can be symbolized as A_i, where the subscript or *index i* is understood to take the sequential values 1, 2, 3. If we have nine quantities, we may employ a double-subscripted notation D_{ij}, where i and j range from 1 to 3 in turn. Later, we will associate these nine components with a higher form of a vector, called a tensor. Further, we may have 27 quantities, C_{ijk}, etc. While i and j range as stated, an exception is made when two subscripts are identical, such as D_{jj}. The *Einstein summation convention* states that a subscript appearing *twice* is *summed* from 1 to 3. No subscript can appear more than twice in a single term. As an example, we have the inner product, (1.5), rewritten as

$$A_i B_i = \sum_{i=1}^{3} A_i B_i \tag{1.15a}$$
$$= A_1 B_1 + A_2 B_2 + A_3 B_3.$$

Also,

$$D_{jj} = D_{11} + D_{22} + D_{33}. \tag{1.15b}$$

It is apparent from the preceding examples that there are two distinct types of indices. The first type appears only once in each term of the equation and ranges from 1 to 3. It is called a *free* index. The second type appears twice in a single term and is summed from 1 to 3. Since it is immaterial which letter is used in this context, a repeated subscript is called a *dummy* index. That is, $D_{ii} = D_{jj} = D_{kk}$. From the preceding discussion, it may be deduced that the number of individual terms represented by a single product is 3^k, where k is the number of free indices. There are some situations where double subscripts occur but the summation convention is not intended. This is indicated by enclosing the subscripts in parentheses [3]. For example, the individual components D_{11}, D_{22}, and D_{33} could be represented by $D_{(ii)} i = 1, 2, 3$. Also in the following sections, the summation convention is applied only to the indices i, j, k unless otherwise stated.

The *product* of the three components of a vector is expressed by the Pi convention:

$$\prod_{i=1}^{3} A_i = A_1 A_2 A_3.$$ (1.16)

Partial differentiation may also be abbreviated using the *comma* convention:

$$\frac{\partial A_i}{\partial x_j} = A_{i,j}.$$ (1.17a)

Since both i and j are free indices, (1.17a) represents $3^2 = 9$ quantities. With repeated indices, the equation becomes

$$\frac{\partial A_i}{\partial x_i} = A_{i,i}$$

$$= \Delta A$$ (1.17b)

which is the divergence as defined in (1.11). Further,

$$\frac{\partial D_{ij}}{\partial x_j} = D_{ij,j}$$

$$= D_{i1,1} + D_{i2,2} + D_{i3,3}$$ (1.17c)

that takes on $3^1 = 3$ values for each $i = 1, 2, 3$. This example combines the summation and comma conventions.

1.5   Coordinate Rotation

In Fig. 1.1 (see [2]), we show a position vector to point P, $\mathbf{r}$, resolved into components with respect to *two* different Cartesian systems, x_i and x_i', having a common origin. The unit vectors in the x_i' system are indicated as $\mathbf{e}_i'$ in the figure.

First, we consider the point P with coordinates $P(x_1, x_2, x_3) = P(x_i)$ in the unprimed system and $P(x_1', x_2', x_3') = P(x_i')$ in the primed system. The linear transformation between the coordinates of P is given by

$$\begin{aligned} x_1' &= \alpha_{11}x_1 + \alpha_{12}x_2 + \alpha_{13}x_3 \\ x_2' &= \alpha_{21}x_1 + \alpha_{22}x_2 + \alpha_{23}x_3 \\ x_3' &= \alpha_{31}x_1 + \alpha_{32}x_2 + \alpha_{33}x_3 \end{aligned}$$ (1.18a)

or

$$x'_i = \alpha_{ij} x_j \tag{1.18b}$$

using the summation convention. Each of the nine quantities α_{ij} is the cosine of the angle between the i th primed and the jth unprimed axis, that is,

$$\alpha_{ij} = \cos\left(x'_i, x_j\right) = \frac{\partial x_j}{\partial x'_i} = \mathbf{e}'_i \cdot \mathbf{e}_j = \cos\left(\mathbf{e}'_i, \mathbf{e}_j\right). \tag{1.19}$$

The α_{ij}'s are known as *direction cosines* and are conveniently arranged in tabular form for computation:

$$
\begin{array}{c|ccc}
 & x_1 & x_2 & x_3 \\
\hline
x'_1 & \alpha_{11} & \alpha_{12} & \alpha_{13} \\
x'_2 & \alpha_{21} & \alpha_{22} & \alpha_{23} \\
x'_3 & \alpha_{31} & \alpha_{32} & \alpha_{33}
\end{array}
\tag{1.20}
$$

It is emphasized that, in general, $\alpha_{ij} \neq \alpha_{ji}$. From a computational standpoint, (1.19) indicates that expressing the unit vectors in the x'_i coordinate system, $\mathbf{e}'_i$, in terms of those in the x_i system, $\mathbf{e}_i$, is tantamount to evaluating the corresponding α_{ij} terms. A numerical example is given in Sect. 2.2.6 and 2.4.3.

We next consider the position vector $\mathbf{r}$ and recognize that the *components* are related by (1.18a) and (1.18b). Conversely, any quantity which obeys this transformation law is a vector. This somewhat indirect definition of a vector proves to be convenient for defining higher-order quantities, *Cartesian tensors*.

From a computational standpoint, it is often convenient to carry out the transformations indicated in (1.18a) and (1.18b) in matrix form as

$$x' = Rx \tag{1.21}$$

in which

$$x' = \{x'_1, x'_2, x'_3\} \tag{1.22a}$$

$$x = \{x_1, x_2, x_3\} \tag{1.22b}$$

$$R = \begin{bmatrix} \alpha_{11} & \alpha_{12} & \alpha_{13} \\ \alpha_{21} & \alpha_{22} & \alpha_{23} \\ \alpha_{31} & \alpha_{32} & \alpha_{33} \end{bmatrix}. \tag{1.22c}$$

The brackets { } indicate a column vector, the transpose { }^{T} or $\lfloor \quad \rfloor$ is a row matrix, and R may be called a *rotation* matrix.

1.6   Cartesian Tensors

A tensor of order n is a set of 3^n quantities which transform from one coordinate system, x_i, to another, x_i', by a specified law, as follows:

n	Order	Transformation law
0	Zero (scalar)	$A_i' = A_i$
1	One (vector)	$A_i' = \alpha_{ij} A_j$
2	Two (dyadic)	$A_{ij}' = \alpha_{ik} \alpha_{jl} A_{kl}$
3	Three	$A_{ijk}' = \alpha_{il} \alpha_{jm} \alpha_{kn} A_{lmn}$
4	Four	$A_{ijkl}' = \alpha_{im} \alpha_{jn} \alpha_{kp} \alpha_{lq} A_{mnpq}$

Order zero and order one tensors are familiar physical quantities, whereas the higher-order tensors are useful to describe physical quantities with a corresponding number of associated directions. Second-order tensors (dyadics) are particularly prevalent in elasticity, and the transformation may be carried out in a matrix format, analogous to (1.21), as

$$A' = RAR^T, \tag{1.23}$$

in which

$$A' = \begin{bmatrix} A_{11}' & A_{12}' & A_{13}' \\ A_{21}' & A_{22}' & A_{23}' \\ A_{31}' & A_{32}' & A_{33}' \end{bmatrix} \tag{1.24}$$

and A is similar.

It may be helpful to visualize a tensor of order n as having n unit vectors or directions associated with each component. Thus, a scalar has no directional association (isotropic), and a vector is directed in one direction. A second-order tensor has two associated directions, perhaps one direction *in* which it acts and another defining the surface *on* which it is acting.

1.7   Algebra of Cartesian Tensors

Tensor arithmetic and algebra are similar to matrix operations in regard to addition, subtraction, equality, and scalar multiplication. Multiplication of two tensors of order n and m produces a new tensor of order $n + m$. For example,

$$A_i B_{jk} = C_{ijk}. \tag{1.25}$$

For two repeated indices, the summation convention holds, as shown in (1.15b).

1.8 Operational Tensors

Additional tensor operations are facilitated by the use of the *Kronecker delta* δ_{ij} defined such that

$$\begin{aligned} \delta_{ij} &= 1 \quad \text{if} \quad i = j \\ \delta_{ij} &= 0 \quad \text{if} \quad i \neq j \end{aligned} \tag{1.26}$$

and the *permutation tensor* ε_{ijk} defined such that

$$\varepsilon_{ijk} = \frac{1}{2}(i-j)(j-k)(k-i). \tag{1.27}$$

Thus,

$\varepsilon_{ijk} = 0$	If any two of i, j, k are equal
$\varepsilon_{ijk} = 1$	For an even permutation (forward on the number line 1, 2, 3, 1, 2, 3, ...)
$\varepsilon_{ijk} = -1$	For an odd permutation (backward on the number line 3, 2, 1, 3, 2, 1, ...)

Hence, $\varepsilon_{112} = 0$, $\varepsilon_{231} = +1$, and $\varepsilon_{321} = -1$.

The Kronecker delta δ_{ij} is used to change the subscripts in a tensor by multiplication, as illustrated in the following:

$$\begin{aligned} \delta_{ij}A_i &= \delta_{1j}A_1 + \delta_{2j}A_2 + \delta_{3j}A_3 = A_j \\ \delta_{ij}D_{jk} &= \delta_{i1}D_{1k} + \delta_{i2}D_{2k} + \delta_{i3}D_{3k} = D_{ik} \\ \delta_{ij}C_{ijk} &= \delta_{11}C_{11k} + \delta_{22}C_{22k} + \delta_{33}C_{33k} = C_{iik} = A_k. \end{aligned} \tag{1.28}$$

The last illustration, in which two subscripts in C_{ijk} were made identical by δ_{ij}, results in C_{ijk} being changed from a third- to a first-order tensor. This is known as *contraction* and generally reduces the order of the original tensor by two.

The Kronecker delta δ_{ij} is also useful in vector algebra and for coordinate transformations. Starting with the dot product of two unit vectors

$$\mathbf{e}_i \cdot \mathbf{e}_j = \delta_{ij} \tag{1.29}$$

and seeking the component of a vector $\mathbf{A} = A_i\mathbf{e}_i$ in the j-direction, A_j, we do the following:

$$\begin{aligned} \mathbf{e}_j \cdot \mathbf{A} &= \left(\mathbf{e}_j \cdot \mathbf{e}_i\right)A_i \\ &= \delta_{ji}A_i \\ &= A_j. \end{aligned} \tag{1.30}$$

Next, we consider (1.29) for transformed coordinates:

$$
\begin{aligned}
\delta_{ij} &= \mathbf{e}'_i \cdot \mathbf{e}'_j \\
&= (\alpha_{ik}\mathbf{e}_k) \cdot (\alpha_{jl}\mathbf{e}_l) \\
&= \alpha_{ik}\alpha_{jl}\mathbf{e}_k \cdot \mathbf{e}_l \\
&= \alpha_{ik}\alpha_{jl}\delta_{kl} \\
&= \alpha_{ik}\alpha_{jk}.
\end{aligned}
\tag{1.31}
$$

Equation (1.31) is useful for demonstrating some important properties of direction cosines.

First, taking $i = j$ and summing only on k, we get

$$
\begin{aligned}
1 &= \alpha_{(i)k}\alpha_{(i)k} \\
&= \alpha_{i1}^2 + \alpha_{i2}^2 + \alpha_{i3}^2.
\end{aligned}
\tag{1.32}
$$

Expanding (1.32), we have

$$
\begin{aligned}
i = 1: \quad & \alpha_{11}^2 + \alpha_{12}^2 + \alpha_{13}^2 = 1 \\
i = 2: \quad & \alpha_{21}^2 + \alpha_{22}^2 + \alpha_{23}^2 = 1 \\
i = 3: \quad & \alpha_{31}^2 + \alpha_{32}^2 + \alpha_{33}^2 = 1,
\end{aligned}
$$

which is the *normality* property of direction cosines.

Then, taking $i \neq j$, we get

$$
\begin{aligned}
0 &= \alpha_{ik}\alpha_{jk} \\
&= \alpha_{i1}\alpha_{j1} + \alpha_{i2}\alpha_{j2} + \alpha_{i3}\alpha_{j3}.
\end{aligned}
\tag{1.33}
$$

Expanding (1.33), we have

$$
\begin{aligned}
i = 1, j = 2 &: \alpha_{11}\alpha_{21} + \alpha_{12}\alpha_{22} + \alpha_{13}\alpha_{23} = 0 \\
i = 1, j = 3 &: \alpha_{11}\alpha_{31} + \alpha_{12}\alpha_{32} + \alpha_{13}\alpha_{33} = 0 \\
i = 2, j = 1 &: \alpha_{21}\alpha_{11} + \alpha_{22}\alpha_{12} + \alpha_{23}\alpha_{13} = 0 \\
i = 2, j = 3 &: \alpha_{21}\alpha_{31} + \alpha_{22}\alpha_{32} + \alpha_{23}\alpha_{33} = 0 \\
i = 3, j = 1 &: \alpha_{31}\alpha_{11} + \alpha_{32}\alpha_{12} + \alpha_{33}\alpha_{13} = 0 \\
i = 3, j = 2 &: \alpha_{31}\alpha_{21} + \alpha_{32}\alpha_{22} + \alpha_{33}\alpha_{23} = 0,
\end{aligned}
$$

which is the *orthogonality* property of direction cosines.

The permutation tensor ε_{ijk} is useful for vector cross-product operations. If we take

$$
\begin{aligned}
\varepsilon_{ijk}A_jB_k\mathbf{e}_i &= (\varepsilon_{123}A_2B_3 + \varepsilon_{132}A_3B_2)\mathbf{e}_1 + (\varepsilon_{213}A_1B_3 + \varepsilon_{231}A_3B_2)\mathbf{e}_2 \\
&\quad + (\varepsilon_{312}A_1B_2 + \varepsilon_{321}A_2B_1)\mathbf{e}_3 \\
\varepsilon_{ijk}A_jB_k\mathbf{e}_i &= (A_2B_3 - A_3B_2)\mathbf{e}_1 + (A_3B_1 - A_1B_3)\mathbf{e}_2 \\
&\quad + (A_1B_2 - A_2B_1)\mathbf{e}_3,
\end{aligned}
\tag{1.34}
$$

we obtain an expression which is identical to (1.6a) and (1.6b). Thus, $\varepsilon_{ijk}A_jB_k$ gives the components of $\mathbf{A} \times \mathbf{B}$.

1.9   Computational Examples

To illustrate computations and manipulations using Cartesian tensors, we present the following illustrations:

1. Show that $\delta_{ij}\delta_{jk} = \delta_{ik}$:

   Expanding $\delta_{ij}\delta_{jk}$, we get

   $$
   \begin{aligned}
   \delta_{ij}\delta_{jk} &= \delta_{i1}\delta_{1k} + \delta_{i2}\delta_{2k} + \delta_{i3}\delta_{3k} = 1 \times \delta_{ik} \text{ for a selected} \\
   i &= 1, 2, \text{ or } 3
   \end{aligned}
   $$

2. Show that $\varepsilon_{ijk}A_jA_k = 0$:

   Expanding $\varepsilon_{ijk}A_jA_k$, we find

   $$
   \begin{aligned}
   \varepsilon_{ijk}A_jA_k &= \varepsilon_{123}A_2A_3 + \varepsilon_{132}A_3A_2 + \varepsilon_{231}A_3A_1 + \varepsilon_{213}A_1A_3 \\
   &\quad + \varepsilon_{312}A_1A_2 + \varepsilon_{321}A_2A_1 \\
   &= +1 - 1 + 1 - 1 + 1 - 1 \\
   &= 0.
   \end{aligned}
   $$

3. Prove that the product of two first-order tensors is a second-order tensor:

   Let A_i and B_j be two first-order tensors and C_{ij} be their product; then

   $$
   \begin{aligned}
   A_i &= \alpha_{ik}A_k \\
   B_j &= \alpha_{jl}B_l \\
   A_iB_j &= \alpha_{ik}\alpha_{jl}A_kB_l = \alpha_{ik}\alpha_{jl}C_{kl} = C_{ij}
   \end{aligned}
   $$

   so that C_{ij} transforms as a second-order tensor.

1.10 Fundamentals of MATLAB

The core ideas of MATLAB are matrices or arrays that are the basic computation units. As a computational tool, MATLAB has been widely used in the field of scientific computation. In this book, we use MATLAB to demonstrate some computational applications to elasticity. Several examples solved analytically will be solved using MATLAB. As a useful tool to implement and test algorithms, codes written in MATLAB can be translated to other codes such as Fortran or C.

In the computational framework, the components of the stress tensor σ, defined in Section 2.2.2, can be represented in a 3×3 matrix $\sigma = \begin{bmatrix} \sigma_{11} & \sigma_{12} & \sigma_{13} \\ \sigma_{21} & \sigma_{22} & \sigma_{23} \\ \sigma_{31} & \sigma_{32} & \sigma_{33} \end{bmatrix}$. The matrix notation of the stress tensor makes it easy for computational implementation. Similarly, vectors such as the normal n can be denoted as a 3×1 matrix $n = \{n_1, n_2, n_3\}$.

MATLAB offers a very intuitive way of data representation. The stress tensor can be input as

```
sig = [sig_11, sig_12, sig_13;
       sig_21, sig_22, sig_23;
       sig_31, sig_32, sig_33];
```

where each element is the corresponding component of the stress tensor. The normal vector can be input as

```
n = [n_1; n_2; n_3];
```

where n_1 to n_3 are the corresponding vector components.

Frequently used operations are the matrix product and dot product. MATLAB uses "*" for the matrix product. For example, A*B=C can be input as C = A * B. The dot product between the two vectors, e.g., v1*v2 = v3, can be calculated by v3 = v1 * v2. Therefore, in a state of stress with stress tensor sig, the traction force on the plane with a normal vector n can be calculated by sig*n. For other command usage, please refer to Appendix I.

1.11 Introduction to MuPAD

Many elasticity problems involve presenting analytical solutions. Although analytical solutions are not obtainable for all problems, they provide a useful way to verify computational results.

There are two ways to obtain analytical solutions in MATLAB. One is to use the "symbolic toolbox" and the other is to use the embedded package "MuPAD."

Although the commands are different between the two, both tools interface within MATLAB and can exchange data and variables via the MATLAB workspace.

In the command window, simply type in "mupad" and the MuPAD interface window will pop out. An important feature of the MuPAD is the visualization of the symbols and equations. This is one of the major advantages of the "symbolic toolbox." For example, create a symbol "σ" and use "sig" for representation in the scripts.

```
sig := Symbol::sigma
    σ
```

In this case, we can use sig for all the symbolic calculations, but in the final results, symbol "σ" will be shown. Therefore, if we use "sig" for computation, the command prompt output will show symbol "σ" rather than "sig". This is useful for visualizing all the equations. For other command usage, please refer to Appendix II.

In the following chapters, we demonstrate the use of MuPAD using typical examples in elasticity.

1.12   Exercises

1.1 Show that (after [2]):

    (a) $\delta_{ij}\delta_{ij} = 3$
    (b) $\varepsilon_{ijk}\varepsilon_{kji} = -6$
    (c) $\alpha_{ij}b_{jk} = \alpha_{in}b_{nk}$
    (d) $\delta_{ij}\delta_{jk}\delta_{ki} = 3$
    (e) $\delta_{ij}C_{jkl} = C_{ikl}$

1.2 Prove that if $\delta_{ij} = \delta'_{ij}$, then δ_{ij} is a second-order tensor (after [2]).

1.3 Prove that the product of a first- and a second-order tensor is a third-order tensor.

1.4 Two first-order tensors are related by

$$A_i = C_{ij}B_j.$$

    Prove that C_{ij} is a second-order tensor.

1.5 Show that if B_i is a first-order tensor, $B_{i,j}$ is a second-order tensor [2].

1.6 If a square matrix **C** has the property

$$C^{-1} = C^T,$$

    it is said to be orthogonal. This property is used in the derivation of (1.23). Show that the matrix of direction cosines **R** as defined in (1.22c) is orthogonal.

1.7 Write the operator $\nabla^2(\ )$ defined in (1.10b) in indicial notation.

1.8 Write the divergence theorem defined in (1.14) in indicial notation.

1.9 Show that **curl A** $= \nabla \times \mathbf{A} = \varepsilon_{ijk} A_{j,i} \mathbf{e}_k$.

References

1. Westergaard HM (1964) Theory of elasticity and plasticity. Dover Publications, Inc., New York
2. Tauchert TR (1974) Energy principles in structural mechanics. McGraw-Hill Book Company Inc., New York
3. Ma Y, Desai CS (1990) Alternative definition of finite strains. J Eng Mech, ASCE 116 (4):901–919

Chapter 2
Traction, Stress, and Equilibrium

Abstract An approach to the solution of problems in solid mechanics is to establish relationships first between applied loads and internal stresses and, subsequently, to consider deformations. Another way is to examine deformations initially and then proceed to the stresses and applied loads. Regardless of the eventual solution path selected, it is necessary to derive the component relationships individually. In this chapter, the first set of equations describing equilibrium between external and internal forces and stresses is derived from the principles of linear and angular momentum. The stresses are transformed into principal coordinates, and several special stress states are defined.

2.1 Introduction

A popular approach to the solution of problems in solid mechanics is to establish relationships first between applied loads and internal stresses and, subsequently, to consider deformations. Another way is to examine deformations initially and then proceed to the stresses and applied loads. Regardless of the eventual solution path selected, it is necessary to derive the component relationships individually. In this chapter, the first set of equations, which describes equilibrium between external and internal forces and stresses, is derived.

2.2 State of Stress

2.2.1 Traction and Couple-Stress Vectors

A deformable body subject to external loading is shown in Fig. 2.1. There may be loads applied over the exterior, properly called *surface forces*, and loads distributed within the interior, known as *body forces*. An example of the latter is the effect of gravity which produces the *self-weight* of the body.

© Springer International Publishing AG, part of Springer Nature 2018
P. L. Gould, Y. Feng, *Introduction to Linear Elasticity*,
https://doi.org/10.1007/978-3-319-73885-7_2

Fig. 2.1 Deformable body under external loading (After Tauchert [1]). Reproduced with permission

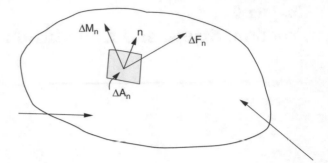

Focusing on an element with an area ΔA_n on or within the body and oriented as specified by the unit normal **n**, we accumulate the resultant force $\Delta \mathbf{F}_n$ and moment $\Delta \mathbf{M}_n$. Both are vector quantities and are *not*, in general, parallel to **n**. Next, we represent the intensity of the resultants on the area ΔA_n in the form [1]

$$\lim_{\Delta A_n \to 0} \frac{\Delta \mathbf{F}_n}{\Delta A_n} = \frac{d\mathbf{F}_n}{dA_n} = \mathbf{T}_n \tag{2.1a}$$

$$\lim_{\Delta A_n \to 0} \frac{\Delta \mathbf{M}_n}{\Delta A_n} = \frac{d\mathbf{M}_n}{dA_n} = \mathbf{C}_n, \tag{2.1b}$$

where $\mathbf{T}_n$ is known as the stress vector or *traction*, and $\mathbf{C}_n$ is called the *couple-stress vector*.

The elementary theory of elasticity proceeds on the assumption that $\mathbf{C}_n = 0$ [2], while the traction $\mathbf{T}_n$ is the stress intensity at the point for the particular orientation of the area element specified by **n**. A complete description at the point requires that the state of stress be known for all directions so that $\mathbf{T}_n$ itself is necessary but not sufficient for this purpose.

2.3   Components of Stress

We now study an infinitesimal rectangular parallelepiped at the point in question and erect a set of Cartesian coordinates x_i parallel to the sides, as shown in Fig. 2.2 [1]. Corresponding to each coordinate axis is a unit vector $\mathbf{e}_i$. Shown in the figure are the tractions $\mathbf{T}_i$ acting on each face i, with the subscript chosen corresponding to the face normal $\mathbf{e}_i$. It is again emphasized that, in general, $\mathbf{T}_i$ is *not* parallel to $\mathbf{e}_i$, which is perpendicular to the face of the parallelepiped.

Each traction may be written in terms of its Cartesian components in the form

$$\mathbf{T}_i = \sigma_{ij}\mathbf{e}_j, \tag{2.2}$$

which is expanded explicitly into the three equations:

Fig. 2.2 Components of
stress (After Tauchert [1]).
Reproduced with
permission

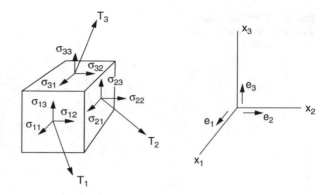

$$\mathbf{T}_1 = \sigma_{11}\mathbf{e}_1 + \sigma_{12}\mathbf{e}_2 + \sigma_{13}\mathbf{e}_3 \tag{2.3a}$$

$$\mathbf{T}_2 = \sigma_{21}\mathbf{e}_1 + \sigma_{22}\mathbf{e}_2 + \sigma_{23}\mathbf{e}_3 \tag{2.3b}$$

$$\mathbf{T}_3 = \sigma_{31}\mathbf{e}_1 + \sigma_{32}\mathbf{e}_2 + \sigma_{33}\mathbf{e}_3. \tag{2.3c}$$

The coefficients σ_{11}, σ_{12}, ..., σ_{33} are known as stresses, while the entire array forms the Cauchy stress tensor when the appropriate transformation rule is verified. The subscript and sign conventions for components of stress σ_{ij} are as follows:

1. The *first* subscript i refers to the *normal* $\mathbf{e}_i$ which denotes the face *on* which $\mathbf{T}_i$ acts.
2. The *second* subscript j corresponds to the *direction* $\mathbf{e}_j$ in which the stress acts.
3. The so-called *normal* or *extensional* components $\sigma_{(ii)}$ are positive if they produce tension and negative if they produce compression. The *shearing* components σ_{ij} $(i \neq j)$ are positive if directed in the positive x_j-direction while acting on the face with the unit normal $+\mathbf{e}_i$ or if directed in the negative x_j-direction while acting on the face with unit normal $-\mathbf{e}_i$.

While it is generally vital to distinguish between tension and compression, the difference between the positive and negative shear directions is quite arbitrary for most materials. See Sects. 12.4.3 and 12.4.4.

2.4   Stress at a Point

We are now in a position to pursue the main thrust of this section and thus establish sufficient conditions to completely describe the state of stress at a point. We will show that this may be accomplished by specifying the tractions $\mathbf{T}_i$ on each of the three planes $\mathbf{e}_i$ which, by (2.3a), (2.3b), and (2.3c), is equivalent to specifying the nine components of stress σ_{ij}. Then, if the traction $\mathbf{T}_n$ acting on any *arbitrary* element of surface defined by an appropriate $\mathbf{n}$ can be evaluated, the proposition is proved, and the stress tensor σ_{ij}, referred to any convenient Cartesian system, completely specifies the state of stress at the point.

Fig. 2.3 Tractions (After
Tauchert [1]). Reproduced
with permission

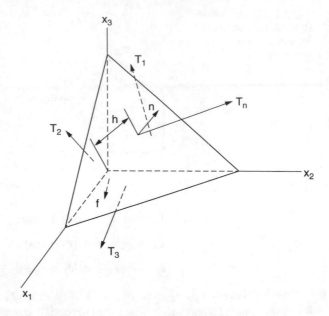

The differential tetrahedron in Fig. 2.3 [2] shows the traction $\mathbf{T}_n$ acting on the
plane identified by $\mathbf{n}$, along with tractions on the faces indicated by $\mathbf{e}_i$ and the body
force per unit volume $\mathbf{f}$ [1]. The force on the sloping face is $\mathbf{T}_n dA_n$, while the force on
each of the other faces is $-\mathbf{T}_i dA_i$, $i = 1, 2, 3$, since they have unit normals in the
negative $\mathbf{e}_i$-directions. The areas of the planes are related by [3]

$$dA_i = dA_n \cos(\mathbf{n}, \mathbf{e}_i) = dA_n \mathbf{n} \cdot \mathbf{e}_i \tag{2.4}$$

so that

$$dA_n = \frac{dA_i}{\mathbf{n} \cdot \mathbf{e}_i} = \frac{dA_i}{n_i}, \tag{2.5}$$

where

$$n_i = \mathbf{n} \cdot \mathbf{e}_i = \cos(\mathbf{n}, \mathbf{e}_i) \tag{2.6}$$

is the component of $\mathbf{n}$ in the $\mathbf{e}_i$-direction and also a direction cosine.

Force equilibrium for the tetrahedron gives

$$\mathbf{T}_n dA_n - \mathbf{T}_1 dA_1 - \mathbf{T}_2 dA_2 - \mathbf{T}_3 dA_3 + \mathbf{f}\left(\frac{1}{3} h dA_n\right) = 0, \tag{2.7}$$

where h is the height of the tetrahedron. Using (2.5), (2.7) becomes

$$\left(\mathbf{T}_n - \mathbf{T}_i n_i + \mathbf{f}\frac{h}{3}\right) dA_n = 0. \tag{2.8}$$

Resolving $\mathbf{T}_n$ into Cartesian components $T_i \mathbf{e}_i$ and taking the limit as $h \to 0$, equilibrium is satisfied if

$$T_i \mathbf{e}_i = \mathbf{T}_i n_i. \tag{2.9}$$

The next step is to write $\mathbf{T}_i$ in terms of the stress components using (2.2). However, it is convenient first to change the dummy index on the r.h.s. of (2.9) from i to j; thus from (2.2)

$$\mathbf{T}_i n_i = \mathbf{T}_j n_j = \sigma_{ji} n_j \mathbf{e}_i, \tag{2.10}$$

which permits coefficients of $\mathbf{e}_i$ in (2.9) and (2.10) to be equated yielding

$$T_i = \sigma_{ji} n_j. \tag{2.11}$$

Conversely, if the components T_i are known, the magnitude of $\mathbf{T}_n$ may be evaluated as

$$T_n = |\mathbf{T}_n| = (T_i T_i)^{1/2}. \tag{2.12}$$

Since T_n represents a component of traction acting on any arbitrary plane as defined by $\mathbf{n}$, knowledge of the stress components referred to the Cartesian coordinates is indeed sufficient to completely specify the state of stress at the point. In (2.11) T_i and n_j are both components of vectors (order 1 tensors) so that the σ_{ji} are components of an order 2 tensor $\boldsymbol{\sigma}$. Therefore, if the stress components are known in one coordinate system, say the x_i system, they may be evaluated for another coordinate system, say the x_i' system, by the transformation law for second-order tensors

$$\sigma_{ij}' = \alpha_{ik} \alpha_{jl} \sigma_{kl}, \tag{2.13}$$

where each direction cosine

$$\alpha_{ij} = \cos\left(x_i', x_j\right), \tag{2.14}$$

as introduced in Sect. 1.5, represents the cosine of the angle between the x_i'- and x_j-axes.

Transformation rules play an important role in the theory of elasticity, and it is worth restating that $\alpha_{ij} \neq \alpha_{ji}$, that is, the direction cosines are *not* symmetric.

2.4.1   Stress on a Normal Plane

It is sometimes useful to resolve $\mathbf{T}_n$ into components that are normal and tangential to the differential surface element dA_n, as shown in Fig. 2.4. Double subscripts on n are not summed here. The normal component σ_{nn} is calculated by

Fig. 2.4 Differential
surface element

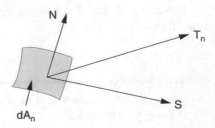

$$\sigma_{nn} = |\mathbf{N}| = \mathbf{T}_n \cdot \mathbf{n}$$
$$= T_i \mathbf{e}_i \cdot \mathbf{n} \tag{2.15}$$
$$= T_i n_i$$

or from (2.11),

$$\sigma_{nn} = \sigma_{ji} n_j n_i. \tag{2.16}$$

The tangential component σ_{ns} is

$$\sigma_{ns} = |\mathbf{S}| = \mathbf{T}_n \cdot \mathbf{s}$$
$$= T_i \mathbf{e}_i \cdot \mathbf{s}$$
$$= T_i s_i \tag{2.17}$$
$$= \sigma_{ji} n_j s_i,$$

where

$$s_i = \mathbf{e}_i \cdot \mathbf{s} \tag{2.18}$$

similar to (2.6). The unit vector $\mathbf{s}$ is coplanar with $\mathbf{n}$.

It is often expedient to calculate σ_{ns} using the Pythagorean theorem as

$$\sigma_{ns} = \left(T_i T_i - \sigma_{nn}^2\right)^{1/2}. \tag{2.19}$$

Carrying the resolution one step further, the Cartesian components of $\mathbf{N}$ and $\mathbf{S}$ may be evaluated using (k) to designate the axes:

$$\sigma_{nn(k)} = \mathbf{N} \cdot \mathbf{e}_k = \sigma_{nn} \mathbf{n} \cdot \mathbf{e}_k$$
$$= \sigma_{nn} n_k \tag{2.20}$$
$$= \sigma_{ji} n_j n_i n_k, \quad \text{where } k = 1, 2, 3$$

from (2.16). For σ_{ns} simple subtraction gives

$$\sigma_{ns(k)} = T_k - \sigma_{nn(k)} \quad k = 1, 2, 3, \tag{2.21a}$$

where T_k are the Cartesian components of $\mathbf{T}$, as given by (2.11). The expression may be written explicitly as

$$
\begin{aligned}
\sigma_{\mathrm{ns}(k)} &= \sigma_{jk} n_j - \sigma_{ji} n_j n_i n_k \\
&= (\delta_{ki} - n_i n_k) \sigma_{ji} n_j .
\end{aligned}
\tag{2.21b}
$$

An application of this resolution is presented in Sect. 12.2.3.

2.4.2   Dyadic Representation of Stress

Conceptually, it may be helpful to view the stress tensor as a vector-like quantity having a magnitude and associated direction(s) specified by unit vectors. The *dyadic*, attributed to the mathematician J. Willard Gibbs [4], is such a representation. We write the stress tensor or stress dyadic as

$$
\begin{aligned}
\boldsymbol{\sigma} &= \sigma_{ij} \mathbf{e}_i \mathbf{e}_j \\
&= \sigma_{11}\mathbf{e}_1\mathbf{e}_1 + \sigma_{12}\mathbf{e}_1\mathbf{e}_2 + \sigma_{13}\mathbf{e}_1\mathbf{e}_3 + \sigma_{21}\mathbf{e}_2\mathbf{e}_1 \\
&\quad + \sigma_{22}\mathbf{e}_2\mathbf{e}_2 + \sigma_{23}\mathbf{e}_2\mathbf{e}_3 + \sigma_{31}\mathbf{e}_3\mathbf{e}_1 + \sigma_{32}\mathbf{e}_3\mathbf{e}_2 + \sigma_{33}\mathbf{e}_3\mathbf{e}_3 ,
\end{aligned}
\tag{2.22}
$$

where the juxtaposed double vectors are termed *dyads*. The corresponding tractions are evaluated by an operation analogous to the scalar or dot product operation in vector arithmetic:

$$
\mathbf{T}_i = \boldsymbol{\sigma} \cdot \mathbf{e}_i = \sigma_{ij}\mathbf{e}_j .
\tag{2.23}
$$

The dot ($\cdot$) operation of $\mathbf{e}_i$ on $\boldsymbol{\sigma}$ selects the components with the second vector of the dyad equal to $\mathbf{e}_i$ since $\mathbf{e}_i \cdot \mathbf{e}_j = \delta_{ij}$. Equation (2.23) is identical to (2.2).

Similarly, the normal and tangential components of the traction $\mathbf{T}_n$ on a plane defined by normal $\mathbf{n}$ are

$$
\begin{aligned}
\sigma_{\mathrm{nn}} &= \boldsymbol{\sigma} \cdot \mathbf{n} \cdot \mathbf{n} \\
&= \mathbf{T}_n \cdot \mathbf{n} \\
&= \sigma_{ij} n_i n_j
\end{aligned}
\tag{2.24}
$$

and

$$
\begin{aligned}
\sigma_{\mathrm{ns}} &= \boldsymbol{\sigma} \cdot \mathbf{n} \cdot \mathbf{s} \\
&= \mathbf{T}_n \cdot \mathbf{s} \\
&= \sigma_{ij} n_i s_j
\end{aligned}
\tag{2.25}
$$

as previously found in (2.16) and (2.17), respectively.

2.4.3   Computational Example

The problem statement follows:

1. The components of stress at a point in Cartesian coordinates are given by
   $\sigma_{xx} = 500$; $\sigma_{xy} = \sigma_{yx} = 500$; $\sigma_{yy} = 1000$; $\sigma_{yz} = \sigma_{zy} = -750$; $\sigma_{zx} = \sigma_{xz} = 800$;
   and $\sigma_{zz} = -300$.
2. A plane is defined by the unit vector

$$\mathbf{n} = \frac{1}{2}\mathbf{e}_x + \frac{1}{2}\mathbf{e}_y + \frac{1}{\sqrt{2}}\mathbf{e}_z.$$

3. It is desired to compute the traction and the normal and tangential components on the plane.
4. Note: at this point in the development, the given state of stress cannot be verified as correct or admissible. The values are chosen in order to illustrate a computational format for the equations developed thus far.
5. Let $x, y, z = x_1, x_2, x_3$.

The solution in component form is as follows:

1. Evaluate n_i by using (2.6):

$$n_i = \mathbf{n} \cdot \mathbf{e}_i,$$
$$n_1 = 1/2; \quad n_2 = 1/2; \quad n_3 = 1/\sqrt{2}.$$

2. Evaluate T_i by using (2.11):

$$T_i = \sigma_{ji}n_j;$$
$$T_1 = \sigma_{11}n_1 + \sigma_{21}n_2 + \sigma_{31}n_3$$
$$= (500)(1/2) + (500)(1/2) + (800)(1/\sqrt{2})$$
$$= 1,066;$$
$$T_2 = \sigma_{12}n_1 + \sigma_{22}n_2 + \sigma_{32}n_3$$
$$= (500)(1/2) + (1,000)(1/2) + (-750)(1/\sqrt{2})$$
$$= 220;$$
$$T_3 = \sigma_{13}n_1 + \sigma_{23}n_2 + \sigma_{33}n_3$$
$$= (800)(1/2) + (-750)(1/2) + (-300)(1/\sqrt{2})$$
$$= -187.$$

3. Evaluate $|\mathbf{T}_n|$ by using (2.12):

$$|\mathbf{T}_n| = (T_i T_i)^{1/2}$$
$$= \left[(1,066)^2 + (220)^2 + (-187)^2 \right]^{1/2}$$
$$= 1,104.$$

4. Evaluate σ_{nn} and σ_{ns} by using (2.15) or (2.16) and (2.19):

$$\sigma_{nn} = T_i n_i$$
$$= (1,066)(1/2) + (220)(1/2) + (-187)(1/\sqrt{2}) = 511,$$

or using (2.16)

$$\sigma_{nn} = \sigma_{ji} n_j n_i,$$

where

j	i	$(+)$	$(-)$
1	1	$(500)(1/2)(1/2)$	
1	2	$(500)(1/2)(1/2)$	
1	3	$(800)(1/2)\ (1/\sqrt{2})$	
2	1	$(500)(1/2)(1/2)$	
2	2	$(1000)(1/2)(1/2)$	
2	3		$(750)(1/2)\ (1/\sqrt{2})$
3	1	$(800)(1/2)\ (1/\sqrt{2})$	
3	2		$(750)\ (1/\sqrt{2})\ (1/2)$
3	3		$(300)\ (1/\sqrt{2})\ (1/\sqrt{2})$
		1191	680

$$\sigma_{nn} = 1,191 - 680 = 511.$$

From (2.19),

$$\sigma_{ns} = \left(T_i T_i - \sigma_{nn}^2 \right)^{1/2};$$
$$\sigma_{ns} = \left[(1,104)^2 - (511)^2 \right]^{1/2} = 979.$$

Finally the Cartesian components are computed from (2.20) as

$$\sigma_{nn(1)} = \sigma_{nn} n_1 = 511(1/2) = 256$$
$$\sigma_{nn(2)} = \sigma_{nn} n_2 = 511(1/2) = 256$$
$$\sigma_{nn(3)} = \sigma_{nn} n_3 = 511(1/\sqrt{2}) = 361$$

and (2.21a) as

$$\sigma_{ns(1)} = T_1 - \sigma_{nn(1)} = 1,066 - 256 = 810$$
$$\sigma_{ns(2)} = T_2 - \sigma_{nn(2)} = 220 - 256 = -36$$
$$\sigma_{ns(3)} = T_3 - \sigma_{nn(3)} = -187 - 361 = -548.$$

To check,

$$\sigma_{ns} = \left[(810)^2 + (-36)^2 + (-548)^2\right]^{1/2} = 979.$$

The solution in dyadic form is as follows:

1. The stress dyadic $\boldsymbol{\sigma}$ from (2.22)

$$\boldsymbol{\sigma} = \sigma_{ij}\mathbf{e}_i\mathbf{e}_j$$
$$= 500\mathbf{e}_1\mathbf{e}_1 + 500\mathbf{e}_1\mathbf{e}_2 + 800\mathbf{e}_1\mathbf{e}_3 + 500\mathbf{e}_2\mathbf{e}_1 + 1,000\mathbf{e}_2\mathbf{e}_2$$
$$- 750\mathbf{e}_2\mathbf{e}_3 + 800\mathbf{e}_3\mathbf{e}_1 - 750\mathbf{e}_3\mathbf{e}_2 - 300\mathbf{e}_3\mathbf{e}_3.$$

2. Evaluate $\mathbf{T}_n$ from (2.23):

$$\mathbf{T}_n = \boldsymbol{\sigma} \cdot \mathbf{n}$$
$$= \left[500(1/2) + 500(1/2) + 800(1/\sqrt{2})\right]\mathbf{e}_1$$
$$+ \left[500(1/2) + 1,000(1/2) - 750(1/\sqrt{2})\right]\mathbf{e}_2$$
$$+ \left[800(1/2) - 750(1/2) - 300(1/\sqrt{2})\right]\mathbf{e}_3$$
$$= 1,066\mathbf{e}_1 + 220\mathbf{e}_2 - 187\mathbf{e}_3;$$
$$|\mathbf{T}_n| = \left[(1,066)^2 + (220)^2 + (-187)^2\right]^{1/2} = 1,104.$$

3. Evaluate σ_{nn} and σ_{ns} from (2.24):

$$\sigma_{nn} = \mathbf{T}_n \cdot \mathbf{n}$$
$$= (1,066)(1/2) + 220(1/2) - 187(1/\sqrt{2}) = 511.$$

From (2.19),

$$\sigma_{ns} = \left(T_i T_i - \sigma_{nn}^2\right)^{1/2}$$
$$= \left[(1,104)^2 - (511)^2\right] = 979.$$

2.5 Equilibrium

2.5.1 Physical and Mathematical Principles

The state of stress at a point in any direction has been shown to be completely determined by the components of the Cartesian stress tensor σ_{ij}. Naturally, the stresses vary within the body. The equations governing the distribution of stresses are known as the equations of equilibrium and are derived from the application of the fundamental physical principles of linear and angular momentum to the region shown in Fig. 2.5, with surface area A and volume V.

The principle of linear momentum is

$$\int_V \mathbf{f} \ dV + \int_A \mathbf{T} \ dA = \int_V \rho \ddot{\mathbf{u}} \ dV \tag{2.26}$$

in which ρ is the mass density; $\mathbf{u}$ is the displacement vector; and the symbol ($\ddot{\ }$) signifies differentiation twice with respect to time.

The principle of angular momentum is

$$\int_V (\mathbf{r} \times \mathbf{f}) \ dV + \int_A (\mathbf{r} \times \mathbf{T}) \ dA = \int_V (\mathbf{r} \times \rho \ddot{\mathbf{u}}) \ dV \tag{2.27}$$

in which $\mathbf{r}$ is the position vector as shown in Fig. 2.5.

Also useful in the derivation is the divergence theorem stated as (1.14)

$$\int_V \nabla \cdot \mathbf{G} \ dV = \int_A \mathbf{n} \cdot \mathbf{G} \ dA. \tag{2.28}$$

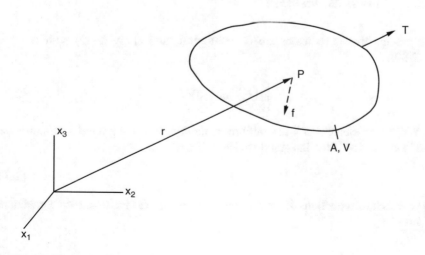

Fig. 2.5 Body in equilibrium in Cartesian space

The preceding equations may be written in component form. The body forces are

$$\mathbf{f} = f_i \mathbf{e}_i \qquad (2.29a)$$

and the tractions are

$$\begin{aligned} \mathbf{T} &= T_i \mathbf{e}_i \\ &= \sigma_{ji} n_j \mathbf{e}_i \end{aligned} \qquad (2.29b)$$

from (2.11). Considering the position vector $\mathbf{r}$ to point $P(x_i)$ in Fig. 2.5,

$$\begin{aligned} \mathbf{r} &= x_j \mathbf{e}_j; \\ \mathbf{r} \times \mathbf{f} &= \varepsilon_{ijk} x_j f_k \mathbf{e}_i; \\ \mathbf{r} \times \mathbf{T} &= \varepsilon_{ijk} x_j T_k \mathbf{e}_i \\ &= \varepsilon_{ijk} x_j \sigma_{lk} n_l \mathbf{e}_i. \end{aligned} \qquad (2.29c)$$

For static problems, the r.h.s. of (2.26) and (2.27) are zero. Substituting (2.29c) into (2.26), (2.27), and (2.28), we have the static equations of linear and angular momentum and the divergence theorem in component form as, respectively,

$$\int_V f_i dV + \int_A \sigma_{ji} n_j dA = 0; \qquad (2.30)$$

$$\int_V \varepsilon_{ijk} x_j f_k \, dV + \int_A \varepsilon_{ijk} x_j \sigma_{lk} n_l \, dA = 0; \qquad (2.31)$$

$$\int_V G_{i,i} \, dV = \int_A n_i G_i \, dA. \qquad (2.32)$$

2.5.2   Linear Momentum

We assume that the stresses are C^1 continuous and apply (2.32) with $G_i = \sigma_{ji}$ to (2.30):

$$\int_V \left(f_i + \sigma_{ji,j} \right) dV = 0. \qquad (2.33)$$

With every element of V in equilibrium, the region of integration is arbitrary, and (2.33) is satisfied if the integrand vanishes. Therefore,

$$\sigma_{ji,j} + f_i = 0 \qquad (2.34)$$

represents three equations of equilibrium in terms of the nine unknown components of stress σ_{ij}.

2.5.3 Angular Momentum

Taking $\varepsilon_{ijk}x_j\sigma_{lk}$ for G_i and n_l for n_i, (2.32) is applied to (2.31) and gives

$$\int_V \left[\varepsilon_{ijk}x_jf_k + \left(\varepsilon_{ijk}x_j\sigma_{lk}\right)_{,l}\right] dV = 0. \tag{2.35}$$

The second term of the integrand is expanded by product differentiation as

$$\left(\varepsilon_{ijk}x_j\sigma_{lk}\right)_{,l} = \varepsilon_{ijk}\sigma_{lk}x_{j,l} + \varepsilon_{ijk}x_j\sigma_{lk,l}. \tag{2.36}$$

Further, it is recognized that x_j, the component of $\mathbf{r}$ in the x_j-direction, changes only in that direction; that is,

$$x_{j,l} = \delta_{jl}. \tag{2.37}$$

Thus (2.35) becomes

$$\int_V \varepsilon_{ijk}\left(x_jf_k + \delta_{jl}\sigma_{lk} + x_j\sigma_{lk,l}\right) dV = 0. \tag{2.38}$$

The first and third terms of the integrand combine into

$$\varepsilon_{ijk}x_j(f_k + \sigma_{lk,l}) = 0 \tag{2.39}$$

from (2.34), and the remaining term becomes

$$\varepsilon_{ijk}\delta_{jl}\sigma_{lk} = \varepsilon_{ijk}\sigma_{jk} \tag{2.40}$$

so that the equation reduces to

$$\int_V \varepsilon_{ijk}\sigma_{jk} dV = 0, \tag{2.41}$$

which is satisfied if

$$\varepsilon_{ijk}\sigma_{jk} = 0. \tag{2.42}$$

Equation (2.42) may be evaluated for $i = 1, 2, 3$. For $i = 1$, $\sigma_{23} - \sigma_{32} = 0$; for $i = 2$, $\sigma_{31} - \sigma_{13} = 0$; for $i = 3$, $\sigma_{12} - \sigma_{21} = 0$; or in general,

$$\sigma_{ij} = \sigma_{ji} \tag{2.43}$$

which is a statement of the *symmetry* of the stress tensor and which furthermore implies that σ_{ij} has but six independent components instead of nine components. Equation (2.43) is very important in the entire field of solid mechanics.

We may now rewrite (2.11) as

$$T_i = \sigma_{ij} n_j \tag{2.44}$$

and (2.34) as

$$\sigma_{ij,j} + f_i = 0, \tag{2.45}$$

which is now a set of three equations in six unknowns. Since they are used repeatedly, it is useful to write the latter equations in explicit form:

$$\sigma_{11,1} + \sigma_{12,2} + \sigma_{13,3} + f_1 = 0 \tag{2.46a}$$
$$\sigma_{21,1} + \sigma_{22,2} + \sigma_{23,3} + f_2 = 0 \tag{2.46b}$$
$$\sigma_{31,1} + \sigma_{32,2} + \sigma_{33,3} + f_3 = 0, \tag{2.46c}$$

which represents a system which is still statically indeterminate.

2.5.4   Computational Example

Although it is not possible to solve the equilibrium equations alone since there are more unknowns than equations, it is possible to test a given set of stresses against the equilibrium equations, (2.46a), (2.46b), and (2.46c).

We take the components of stress for an elastic body to be given by

$$\sigma_{11} = x^2 + y + 3z^2; \quad \sigma_{22} = 2x + y^2 + 2z; \quad \sigma_{33} = -2x + y + z^2$$
$$\sigma_{12} = \sigma_{21} = -xy + z^3; \quad \sigma_{13} = \sigma_{31} = y^2 - xz; \quad \sigma_{23} = \sigma_{32} = x^2 - yz$$
$$f_1 = f_2 = f_3 = 0$$

and show that this state of stress satisfies equilibrium when the values are substituted into (2.46a), (2.46b), and (2.46c), respectively:

$$2x - x - x = 0$$
$$-y + 2y - y = 0$$
$$-z - z + 2z = 0.$$

2.6   Principal Stress

2.6.1   Definition and Derivation

We consider Fig. 2.6a, a normal section through the body. The traction $\mathbf{T}_n$ acts on the plane defined by $\mathbf{n}$ that appears edgewise in the figure. $\mathbf{T}_n$ is not necessarily shown true in this view, but the normal component σ_{nn} is in the plane of the figure.

Fig. 2.6 (a) Normal
section; (b) principal plane

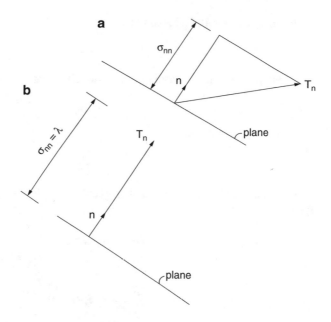

Using the dyadic form,

$$\sigma_{nn} = \mathbf{T}_n \cdot \mathbf{n} = \boldsymbol{\sigma} \cdot \mathbf{n} \cdot \mathbf{n}, \tag{2.47}$$

we seek the orientation of a plane (given by the direction of **n**) such that σ_{nn} is an extremum (maximum or minimum). Such a plane (or planes) is called *principal*.

From differential calculus, the extremum is achieved when $\frac{dn}{dn} = 0$. From (2.47), treating the sought-after direction n as a variable,

$$\frac{d\sigma_{nn}}{dn} = \mathbf{n} \cdot \frac{d(\boldsymbol{\sigma} \cdot \mathbf{n})}{dn} + (\boldsymbol{\sigma} \cdot \mathbf{n}) \cdot \frac{dn}{dn}. \tag{2.48}$$

Since $\boldsymbol{\sigma}$ is not a function of **n**, the first term is written as $\mathbf{n} \cdot \boldsymbol{\sigma} \cdot \frac{dn}{dn}$, and (2.48) becomes

$$\begin{aligned}
\frac{d\sigma_{nn}}{dn} &= 2(\boldsymbol{\sigma} \cdot \mathbf{n}) \cdot \frac{dn}{dn} \\
&= 2\mathbf{T}_n \cdot \frac{dn}{dn} \\
&= 0.
\end{aligned} \tag{2.49}$$

If $\mathbf{T}_n \cdot \frac{dn}{dn} = 0$, $\mathbf{T}_n$ is normal to $\frac{dn}{dn}$. Furthermore $\frac{dn}{dn}$ is itself normal to **n**. Therefore, $\mathbf{T}_n$ must be *parallel* to $\frac{dn}{dn}$ for (2.49) to hold. The condition corresponding to the extremum is shown in Fig. 2.6b. Obviously, the tangential component $\sigma_{ns} = 0$, and the normal component for this case is designated as

$$\sigma_{nn} = |\mathbf{T}_n| = \lambda. \tag{2.50}$$

We may express the preceding by

$$\mathbf{T}_n = \lambda\mathbf{n} \tag{2.51}$$

or in component form by dotting each side with $\mathbf{e}_i$ as

$$T_i = \lambda n_i. \tag{2.52}$$

Also, from (2.11), we have

$$T_i = \sigma_{ji}n_j. \tag{2.53}$$

Equating (2.52) and (2.53) and employing the index-changing property of the Kronecker delta,

$$\lambda n_i = \lambda n_j \delta_{ij} = \sigma_{ji}n_j \tag{2.54}$$

or

$$\left(\sigma_{ji} - \lambda\delta_{ij}\right)n_j = 0. \tag{2.55}$$

Equation (2.55) is a set of three homogeneous algebraic equations in four unknowns, n_j with $j = 1, 2, 3$, and λ. The components n_j determine the orientation of the *principal plane*, and λ is called the *principal stress*. This is known as a linear eigenvalue problem, and using Cramer's rule, it may be shown that a nontrivial solution may be found if the determinant of the coefficients of n_j vanishes, that is,

$$\left|\sigma_{ji} - \lambda\delta_{ij}\right| = 0 \tag{2.56a}$$

or in expanded form

$$\begin{vmatrix} \sigma_{11} - \lambda & \sigma_{12} & \sigma_{13} \\ \sigma_{21} & \sigma_{22} - \lambda & \sigma_{23} \\ \sigma_{31} & \sigma_{32} & \sigma_{33} - \lambda \end{vmatrix} = 0. \tag{2.56b}$$

Equation (2.56b) expands into a cubic equation for λ indicating that three principal stresses, $\lambda_1 = \sigma^{(1)}$, $\lambda_2 = \sigma^{(2)}$, and $\lambda_3 = \sigma^{(3)}$, exist. Corresponding to each principal stress is a distinct orientation, evaluated by substituting, in turn, a known value of $\sigma^{(k)}$, that is, $\sigma^{(1)}$, $\sigma^{(2)}$, and $\sigma^{(3)}$, into (2.55) and solving for the corresponding components of $n_j^{(k)}$, that is, $n_j^{(1)}$, $n_j^{(2)}$, and $n_j^{(3)}$. Since (2.55) originally represented three equations in four unknowns, it is a linearly dependent set that must be supplemented by an additional relationship:

$$n_j^{(k)} n_j^{(k)} = 1 \tag{2.57}$$

expressing the "length" of the unit normal with $k = 1, 2, 3$ in turn. As a matter of mathematical completeness, it is instructive to demonstrate that all three roots of the cubic equation for the principal stresses are real. This has been addressed and confirmed by Chou and Pagano [7].

Since the principal stresses at a point represent the greatest magnitudes of tension and compression that can exist under a particular loading case, they are of great importance in engineering design. In the strength of materials, a graphical representation of principal stresses in two dimensions, Mohr's circle, is widely used for stress analysis [7].

2.6.2 Computational Format, Stress Invariants, and Principal Coordinates

It is often expedient to substitute the numerical values of the stresses, which have been calculated with respect to a selected Cartesian coordinate system, into (2.56a) and (2.56b) and to expand, simplify, and solve the determinant as an algebraic equation.

It may be shown [2] that a general expansion of (2.56a) and (2.56b) is

$$\lambda^3 - Q_1\lambda^2 + Q_2\lambda - Q_3 = 0, \tag{2.58}$$

in which

$$Q_1 = \sigma_{11} + \sigma_{22} + \sigma_{33}$$
$$= \sigma_{ii}, \quad \text{the sum of the main diagonal of } [\sigma_{ij}]; \tag{2.59a}$$

$$Q_2 = \begin{vmatrix} \sigma_{11} & \sigma_{21} \\ \sigma_{12} & \sigma_{22} \end{vmatrix} + \begin{vmatrix} \sigma_{11} & \sigma_{31} \\ \sigma_{13} & \sigma_{33} \end{vmatrix} + \begin{vmatrix} \sigma_{22} & \sigma_{32} \\ \sigma_{23} & \sigma_{33} \end{vmatrix}$$
$$= \frac{1}{2}\varepsilon_{mik}\varepsilon_{mjl}\sigma_{ij}\sigma_{kl}, \quad \text{the sum of the minors of the main diagonal of } [\sigma_{ij}];$$
$$\tag{2.59b}$$

$$Q_3 = \begin{vmatrix} \sigma_{11} & \sigma_{12} & \sigma_{13} \\ \sigma_{21} & \sigma_{22} & \sigma_{23} \\ \sigma_{31} & \sigma_{32} & \sigma_{33} \end{vmatrix}$$
$$\tag{2.59c}$$
$$= \frac{1}{6}\varepsilon_{mik}\varepsilon_{nil}\sigma_{ij}\sigma_{kl}\sigma_{mn}, \quad \text{the determinant of } [\sigma_{ij}].$$

It is instructive to examine the coefficients of (2.58) in terms of scalar quantities that can be constructed out of the tensor σ_{ij}, that is,

$$P_1 = \sigma_{ii};$$
$$P_2 = \sigma_{ij}\sigma_{ij}; \tag{2.60}$$
$$P_3 = \sigma_{ij}\sigma_{jk}\sigma_{ki}.$$

Such scalar quantities (no free indices) constructed from a tensor are obviously independent of any particular coordinate system and are therefore known as *invariants*. For a principal coordinate system that coincides with the directions of the principal stresses and is subsequently shown to be orthogonal, all the σ_{ij}, with $i \neq j$, terms vanish so that $P_1, P_2,$ and P_3 may be written in terms of the principal stresses as

$$
\begin{aligned}
P_1 &= \sigma^{(1)} + \sigma^{(2)} + \sigma^{(3)}; \\
P_2 &= \sigma^{(1)^2} + \sigma^{(2)^2} + \sigma^{(3)^2}; \\
P_3 &= \sigma^{(1)^3} + \sigma^{(2)^3} + \sigma^{(3)^3}.
\end{aligned}
\tag{2.61}
$$

Finally, it may be shown that [2]

$$
\begin{aligned}
Q_1 &= P_1; \\
Q_2 &= \frac{1}{2}(P_1^2 - P_2); \\
Q_3 &= \frac{1}{6}(P_1^3 - 3P_1P_2 + 2P_3)
\end{aligned}
\tag{2.62}
$$

so that $Q_1, Q_2,$ and Q_3 are likewise *invariants*. Invariants are quite important in many applications of the theory of elasticity. For example, see Herrmann et al. [5]. In the literature, they are commonly denoted as

$$
\begin{aligned}
J_1 &= P_1 = \sigma_{ii} \\
J_2 &= \frac{1}{2}P_2 = \frac{1}{2}\sigma_{ij}\sigma_{ij} \\
J_3 &= \frac{1}{3}P_3 = \frac{1}{3}\sigma_{ij}\sigma_{jk}\sigma_{ki}.
\end{aligned}
\tag{2.63}
$$

We next take up the problem of evaluating the components of the normals to the principal planes $n_j^{(k)}$. This is carried out by back-substitutions of each computed $\sigma^{(k)}$ into two permutations of (2.55), along with the use of (2.57). Care should be taken to insure that the two equations extracted from (2.55) are linearly independent. As an illustration, we take $i = 1, 2$ in (2.55) giving

$$
i = 1: \quad \left[\sigma_{11} - \sigma^{(k)}\right]n_1^{(k)} + \sigma_{21}n_2^{(k)} = -\sigma_{31}n_3^{(k)};
\tag{2.64a}
$$

$$
i = 2: \quad \sigma_{12}n_1^{(k)} + \left[\sigma_{22} - \sigma^{(k)}\right]n_2^{(k)} = -\sigma_{32}n_3^{(k)};
\tag{2.64b}
$$

$$
n_1^{(k)^2} + n_2^{(k)^2} + n_3^{(k)^2} = 1.
\tag{2.64c}
$$

Considering (a) and (b), we solve for $n_1^{(k)}$ and $n_2^{(k)}$ in terms of $n_3^{(k)}$ using Cramer's rule:

$$
n_1^{(k)} = n_3^{(k)}\frac{D_1}{D},
\tag{2.65a}
$$

$$
n_2^{(k)} = n_3^{(k)}\frac{D_2}{D},
\tag{2.65b}
$$

where

$$D_1 = \begin{vmatrix} -\sigma_{31} & \sigma_{21} \\ -\sigma_{32} & \sigma_{22} - \sigma^{(k)} \end{vmatrix}, \tag{2.66a}$$

$$D_2 = \begin{vmatrix} \sigma_{11} - \sigma^{(k)} & -\sigma_{31} \\ \sigma_{12} & -\sigma_{32} \end{vmatrix}, \tag{2.66b}$$

$$D = \begin{vmatrix} \sigma_{11} - \sigma^{(k)} & \sigma_{21} \\ \sigma_{12} & \sigma_{22} - \sigma^{(k)} \end{vmatrix}. \tag{2.66c}$$

Then, substituting into (2.64c) gives

$$n_3^{(k)^2} \left[\left(\frac{D_1}{D} \right)^2 + \left(\frac{D_2}{D} \right)^2 + 1 \right] = 1. \tag{2.67}$$

With $\sigma^{(k)}$ real, $n_3^{(k)}$ will be real, and $n_1^{(k)}$ and $n_2^{(k)}$ are found from (2.65a) and (2.65b). Finally,

$$\mathbf{n}^{(k)} = n_1^{(k)} \mathbf{e}_1 + n_2^{(k)} \mathbf{e}_2 + n_3^{(k)} \mathbf{e}_3. \tag{2.68}$$

The procedure is carried out for each $\sigma^{(k)}$, $k = 1, 2, 3$, in turn.

It now remains to show that the calculated $\mathbf{n}^{(k)}$ corresponding to the principal directions are orthogonal. We take $k = 1$ and $k = 2$, respectively, and form

$$\sigma_{ij} n_j^{(1)} = \sigma^{(1)} n_i^{(1)} \tag{2.69a}$$

$$\sigma_{ij} n_j^{(2)} = \sigma^{(2)} n_i^{(2)}. \tag{2.69b}$$

We multiply (2.69a) by $n_i^{(2)}$ and (2.69b) by $n_i^{(1)}$ giving

$$\sigma_{ij} n_j^{(1)} n_i^{(2)} = \sigma^{(1)} n_i^{(1)} n_i^{(2)} \tag{2.70a}$$

$$\sigma_{ij} n_j^{(2)} n_i^{(1)} = \sigma^{(2)} n_i^{(2)} n_i^{(1)}. \tag{2.70b}$$

Since i and j are repeated indices, the l.h.s. of (2.70a) and (2.70b) are equal; hence, equating the r.h.s.,

$$\left(\sigma^{(1)} - \sigma^{(2)} \right) n_i^{(1)} n_i^{(2)} = 0, \tag{2.71}$$

and if $\sigma^{(1)}$ and $\sigma^{(2)}$ are different, $n_i^{(1)}$ and $n_i^{(2)}$ must be orthogonal.

Once the principal stresses are evaluated and the principal planes are located, it is generally convenient to refer further computations to the *principal coordinates* that have now been shown to be orthogonal. Referring to Fig. 2.7 and designating the principal coordinates by x_i^*, it is obvious that the state of stress is relatively simple

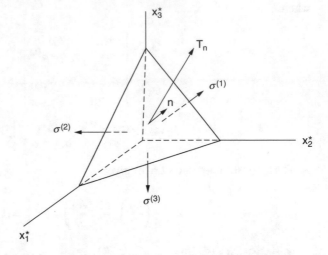

Fig. 2.7 Principal coordinates and stresses (After Tauchert [1]). Reproduced with permission

with only normal stresses and no shearing stresses acting. The components of the traction $\mathbf{T}_n$ in the principal directions are found from (2.11) as

$$T_1^* = \sigma^{(1)} n_1^* \tag{2.72a}$$

$$T_2^* = \sigma^{(2)} n_2^* \tag{2.72b}$$

$$T_3^* = \sigma^{(3)} n_3^*, \tag{2.72c}$$

where

$$n_i^* = \mathbf{n} \cdot \mathbf{e}_i^*. \tag{2.72d}$$

A visual representation of the traction vector $\mathbf{T}_n$ and components T_i^* is shown in Sect. 2.5, and computational illustrations using principal coordinate solutions are given in Sect. 2.6.

2.6.3 Computational Example

As a continuation of the example in Sect. 2.4.3, we now seek the principal stresses and directions for the state of stress previously enumerated. Equation (2.56b) becomes

$$\begin{vmatrix} 500 - \lambda & 500 & 800 \\ 500 & 1,000 - \lambda & -750 \\ 800 & -750 & -300 - \lambda \end{vmatrix} = 0. \tag{2.73}$$

It is convenient to work with the stress $\times\ 10^{-3}$. Expanding the determinant directly gives

$$\bar{\lambda}^3 - 1.2\bar{\lambda}^2 - 1.4025\bar{\lambda} + 1.59625 = 0, \qquad (2.74)$$

where $\bar{\lambda} = \lambda \times 10^{-3}$.

We may also compute the coefficients by using the invariants as described in (2.60) and (2.62) and compare the values to those in (2.74):

$$P_1 = \sigma_{ii} = 0.500 + 1.000 - 0.300 = 1.2$$
$$-Q_1 = -P_1 = 1.2 \ \surd$$
$$P_2 = \sigma_{ij}\sigma_{ij}$$
$$= (0.500)^2 + 2(0.500)^2 + 2(0.800)^2 + (1.00)^2 + 2(-0.750)^2 + (0.300)^2$$
$$= 4.245$$
$$Q_2 = \frac{1}{2}\Big[(1.2)^2 - 4.245\Big] = -1.4025 \ \surd$$

or, using the sum of the minors of the main diagonal of σ_{ij},

$$Q_2 = [(1.00)(-0.300) - (-0.750)(-0.750)]$$
$$+ [(0.500)(-0.300) - (0.800)(0.800)]$$
$$+ [(0.500)(1.000) - (0.500)(0.500)]$$
$$= -1.4025 \ \surd$$
$$Q_3 = \det[\sigma_{ij}]$$
$$= (0.500)(1.00)(-0.300) + 2(0.800)(0.500)(-0.750)$$
$$- (0.800)(0.800)(1.000) - (-0.750)(-0.750)(0.500)$$
$$- (0.500)(0.500)(-0.300)$$
$$= -1.59625$$
$$-Q_3 = +1.59625 \ \surd.$$

Solving a cubic equation may be tedious, but straightforward procedures are found in standard mathematics texts and on programmable calculators. The solution to (2.74) may be verified as

$$\bar{\lambda}_1 = -1.168; \quad \bar{\lambda}_2 = 1.380; \quad \bar{\lambda}_3 = 0.988$$

so that the principal stresses are

$$\sigma^{(1)} = -1,168$$
$$\sigma^{(2)} = 1,380 \qquad (2.75)$$
$$\sigma^{(3)} = 988 \quad .$$

From (2.64a), (2.64b), (2.64c) and (2.68), the corresponding unit normal vectors are

$$\mathbf{n}^{(1)} = -0.4903\mathbf{e}_1 + 0.3838\mathbf{e}_2 + 0.7825\mathbf{e}_3 \qquad (2.76\text{a})$$

$$\mathbf{n}^{(2)} = -0.2514\mathbf{e}_1 - 0.9207\mathbf{e}_2 + 0.2988\mathbf{e}_3 \qquad (2.76\text{b})$$

$$\mathbf{n}^{(3)} = 0.8298\mathbf{e}_1 - 0.0749\mathbf{e}_2 + 0.5530\mathbf{e}_3 \qquad (2.76\text{c})$$

or, in general,

$$\mathbf{n}^{(j)} = n_{ji}\mathbf{e}_i. \qquad (2.76\text{d})$$

The unit normals $\mathbf{n}^{(j)}$ now become the basis vectors $\mathbf{e}_j^*$ for the principal coordinate system x_j^*, and the components n_{ji} are the direction cosines for the transformation between the x_j^*- and x_i-axes.

2.6.4   Stress Ellipsoid

Gabriel Lamé is credited by several authors for creating the visual representation of the components of the traction $\mathbf{T_n}$ shown in Fig. 2.7 as the normal $\mathbf{n}$ defining the plane of interest changes orientation.

If we take the direction of a particular $\mathbf{n}$ referred to the principal coordinates as n_i^* as defined in (2.72d), we evaluate (2.57) as

$$\left(n_1^*\right)^2 + \left(n_2^*\right)^2 + \left(n_3^*\right)^2 = 1. \qquad (2.77)$$

Then we substitute n_1^*, n_2^*, and n_3^* from (2.72a), (2.72b), and (2.72c) into (2.77) to get

$$\left(\frac{T_1^*}{\sigma^{(1)}}\right)^2 + \left(\frac{T_2^*}{\sigma^{(2)}}\right)^2 + \left(\frac{T_3^*}{\sigma^{(3)}}\right)^2 = 1 \qquad (2.78)$$

which is the equation of an ellipsoid surface, the so-called stress ellipsoid.

Referring to Fig. 2.8, adapted from Chou and Pagano [7], the traction vector $\mathbf{T_n}$ pierces the surface at point p. Thus $\mathbf{T_n}$, with components T_i, defines a surface where the vector extends from the origin of the coordinate system to the ellipsoid surface and the semiaxes represent the principal stresses. The longest semiaxis gives the maximum principal stress and the shortest the minimum. Each radius vector of the stress ellipsoid represents to a certain scale the stress state on a corresponding plane through the center of the ellipsoid. The plane corresponding to the traction vector piercing the surface at particular point $p(x_i^*)$ can be found by the *stress director surface* [8] but this calculation is not pursued here.

The general concept of the stress ellipsoid may be progressively simplified as follows [7]:

1. If two of the principal stresses are numerically equal, the stress ellipsoid becomes an ellipsoid of revolution. If they are further of the same sign, the resultant

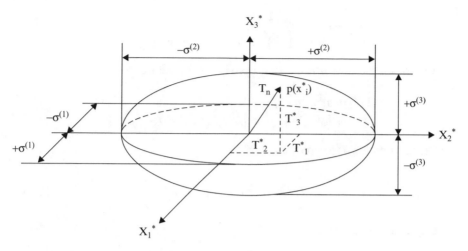

Fig. 2.8 Stress ellipsoid (After Chou and Pagano [6])

stresses on all of the planes passing through the axis of symmetry of the ellipsoid will be equal and perpendicular to the planes on which they act. The stresses on any two perpendicular planes through the axis of symmetry are principal stresses.

2. If all three principal stresses are equal and of the same sign, the stress ellipsoid contracts to a sphere, and any three perpendicular directions may be taken as principal. This is the case of hydrostatic stress to be discussed in Sect. 2.8.5. When one of the principal stresses is zero, the stress ellipsoid reduces to an ellipse, and the vectors representing the stresses on all of the planes through the point lie in the same plane. This is the state of plane stress to be discussed in Sect. 2.8.2.

3. Finally if two principal stresses are zero, we have simple uniaxial tension or compression.

An example plot using the very similar strain ellipse is given in Sect. 3.9, Fig. 3.3b.

2.7 Stresses in Principal Coordinates

2.7.1 Stresses on an Oblique Plane

It was noted in Sect. 2.4.2 that the principal coordinates, once established, provide a convenient basis for further analysis. We first consider a plane oblique to the x_i^*-axes as shown in Fig. 2.7, with direction cosines n_i^*. From (2.16), with $\sigma_{(ii)} = \sigma^{(i)}$ and $\sigma_{ij}(i \neq j) = 0$, we may write the normal component of the traction as

$$\sigma_{nn} = \sigma^{(i)} n_i^* n_i^*$$
$$= \sigma^{(1)} \left(n_1^* \right)^2 + \sigma^{(2)} \left(n_2^* \right)^2 + \sigma^{(3)} \left(n_3^* \right)^2. \tag{2.79}$$

The shearing component is found from (2.19) by taking

$$T_i T_i = \sigma^{(i)} n_i^* \sigma^{(i)} n_i^*$$
$$= \left(\sigma^{(1)} \right)^2 \left(n_1^* \right)^2 + \left(\sigma^{(2)} \right)^2 \left(n_2^* \right)^2 + \left(\sigma^{(3)} \right)^2 \left(n_3^* \right)^2 \tag{2.80}$$

and subtracting $(\sigma_{nn})^2$ as given by (2.79)

$$\sigma_{ns} = \left\{ \sigma^{(i)} n_i^* \sigma^{(i)} n_i^* - \left[\sigma^{(i)} n_i^* n_i^* \right]^2 \right\}^{1/2} \tag{2.81a}$$

Expanding, we have

$$\sigma_{ns} = \left\{ \left(\sigma^{(1)} \right)^2 \left(n_1^* \right)^2 + \left(\sigma^{(2)} \right)^2 \left(n_2^* \right)^2 + \left(\sigma^{(3)} \right)^2 \left(n_3^* \right)^2 - \left(\sigma^{(1)} \right)^2 \left(n_1^* \right)^4 \right.$$
$$- \left(\sigma^{(2)} \right)^2 \left(n_2^* \right)^4 - \left(\sigma^{(3)} \right)^2 \left(n_3^* \right)^4$$
$$- 2\sigma^{(1)} \sigma^{(2)} \left(n_1^* \right)^2 \left(n_2^* \right)^2 - 2\sigma^{(1)} \sigma^{(3)} \left(n_1^* \right)^2 \left(n_3^* \right)^2 \tag{2.81b}$$
$$\left. - 2\sigma^{(2)} \sigma^{(3)} \left(n_2^* \right)^2 \left(n_3^* \right)^2 \right\}^{1/2}.$$

The first three terms in the expanded Eq. (2.81b) come from the summation on the first term of (2.81a), while the last six terms come from first summing the second term of (2.81a) and then squaring the sum. Equation (2.81b) may be simplified if (1.32), the normality of the direction cosines,

$$\left(n_i^* n_i^* \right)^2 = 1, \tag{2.82}$$

is introduced. Replacing $\left(n_1^* \right)^4$ with $\left(n_1^* \right)^2 \left[1 - \left(n_2^* \right)^2 - \left(n_3^* \right)^2 \right]$, etc., this allows the relatively compact form [6]

$$\sigma_{ns} = \left\{ \left[\sigma^{(1)} - \sigma^{(2)} \right]^2 \left(n_1^* \right)^2 \left(n_2^* \right)^2 + \left[\sigma^{(2)} - \sigma^{(3)} \right]^2 \left(n_2^* \right)^2 \left(n_3^* \right)^2 \right.$$
$$\left. + \left[\sigma^{(3)} - \sigma^{(1)} \right]^2 \left(n_3^* \right)^2 \left(n_1^* \right)^2 \right\}^{1/2} \tag{2.83}$$

to be derived. Equation (2.83) reveals that if all of the principal stresses are equal, the shearing stress is zero on any plane.

2.7.2 Stresses on Octahedral Planes

Of particular interest in the theory of material failure, discussed in Chap. 12, is the state of stress on the octahedral planes, which are defined as planes having equal

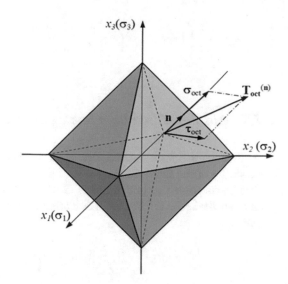

Fig. 2.9 Octahedral planes (Reproduced under the terms of the GNU Free Documentation License, Version 1.2 or any later version published by the Free Software Foundation [9])

direction cosines with each of the principal planes. A general diagram of the octahedral planes is shown on Fig. 2.9 reproduced from Wikipedia [9].

There are obviously eight such planes, one in each octant of the coordinate system. Referring to Fig. 2.9 with x_1, x_2, and x_3 taken as the principal directions, the octahedral plane in the $+x_1^*$, $+x_2^*$, $+x_3^*$ octant may be represented by the triangular surface in Fig. 2.7 where the normal to the surface

$$\mathbf{n}^* = \left(1/\sqrt{3}\right)\left(\mathbf{n}^{(1)} + \mathbf{n}^{(2)} + \mathbf{n}^{(3)}\right),$$ (2.84)

and the respective direction cosines are

$$n_1^* = n_2^* = n_3^* = 1/\sqrt{3}.$$ (2.85)

Substituting (2.85) into (2.79) and (2.83) gives the octahedral normal stress, shown as

σ_{oct} on Fig.2.9,

$$\sigma_{nn}^{\text{oct}} = \frac{1}{3}\left[\sigma^{(1)} + \sigma^{(2)} + \sigma^{(3)}\right]$$
$$= \frac{1}{3}\sigma_{ii},$$ (2.86a)

which is simply the *average* of the principal stresses and the octahedral shearing stress,

shown as T_{oct} on Fig. 2.9,

$$\sigma_{ns}^{oct} = \frac{1}{3}\left\{\left[\sigma^{(1)} - \sigma^{(2)}\right]^2 + \left[\sigma^{(2)} - \sigma^{(3)}\right]^2 + \left[\sigma^{(3)} - \sigma^{(1)}\right]^2\right\}^{1/2}. \qquad (2.86b)$$

2.7.3   Absolute Maximum Shearing Stress

The absolute maximum shearing stress at a point may also be of interest in regard to the capacity of the material.

We first observe that both terms of (2.81a) are positive. Therefore, for σ_{ns} to be a maximum, the second term should be a minimum. This is true if the orientation of σ_{ns} is such that

$$\frac{\partial}{\partial n_i^*}[\sigma^{(i)} n_i^* n_i^*]^2 = 0$$

$$\quad or \qquad\qquad (2.87)$$

$$4[\sigma^{(i)^2} n_i^* n_i^*]n_k^* = 0,$$

where the last subscript is changed to k to avoid a triple repeated index. Equation (2.87) is satisfied only if $n_k^* = 0$, indicating that the planes for which σ_{ns} is maximum will be *parallel* to one of the principal axes.

Now, examining (2.83), two of the three terms will vanish due to $n_k^* = 0$. The nonzero term which makes σ_{ns} maximum is the one that contains the bracketed term with the largest principal stress *difference*. Since this is the only term not containing n_k^*, the maximum shearing stress acts on plane(s) parallel to n_k^*. For example, if $\sigma^{(1)}$ and $\sigma^{(3)}$ are the maximum and minimum principal stresses, then $n_2^* = 0$. In general, with i and j indicating the directions of the maximum and minimum principal stresses, (2.83) gives

$$\left(\sigma_{ns}^{max}\right)^2 = \left[\sigma^{(i)} - \sigma^{(j)}\right]^2 \left(n_i^*\right)^2 \left(n_j^*\right)^2. \qquad (2.88)$$

Since $\left(n_j^*\right)^2 = 1 - \left(n_i^*\right)^2$, the maximum value of the r.h.s. of (2.88) occurs when $\left(n_i^*\right)^2 = \frac{1}{2} = \left(n_j^*\right)^2$ [6], and the absolute maximum shearing stress is

$$\sigma_{ns}^{max} = \frac{1}{2}\left[\sigma^{(i)} - \sigma^{(j)}\right] \qquad (2.89)$$

acting on the planes *bisecting* the planes of maximum and minimum principal stress. For example, if $i = 1$ and $j = 3$,

$$\sigma_{ns}^{max} = \frac{1}{2}\left[\sigma^{(1)} - \sigma^{(3)}\right], \qquad (2.90)$$

and the corresponding planes are shown in Fig. 2.10 [6].

Fig. 2.10 Planes with
maximum shearing stresses
(After Ugural and Fenster,
*Advanced Strength and
Applied Elasticity*, Elsevier,
1975). Reproduced with
permission

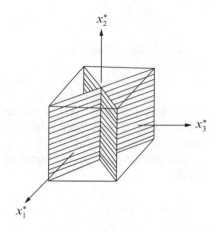

2.7.4   Computational Example

Using the principal stresses found in Sect. 2.4.3, we may evaluate the stresses on the octahedral planes from (2.86a) and (2.86b) as

$$\sigma_{nn}^{oct} = \frac{1}{3}(-1,168 + 1,380 + 988) = 400$$

$$\sigma_{ns}^{oct} = \frac{1}{3}\left\{[-1,168 - 1,380]^2 + [1,380 - 988]^2 + [988 - (-1,168)]^2\right\}^{1/2}$$
$$= 1,120.$$

The maximum shearing stress is found from (2.89), with $i = 2$ and $j = 1$,

$$\sigma_{ns}^{max} = \frac{1}{2}[1,380 - (-1,168)] = 1,274,$$

and acts on planes bisecting the x_1^*- and x_2^*-axes and parallel to x_3^*. These planes are easy to describe in the principal coordinate system as $\mathbf{e}_{Smax} = \pm\cos 45\,\mathbf{e}_1^* \pm \sin 45\,\mathbf{e}_2^*$.

2.8   Properties and Special States of Stress

2.8.1   Projection Theorem

We focus on any point, say P, of the continuum shown in Fig. 2.5. It has been established that any number of tractions may be calculated at P, each associated with a single plane passing through P and identified by the corresponding unit normal. The projection theorem is a relationship between any two tractions at P, say $\mathbf{T}^{(a)}$ and $\mathbf{T}^{(b)}$, and their associated normals, say $\mathbf{n}^{(a)}$ and $\mathbf{n}^{(b)}$, are stated to be

$$\mathbf{T}^{(a)} \cdot \mathbf{n}^{(b)} = \mathbf{T}^{(b)} \cdot \mathbf{n}^{(a)}, \tag{2.91}$$

or, in component form, as

$$T_i^{(a)} n_i^{(b)} = T_i^{(b)} n_i^{(a)} \tag{2.92}$$

meaning that the projection of the traction acting on the first plane onto the second plane is equal to the projection of the traction acting on the second plane onto the first plane.

The proof is easily demonstrated by substituting (2.11) into both sides of (2.92) to obtain

$$\begin{aligned}
\sigma_{ji} n_j^{(a)} n_i^{(b)} &= \sigma_{ji} n_j^{(b)} n_i^{(a)} \\
&= \sigma_{ij} n_j^{(a)} n_i^{(b)}
\end{aligned} \tag{2.93}$$

since the i and j became dummy indices. As $\sigma_{ij} = \sigma_{ji}$, the expressions are identical. The projection theorem may be used to define several of special states of stress.

2.8.2   Plane Stress

Suppose that on one plane passing through P, there are no stresses, that is, $\mathbf{T} = 0$. The projection theorem states that the traction on any other plane passing through P must be perpendicular to the normal to the stress-free plane; hence, it is parallel to that plane. Conversely, if it is established that the traction on any plane is parallel to another plane, then the latter plane will also be stress-free. This is known as a state of *plane stress* and is of practical use since it allows the elasticity problem to be reduced from three to two dimensions, which greatly facilitates the mathematical solution. It is obvious that if one of the principal stresses is zero, the state of stress is plane.

2.8.3   Linear Stress

If there are two nonparallel planes on which $\mathbf{T} = 0$, the traction on any other plane must be parallel to both of these planes and also to their line of intersection. This is a state of *linear stress* and is essentially a one-dimensional problem. For this state, two principal stresses vanish.

2.8.4   Pure Shear

If, for any coordinate system,

$$\sigma_{11} = \sigma_{22} = \sigma_{33} = 0, \tag{2.94}$$

the state of stress is called *pure shear* since in this coordinate system, only shear stresses σ_{ij}, with $i \neq j$, exist. Also, for pure shear, $\sigma_{ii} = 0$ for any coordinate system.

2.8.5 Hydrostatic Stress

If the principal stresses $\sigma^{(i)}$ are all equal, then the stress state is hydrostatic, and the shearing stresses are zero; examine (2.83).

2.9 Computational Examples with MATLAB

Example 2.1 Traction Force and Components (Sect. 2.2)
The traction force and the corresponding normal and tangential components can be calculated using the dot product. The computational example of Sect. 2.2.6 was used to illustrate the application of MATLAB. We first construct the stress tensor and the normal of the plane using a matrix and a column vector:

```
% stress tensor
sig = [500, 500, 800;
       500, 1000, -750;
       800, -750, -300];
% normal vector of the plane
n = [1/2; 1/2; 1/sqrt(2)];
```

Then we evaluate the traction force and the normal and tangential components (note that **T'** is the transpose of **T**):

```
% evaluate traction force
T = sig * n;
% normal component of the traction force
Tn = T' * n;
% tangential component of the traction force
Ts = sqrt(norm(T)^2 - norm(Tn)^2);
```

To check the results, we can print out the values in the command window using the following commands:

```
% print results
fprintf('Magnitude of the traction force magnitude T = %4.0f\n', ...
norm(T))
fprintf('Normal component Tn = %4.0f\n', Tn)
```

Results will be displayed in the command window as follows:

```
Magnitude of the traction force magnitude T = 1104
Normal component Tn = 510
```
Note:

1. The dot product can also be computed with Tn = dot (T, n).
2. fprintf outputs formatted strings in the command window.

Example 2.2 Principal Stresses and Directions (Sect. 2.4)

Computing principal stresses and principal directions are quite straightforward using embedded MATLAB function "eig." We use the same stress tensor as defined in Example 1 for illustration.

```
% stress tensor
sig = [500, 500, 800;
       500, 1000, -750;
       800, -750, -300];

% princial stresses and directions
[v_principal, sig_principal] = eig(sig);
```

Note that v_principal is the principal direction vectors and sig_principal is the corresponding principal stresses.

We can use the fprintf to print out the results (Note here that the "..." symbol is to continue the command with the following line):

```
% print results
fprintf('1st principal stress: %4.0f, 1st principal direction = ...
[%0.4f, %0.4f, %0.4f],\n',
       sig_principal (1,1), v_principal(:,1));
fprintf('2nd principal stress: %4.0f, 2nd principal direction = ...
[%0.4f, %0.4f, %0.4f],\n',
       sig_principal (2,2), v_principal(:,2));
fprintf('3rd principal stress: %4.0f, 3rd principal direction = ...
[%0.4f, %0.4f, %0.4f],\n',
       sig_principal (3,3), v_principal(:,3));
```

Results will be displayed in the command window as follows:

```
1st principal stress: -1169, 1st principal direction =
[-0.4902, 0.3837, 0.7826],
  2nd principal stress: 992, 2nd principal direction =
[-0.8305, 0.0670, -0.5530],
  3rd principal stress: 1376, 3rd principal direction =
[-0.2646, -0.9210, 0.2858],
```

Example 2.3 Octahedral Shear and Normal Stress (Sect. 2.6.2)

Equation (2.86) shows the computation of octahedral shear and normal stresses. Therefore, it is straightforward to calculate the octahedral shear and normal stresses based on principal stresses. We used the same principal stresses calculated in Example 2.2 for illustration.

```
% octahedreal normal stress
sig_oct_n = 1/3 * (sig_principal(1,1) + sig_principal(2,2) +
sig_principal(3,3));
% octahedreal shear stress
sig_oct_s = 1/3 * sqrt((sig_principal(1,1) - sig_principal(2,2))^2 +
...
(sig_principal(2,2) - sig_principal (3,3))^2 + (sig_principal(3,3)
- sig_principal(1,1))^2);

% print results
fprintf('octahedreal normal stress = %4.0f\n', sig_oct_n)
fprintf('octahedreal shear stress = %4.0f\n', sig_oct_s)
```

Results will be displayed in the command window as follows:

```
octahedral normal stress = 400
octahedral shear stress = 1120
```

2.10 Exercises

2.1 Find the components of the traction on a plane defined by $n_1 = 1/\sqrt{2}$, $n_2 = 1/\sqrt{2}$, and $n_3 = 0$ for the following states of stress:

(a) $\sigma_{11} = \sigma$, $\sigma_{12} = 0$, $\sigma_{13} = 0$,
   $\sigma_{21} = \sigma$, $\sigma_{22} = 0$, $\sigma_{33} = \sigma$.
(b) $\sigma_{11} = \sigma$, $\sigma_{12} = \sigma$, $\sigma_{13} = 0$,
   $\sigma_{22} = \sigma$, $\sigma_{23} = 0$, $\sigma_{33} = 0$.

2.2 The state of stress at a point P in a structure is given by

$$\sigma_{11} = 20,000, \quad \sigma_{22} = -15,000, \quad \sigma_{33} = 3,000$$
$$\sigma_{12} = 2,000, \quad \sigma_{23} = 2,000, \quad \sigma_{31} = 1,000.$$

(a) Compute the scalar components T_1, T_2, and T_3 of the traction **T** on the plane passing through P whose outward normal vector **n** makes equal angles with the coordinate axes x_1, x_2, and x_3 (from [1]).
(b) Compute the normal and tangential components of stress on this plane.
(c) Write the stress dyadic for the state of stress and repeat parts (a) and (b).
(d) Determine the principal stresses.
(e) Determine the direction of maximum principal stress.
(f) Determine the octahedral shear stress.

(g) Determine the maximum shear stress, and show the planes on which the stress acts in the principal coordinates.

2.3 The state of stress at a point is given by the components of stress tensor σ_{ij}. A plane is defined by the direction cosines of the normal $(1/2, 1/2, 1/\sqrt{2})$. State the general conditions for which the traction on the plane has the same direction as the x_2-axis and a magnitude of 1.0.

2.4 Determine the body forces for which the following stress field describes a state of equilibrium (from [6]):

$$
\begin{aligned}
\sigma_{xx} &= -2x^2 - 3y^2 - 5z, & \sigma_{xy} &= z + 4xy - 6, \\
\sigma_{yy} &= -2y^2 + 7, & \sigma_{xz} &= -3x + 2y + 1, \\
\sigma_{zz} &= 4x + y + 3z - 5 & \sigma_{yz} &= 0.
\end{aligned}
$$

2.5 Determine whether the following stress field is admissible in an elastic body when body forces are negligible.

$$
\boldsymbol{\sigma} =
\begin{bmatrix}
yz + 4 & z^2 + 2x & 5y + z \\
\cdot & xz + 3y & 8x^3 \\
\cdot & \cdot & 2xyz
\end{bmatrix}.
$$

2.6 At a given point in a body,

$$
\begin{aligned}
\sigma_{11} &= \sigma_{22} = \sigma_{33} = -p, \\
\sigma_{12} &= \sigma_{23} = \sigma_{31} = 0.
\end{aligned}
$$

Show that the normal stresses are equal to $-p$ and that the shearing stresses vanish for any other Cartesian coordinate system (from [1]).

2.7 Expand the stress transformation law from the tensor form (2.13) for a two-dimensional state of stress by assuming that the stress components in the x_3-direction are zero and by considering a rotation about the x_3-axis. Compare the resulting expressions with Mohr's circle representation [1].

2.8 The stress tensor at a point $P(x, y, z)$ in a solid is

$$
\boldsymbol{\sigma} =
\begin{bmatrix}
3 & 1 & 1 \\
1 & 0 & 2 \\
1 & 2 & 0
\end{bmatrix}.
$$

(a) Show that the principal stresses are 4, 1, and -2.
(b) Find the orientation of the principal planes.
(c) Compute the octahedral stresses at P.
(d) Determine the magnitude and orientation of the absolute maximum shearing stress at P.

2.9 The stress tensor at a point $P(x, y, z)$ in a solid is

$$\boldsymbol{\sigma} = \begin{bmatrix} 120 & 40 & 30 \\ 40 & 150 & 20 \\ 30 & 20 & 100 \end{bmatrix}.$$

(a) Determine the stresses with respect to a set of axes x', y', and z' for which the matrix of direction cosines is

$$[a_{ij}] = \begin{bmatrix} \dfrac{1}{2} & -\dfrac{\sqrt{3}}{2} & 0 \\ \dfrac{\sqrt{3}}{2} & \dfrac{1}{2} & 0 \\ 0 & 0 & 1 \end{bmatrix}.$$

(b) Compute the principal stresses using both the x-, y-, and z- and x'-, y'-, and z'-axes as the basis (from [6]).

(c) Find the orientation of the principal planes with respect to the x-, y-, and z-axes.

(d) Compute the stress invariants for both coordinate systems, and verify that they are equal.

2.10 For an octahedral plane with a normal defined by (2.84), show that the traction on the plane is

$$\mathbf{T}_n^* - \sigma_{nn}^{oct} \mathbf{n}^* + \mathbf{T}_{ns}^*,$$

where σ_{nn}^{oct} is defined by (2.86a) and where the magnitude of the shearing component $\mathbf{T}_{ns}^*$ is σ_{ns}^{oct} (2.86b).

2.11 Show that the expression for the tangential component of stress σ_{ns}, given by (2.19), can be written as [10]

$$\sigma_{ns} = \left[\sigma_{ik} \sigma_{jm} n_k n_m (\delta_{ij} - n_i n_j) \right]^{1/2}.$$

2.12 For the case of pure shear, the stress matrix is given by

$$\begin{bmatrix} 0 & \tau & 0 \\ \tau & 0 & 0 \\ 0 & 0 & 0 \end{bmatrix}.$$

Determine (1) the principal stresses and directions and (2) the octahedral normal and shear stresses.

2.13 The stress components defined w.r.t. x–y-axes at a certain point in a stressed
     body in plane stress are shown. Using the stress transformation laws, find the
     stress components acting on the sides of the rectangular elements shown.

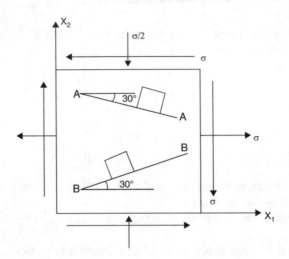

References

 1. Tauchert TR (1974) Energy principles in structural mechanics. McGraw-Hill Book Company
    Inc., New York
 2. Pearson CE (1959) Theoretical elasticity. Harvard monographs in applied science, vol 6. Har-
    vard University Press, Cambridge, MA
 3. Westergaard HM (1964) Theory of elasticity and plasticity. Dover Publications Inc., New York
 4. Davis HF (1961) Hamidi R, Shafigh-Nobari F, Lim CK (1988) Nonlinear behavior of elasto-
    meric bearings. 1: theory. J Eng Mech ASCE 114(11):1811–1830
 5. Ugural AC, Fenster SK (1987) Advanced strength and applied elasticity, 2nd edn. Elsevier,
    New York
 6. Chou PC, Pagano NJ (1992) ELASTICITY tensor, dyadic and engineering approaches. Dover
    Publications, Inc., New York, NY
 7. Timoshenko S, Goodier JN (1951) Theory of elasticity. McGraw-Hill Book Co., New York
 8. Zubelwicz A, Bažant ZP (1987) Constitutive model with rotating active plane and true stress. J
    Eng Mech ASCE 113(3):398–416
 9. Wikipedia
10. Davis HF, Snider AD (1987) Introduction to vector analysis. Allyn & Bacon Inc, Boston

Chapter 3
Deformations

Abstract Displacements with respect to a reference coordinate system may be physically observed, calculated, or measured for a deformed elastic body. Each displacement may be considered to have two components, one of which is due to *relative* movements or *distortions* within the body and the other which is uniform throughout the body, the so-called *rigid body* motion. The relationships between the displacements and the corresponding distortions are known as the *kinematic* or the *strain–displacement* equations of the theory of elasticity and may take several forms, depending on the expected magnitude of the distortions and displacements. Using vector mechanics, the kinematic relations are derived in a general form and then reduced to linear equations. Also nonlinear strain–displacement considerations are introduced. These relationships are important in contemporary fields such as biomechanics.

3.1 Introduction

Displacements with respect to a reference coordinate system may be physically observed, calculated, or measured (at least on the surface) for a deformed elastic body. Each displacement may be considered to have two components, one of which is due to *relative* movements or *distortions* within the body and the other which is uniform throughout the body, the so-called *rigid body* motion. The relationships between the displacements and the corresponding internal distortions are known as the *kinematic* or the *strain–displacement* equations of the theory of elasticity and may take several forms depending on the expected magnitude of the distortions and displacements.

3.2   Strain

We consider the undeformed and the deformed positions of an elastic body shown
in Fig. 3.1. It is convenient to designate two sets of Cartesian coordinates x_i and X_i
with $i = 1, 2, 3$, called *initial* coordinates and *final* coordinates, respectively, which
denote the undeformed and deformed positions of the body [1]. Then we designate a
reference point $P(\bar{x}_i)$ on the undeformed body and $p(\bar{X}_i)$ on the deformed body,
where $\bar{x}_i$ and $\bar{X}_i$ indicate specific values of the coordinates x_i and X_i. Also, we locate
neighboring points $Q(\bar{x}_i + d\bar{x}_i)$ and $q(\bar{X}_i + d\bar{X}_i)$ that are separated from P and
p by differential distances ds and dS, respectively. The latter are given in component
form by

$$(ds)^2 = d\bar{x}_1^2 + d\bar{x}_2^2 + d\bar{x}_3^2 = d\bar{x}_i d\bar{x}_i \tag{3.1a}$$

and

$$(dS)^2 = d\bar{X}_1^2 + d\bar{X}_2^2 + d\bar{X}_3^2 = d\bar{X}_i d\bar{X}_i. \tag{3.1b}$$

In Fig. 3.1, **u** represents the displacement of P to p, and **u** + d**u** is the displacement
of Q to q. From the vector diagram in Fig. 3.1, we note

$$d\mathbf{s} + (\mathbf{u} + d\mathbf{u}) = \mathbf{u} + d\mathbf{S} \tag{3.2}$$

or

$$d\mathbf{u} = d\mathbf{S} - d\mathbf{s}. \tag{3.3}$$

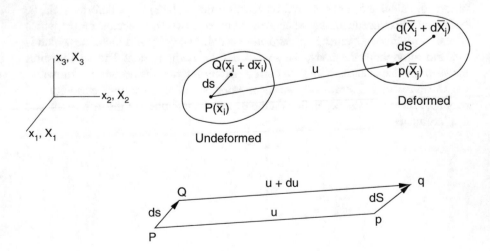

Fig. 3.1 Displacement vector in Cartesian space

We consider this relationship in component form using (3.1a) and (3.1b) as

$$(dS)^2 - (ds)^2 = d\bar{X}_i d\bar{X}_i - d\bar{x}_i d\bar{x}_i. \tag{3.4}$$

We may now select either the initial or the final coordinates as the basis for further computations. First, we choose the initial coordinates x_i so that

$$\bar{X}_i = \bar{X}_i(x_1, x_2, x_3) \tag{3.5a}$$

$$d\bar{X}_i = \frac{\partial X_i}{\partial x_1} d\bar{x}_1 + \frac{\partial X_i}{\partial x_2} d\bar{x}_2 + \frac{\partial X_i}{\partial x_3} d\bar{x}_3 \tag{3.5b}$$

$$= X_{i,j} d\bar{x}_j$$

$$\mathbf{u} = u_1 \mathbf{e}_{x_1} + u_2 \mathbf{e}_{x_2} + u_3 \mathbf{e}_{x_3} = u_i \mathbf{e}_{x_i} \tag{3.5c}$$

$$u_i = X_i - x_i \tag{3.5d}$$

in component form. Substituting (3.5b) and (3.5d) into (3.4), introducing another dummy subscript k, and applying (2.37),

$$
\begin{aligned}
(dS)^2 - (ds)^2 &= X_{i,j} d\bar{x}_j X_{i,k} d\bar{x}_k - d\bar{x}_i d\bar{x}_i \\
&= \left(X_{i,j} X_{i,k} - \delta_{ij} \delta_{ik} \right) d\bar{x}_j d\bar{x}_k \\
&\quad - \left[(x_i + u_i)_{,j} (x_i + u_i)_{,k} - \delta_{jk} \right] d\bar{x}_j d\bar{x}_k \\
&= \left[(\delta_{ij} + u_{i,j})(\delta_{ik} + u_{i,k}) - \delta_{ik} \right] d\bar{x}_j d\bar{x}_k \\
&= \left[\delta_{jk} + u_{j,k} + u_{k,j} + u_{i,j} u_{i,k} - \delta_{jk} \right] d\bar{x}_j d\bar{x}_k \\
&= \left(u_{j,k} + u_{k,j} + u_{l,j} u_{l,k} \right) d\bar{x}_j d\bar{x}_k.
\end{aligned}
\tag{3.6}
$$

It is convenient to interchange the indices i and k so that

$$
\begin{aligned}
(dS)^2 - (ds)^2 &= \left(u_{i,j} + u_{j,i} + u_{k,i} u_{k,j} \right) d\bar{x}_i d\bar{x}_j \\
&= 2\varepsilon_{ij}^L d\bar{x}_i d\bar{x}_j
\end{aligned}
\tag{3.7}
$$

in which

$$\varepsilon_{ij}^L = \frac{1}{2} \left(u_{i,j} + u_{j,i} + u_{k,i} u_{k,j} \right) \tag{3.8}$$

are components of the *Lagrangian*, or *material*, strain tensor $\boldsymbol{\varepsilon}^L$.

We repeat the procedure with the final coordinates X_i as the basis so that

$$\bar{x}_i = \bar{x}_i(X_1, X_2, X_3) \tag{3.9a}$$

$$d\bar{x}_i = \frac{\partial x_i}{\partial X_1} d\bar{X}_1 + \frac{\partial x_i}{\partial X_2} d\bar{X}_2 + \frac{\partial x_i}{\partial X_3} d\bar{X}_3 \tag{3.9b}$$

$$= x_{i,j} d\bar{X}_j$$

$$\mathbf{u} = u_1 \mathbf{e}_{X_1} + u_2 \mathbf{e}_{X_2} + u_3 \mathbf{e}_{X_3} = u_i \mathbf{e}_{X_i} \qquad (3.9c)$$

$$u_i = x_i - X_i. \qquad (3.9d)$$

Substituting (3.9b) and (3.5d) into (3.4) gives

$$
\begin{aligned}
(dS)^2 - (ds)^2 &= d\bar{X}_i d\bar{X}_i - x_{i,j} d\bar{X}_j x_{i,k} d\bar{X}_k \\
&= \left(\delta_{ij} \delta_{ik} - x_{i,j} x_{i,k} \right) d\bar{X}_j d\bar{X}_k \\
&= \left[\delta_{jk} - (X_i - u_i)_{,j} (X_i - u_i)_{,k} \right] d\bar{X}_j d\bar{X}_k \\
&= \left[\delta_{jk} - (\delta_{ij} - u_{i,j})(\delta_{ik} - u_{i,k}) \right] d\bar{X}_j d\bar{X}_k \\
&= \left[\delta_{jk} - \delta_{jk} + u_{k,j} + u_{j,k} - u_{i,j} u_{i,k} \right] d\bar{X}_j d\bar{X}_k \\
&= \left(u_{j,k} + u_{k,j} - u_{i,j} u_{i,k} \right) d\bar{X}_j d\bar{X}_k
\end{aligned}
\qquad (3.10)
$$

or, interchanging i and k,

$$
\begin{aligned}
(dS)^2 - (ds)^2 &= \left(u_{i,j} + u_{j,i} - u_{k,i} u_{k,j} \right) d\bar{X}_i d\bar{X}_j \\
&= 2 \varepsilon_{ij}^E d\bar{X}_i d\bar{X}_j
\end{aligned}
\qquad (3.11)
$$

in which

$$\varepsilon_{ij}^E = \frac{1}{2} \left(u_{i,j} + u_{j,i} - u_{k,i} u_{k,j} \right) \qquad (3.12)$$

are components of the *Eulerian*, or *spatial*, strain tensor $\boldsymbol{\varepsilon}^E$.

If the displacement *gradients* $u_{k,i}$ are *small* in comparison to *unity*, then products of such terms are negligible in (3.8) and (3.12), and they may be dropped. For such cases, we also consider the initial and final coordinate systems to be coincident so that

$$\frac{\partial(\ )}{\partial x_i} = \frac{\partial(\ )}{\partial X_i} \qquad (3.13)$$

and then

$$\varepsilon_{ij}^L = \varepsilon_{ij}^E = \varepsilon_{ij} = \frac{1}{2} \left(u_{i,j} + u_{j,i} \right). \qquad (3.14)$$

The components ε_{ij} refer to the small or *infinitesimal* strain tensor $\boldsymbol{\varepsilon}$. Note that no restriction is placed on the magnitude of the u_i terms but only the gradients. It is thus at least theoretically possible at this point to describe relatively large displacements by infinitesimal strains. We will return to the reduction of (3.8) and (3.12) at the end of Sect. 3.2 and in Sect. 3.8.

The equations represented by (3.14) are a cornerstone of the development of the linear theory of elasticity and are called the *strain–displacement* or *kinematic* equations written in explicit form as

$$\varepsilon_{11} = u_{1,1} = \partial u_1/\partial x_1 \tag{3.15a}$$

$$\varepsilon_{22} = u_{2,2} = \partial u_2/\partial x_2 \tag{3.15b}$$

$$\varepsilon_{33} = u_{3,3} = \partial u_3/\partial x_3 \tag{3.15c}$$

$$\varepsilon_{12} = \varepsilon_{21} = \frac{1}{2}(u_{1,2} + u_{2,1}) = \frac{1}{2}(\partial u_1/\partial x_2 + \partial u_2/\partial x_1) \tag{3.15d}$$

$$\varepsilon_{13} = \varepsilon_{31} = \frac{1}{2}(u_{1,3} + u_{3,1}) = \frac{1}{2}(\partial u_1/\partial x_3 + \partial u_3/\partial x_1) \tag{3.15e}$$

$$\varepsilon_{23} = \varepsilon_{32} = \frac{1}{2}(u_{2,3} + u_{3,2}) = \frac{1}{2}(\partial u_2/\partial x_3 + \partial u_3/\partial x_2). \tag{3.15f}$$

Noting that the u_i terms constitute a vector or a first-order tensor and the $u_{i,j}$ terms represent a tensor of order 2; therefore, $\boldsymbol{\varepsilon}$ is a second-order tensor that transforms just as the stress tensor $\boldsymbol{\sigma}$:

$$\varepsilon'_{ij} = \alpha_{ik}\alpha_{jl}\varepsilon_{kl}. \tag{3.16}$$

3.3 Physical Interpretation of Strain Tensor

The physical definition of strain is familiar from elementary strength of materials. Here, it has been introduced from a mathematical standpoint, and it remains to reconcile the notions.

First, we consider the components $\varepsilon_{(ii)}$, which are called *extensional* strains [1]. For example, consider a fiber parallel to the x_2 coordinate axis of initial length ds and final length dS, for which $dx_1 = 0$, $dx_3 = 0$, and $dx_2 = ds$. Equation (3.7) becomes

$$(dS)^2 - (ds)^2 = 2\varepsilon_{22}ds^2 \tag{3.17}$$

for the case of infinitesimal strain. Considering the l.h.s. of (3.17),

$$(dS)^2 - (ds)^2 = (dS - ds)(dS + ds) \tag{3.18}$$

and

$$\frac{dS - ds}{ds} = \frac{(dS)^2 - (ds)^2}{ds(dS + ds)}. \tag{3.19}$$

Substituting (3.17) into the r.h.s. of (3.19), we have

$$\frac{dS - ds}{ds} = \frac{2\varepsilon_{22}ds^2}{ds(dS + ds)}$$

$$\simeq \frac{2\varepsilon_{22}ds^2}{2ds^2} \tag{3.20}$$

$$= \varepsilon_{22},$$

since the difference between $ds\, dS$ and ds^2 is of higher order for distortions $d\mathbf{u}/\mathbf{u}$, which are small compared to unity. Therefore,

$$\varepsilon_{22} = \frac{dS - ds}{ds} \tag{3.21}$$

which is the *relative elongation* of a fiber parallel to the x_2 coordinate axis. Similar interpretations are true for ε_{11} and ε_{33}.

The components ε_{ij} are called *shearing* strains and may be interpreted by considering intersecting fibers initially parallel to two of the coordinate axes [1]. For example, we select fibers parallel to the x_1- and x_3-axes, as shown in Fig. 3.2, with initial lengths $ds^{(1)}$ and $ds^{(3)}$. These fibers have final lengths of $dS^{(1)}$ and $dS^{(3)}$ and have rotated through angles $\beta^{(1)}$ and $\beta^{(3)}$, respectively. The total change in the initially right angle is

$$\beta^{(1)} + \beta^{(3)} = \beta. \tag{3.22}$$

We consider vectors coincident with the fibers given in component form by

$$d\mathbf{S}^{(1)} = dS_i^{(1)}\mathbf{e}_{x_i} \tag{3.23a}$$

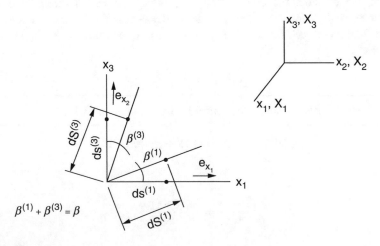

Fig. 3.2 Shearing strains

and

$$dS^{(3)} = dS_i^{(3)} e_{x_i} \qquad (3.23b)$$

in which the components may be written in terms of the initial coordinates as

$$dS_i^{(1)} = \frac{\partial S_i^{(1)}}{\partial x_j} dx_j = S_{i,j}^{(1)} dx_j \qquad (3.24a)$$

and

$$dS_i^{(3)} = \frac{\partial S_i^{(3)}}{\partial x_k} dx_k = S_{i,k}^{(3)} dx_k. \qquad (3.24b)$$

The scalar product of the vectors is

$$dS^{(1)} \cdot dS^{(3)} = dS^{(1)} dS^{(3)} \cos\left(\frac{\pi}{2} - \beta\right). \qquad (3.25)$$

Considering first the l.h.s. of (3.25), in view of (3.23a) and (3.23b), (3.24a) and (3.24b), (3.3), and also (2.37),

$$
\begin{aligned}
dS^{(1)} \cdot dS^{(3)} &= S_{i,j}^{(1)} dx_j S_{i,k}^{(3)} dx_k \\
&= \left(s_i^{(1)} + u_i\right)_{,j} \left(s_i^{(3)} + u_i\right)_{,k} dx_j dx_k \\
&= \left(\delta_{ij} + u_{i,j}\right)\left(\delta_{ik} + u_{i,k}\right) dx_j dx_k \\
&- \left(\delta_{jk} + u_{k,j} + u_{j,k} + u_{i,j} u_{i,k}\right) dx_j dx_k.
\end{aligned}
\qquad (3.26)
$$

Note that $s^{(1)}$ and $s^{(3)}$ are along the original x_1- and x_3-axcs.

The product term is zero for infinitesimal theory, and considering the initial orientation of the fibers, we obtain

$$
\begin{aligned}
dx_1 &= ds^{(1)}, \quad dx_2 = dx_3 = 0; \quad \text{and} \\
dx_3 &= ds^{(3)}, \quad dx_1 = dx_2 = 0.
\end{aligned}
\qquad (3.27)
$$

Therefore, (3.26) has a nonzero value only for $j = 1$ and $k = 3$, for which $\delta_{jk} = 0$. Thus,

$$dS^{(1)} \cdot dS^{(3)} \approx (u_{3,1} + u_{1,3}) ds^{(1)} ds^{(3)}. \qquad (3.28)$$

Now equating the r.h.s. of (3.25) and (3.28),

$$dS^{(1)} dS^{(3)} \cos\left(\frac{\pi}{2} - \beta\right) = (u_{3,1} + u_{1,3}) ds^{(1)} ds^{(3)}. \qquad (3.29)$$

Neglecting the differences between the products of the initial and final lengths of the fibers and assuming small angles of rotation β such that $\cos\left(\frac{\pi}{2} - \beta\right) = \sin\beta \cong \beta$,

$$\beta = (u_{1,3} + u_{3,1}) = 2\varepsilon_{13}. \qquad (3.30)$$

That is, the strain component ε_{13} is equal to *one-half* of the increase in the angle between two fibers initially parallel to the x_1- and x_3-axes. Similar interpretations are obvious for ε_{12} and ε_{23}.

It is important to note that the shearing strains are frequently taken as the *total* angle increase β rather than *one-half* this value [2]. Such a definition is termed an "engineering" strain but is awkward because it violates the tensor character of ε_{ij}. Nevertheless, we should be aware of this form since engineering strains are helpful for physical interpretation and useful in deriving material laws as demonstrated in Sect. 4.3.

Since the strain tensor is formed from the displacement gradients, it is interesting to separate $u_{i,j}$ into two parts:

$$u_{i,j} = \varepsilon_{ij} + \omega_{ij}, \qquad (3.31a)$$

where

$$\omega_{ij} = \frac{1}{2}\left(u_{i,j} - u_{j,i}\right) \qquad (3.31b)$$

by rearranging (3.14). For $i = j$, $\omega_{ij} = 0$, while for $i \neq j$, $\omega_{ij} = -\omega_{ji}$; thus, the ω_{ij} are elements of a skew-symmetric tensor $\boldsymbol{\omega}$. For the previous example,

$$\omega_{13} = \frac{1}{2}\left(u_{1,3} - u_{3,1}\right). \qquad (3.32)$$

From (3.22) and (3.30), we find

$$\beta = \beta^{(1)} + \beta^{(3)} = (u_{1,3} + u_{3,1}) = 2\varepsilon_{13}. \qquad (3.33)$$

By comparison,

$$\omega_{13} = \frac{\beta^{(1)} - \beta^{(3)}}{2} \qquad (3.34)$$

which may be interpreted as the *average* rotation of all fibers around the x_2-axis. For a rigid body rotation, $\varepsilon_{13} = 0$, $\beta^{(1)} = -\beta^{(3)}$, and $\omega_{13} = \beta^{(1)}$. For this reason, $\boldsymbol{\omega}$ is called the rotation tensor [3], which may enter into the computation of displacements.

The physical notions of strain discussed in this section are closely intertwined with the mathematical definitions in the preceding section. This connection extends into the nonlinear Lagrangian and Eulerian strain tensors as well but is understandably more complex than the linear case. A recent unified definition of finite strains by Ma and Desai [4] contains both of the classical nonlinear tensors as special cases and isolates a special case which is essentially a mean of the two. The resulting strain tensor possesses some linear characteristics that assist in physical interpretation and exhibits good computational performances for some classical one- and two-dimensional problems.

Since the notion of strain is strongly supported by observable and/or measurable physical changes in length, shape, and orientation, an example of a nonlinear strain tensor expressed in terms of the physical interpretations of this section is considered in (3.8) with $i = j = 2$:

$$\varepsilon_{22}^L = \frac{1}{2}(2u_{2,2}) + \frac{1}{2}(u_{k,2}u_{k,2})$$

$$= u_{2,2} + \frac{1}{2}(u_{1,2}u_{1,2} + u_{2,2}u_{2,2} + u_{3,2}u_{3,2})$$

$$= u_{2,2} + \frac{1}{2}u_{2,2}u_{2,2} + \frac{1}{2}u_{1,2}u_{1,2} + \frac{1}{2}u_{3,2}u_{3,2}.$$

From (3.31b) and (3.14),

$$\omega_{12} = \frac{1}{2}(u_{1,2} - u_{2,1}) \text{ and } \varepsilon_{12} = \frac{1}{2}(u_{1,2} + u_{2,1})$$

so that

$$\omega_{12} + \varepsilon_{12} = u_{1,2}.$$

Similarly,

$$\omega_{23} = \frac{1}{2}(u_{2,3} - u_{3,2}) \text{ and } \varepsilon_{23} = \frac{1}{2}(u_{2,3} + u_{3,2})$$

giving

$$\varepsilon_{23} - \omega_{23} = u_{3,2}.$$

Then applying (3.15b), ε_{22}^L becomes

$$\varepsilon_{22}^L = \varepsilon_{22} + \frac{1}{2}\varepsilon_{22}^2 + \frac{1}{2}(\varepsilon_{12} + \omega_{12})^2 + \frac{1}{2}(\varepsilon_{23} - \omega_{23})^2. \tag{3.35}$$

The discussion of the nonlinear strain terms is continued in Sect. 3.8.

3.4 Principal Strain

Since we have shown that the strains have the same second-order tensor character as the stresses, we expect to be able to identify a system of principal strains that are the eigenvalues determined from the characteristic equation

$$\left|\varepsilon_{ji} - \lambda\delta_{ji}\right| = 0 \tag{3.36}$$

which is analogous to (2.56a). The three roots λ_1, λ_2, and λ_3 are the principal strains $\varepsilon^{(k)}$, with $k = 1$, 2, and 3.

The corresponding eigenvectors designate the directions associated with each of the principal strains and are computed from

$$\left(\varepsilon_{ji} - \lambda\delta_{ji}\right)n_j = 0 \tag{3.37}$$

along with (2.57). The directions of principal strain are denoted by $n_j^{(k)}$, with $k = 1$, 2, and 3. These directions are mutually perpendicular and, for isotropic elastic materials (to be discussed in Chap. 4), coincide with the directions of the principal stresses. Obviously, the largest extension experienced by any fiber at a point is equal to the largest principal strain [5].

Strains on oblique planes, octahedral strains, and the absolute maximum shearing strain are obtained relative to the principal coordinates in a manner analogous to obtaining corresponding stress quantities, described in Sect. 2.7. Also for completeness, we define the strain invariants R_1, R_2, and R_3 analogous to Q_1, Q_2, and Q_3 in (2.58) and (2.59) with the σ_{ij} replaced by ε_{ij} in each case. Of particular physical significance is

$$R_1 = \varepsilon_{ii} = u_{i,i} \tag{3.38}$$

which is called the cubical strain and is related to the volume change as shown in Sect. 3.5.

3.5 Volume and Shape Change

It is sometimes convenient to separate the components of strain into those that cause changes in the *volume* and those that cause changes in the *shape* of a differential element.

Consider a volume element oriented with the *principal* directions, similar to that shown in Fig. 2.1, but with the shearing stresses equal to zero. The original length of each side is taken as $dx^{(k)}$ and the final length as $dX^{(k)}$ where

$$dX^{(k)} = dx^{(k)}\left(1 + \varepsilon^{(k)}\right). \tag{3.39}$$

The original volume is found as

$$dv = \prod_{k=1}^{3} dx^{(k)} \tag{3.40a}$$

and the final volume by

$$dV = \prod_{k=1}^{3} dX^{(k)}$$

$$= \prod_{k=1}^{3} \left(1 + \varepsilon^{(k)}\right) dx^{(k)} \qquad (3.40b)$$

$$= dv \prod_{k=1}^{3} \left(1 + \varepsilon^{(k)}\right).$$

Now, the *dilatation* is defined as the relative change in volume [6],

$$\Delta = \frac{dV - dv}{dv}$$

$$= \frac{dv \left\{ \left[\prod_{k=1}^{3} \left(1 + \varepsilon^{(k)}\right) \right] - 1 \right\}}{dv} \qquad (3.41)$$

$$= \left[\left(1 + \varepsilon^{(1)}\right) \left(1 + \varepsilon^{(2)}\right) \left(1 + \varepsilon^{(3)}\right) \right] - 1$$

$$= \varepsilon^{(1)} + \varepsilon^{(2)} + \varepsilon^{(3)} + \text{higher-order terms}$$

$$\cong \sum_{k=1}^{3} \varepsilon^{(k)},$$

which is simply the sum of the diagonal terms of the principal strain tensor.

The dilatation Δ is also equal to the sum of the diagonal strains ε_{ii}, which is the first strain invariant R_1, and the divergence of the displacement vector $\nabla \cdot \boldsymbol{u} = u_{i,i}$, where $\boldsymbol{u}$ is defined by (3.5c).

Changes in shape of the volume element are denoted by the shearing strains ε_{ij} with $i \neq j$, sometimes called the *detrusions* [2].

The preceding separation of the strains into dilatational and detrusional components may be formalized by the resolution of the strain tensor into two terms [3]:

$$\boldsymbol{\varepsilon} = \boldsymbol{\varepsilon}_M + \boldsymbol{\varepsilon}_D \qquad (3.42)$$

in which

$$\boldsymbol{\varepsilon}_M = \begin{bmatrix} \varepsilon_M & 0 & 0 \\ 0 & \varepsilon_M & 0 \\ 0 & 0 & \varepsilon_M \end{bmatrix} = \text{mean normal strain} \qquad (3.43)$$

$$\boldsymbol{\varepsilon}_D = \begin{bmatrix} \varepsilon_{11} - \varepsilon_M & \varepsilon_{12} & \varepsilon_{13} \\ \varepsilon_{12} & \varepsilon_{22} - \varepsilon_M & \varepsilon_{23} \\ \varepsilon_{13} & \varepsilon_{23} & \varepsilon_{33} - \varepsilon_M \end{bmatrix} = \text{strain deviator} \qquad (3.44)$$

and where

$$\varepsilon_M = \frac{1}{3}(\varepsilon_{11} + \varepsilon_{22} + \varepsilon_{33})$$

$$= \frac{1}{3}\left(\varepsilon^{(1)} + \varepsilon^{(2)} + \varepsilon^{(3)}\right) \tag{3.45}$$

$$= \frac{1}{3}R_1.$$

The mean normal strain ε_M corresponds to a state of equal elongation in all directions for an element at a given point. The element would remain similar to the original shape but change in volume with a dilatation of ε_{ii}. For this reason, ε_M may be termed the *volumetric* component of strain.

The strain deviator, or *deviatoric* component, ε_D, has a dilatation calculated from (3.41) of

$$\Delta = \sum_{i=1}^{3} \varepsilon_D{}^{(i)}$$

$$= \varepsilon_{D_{ii}}$$

$$= (\varepsilon_{11} - \varepsilon_M) + (\varepsilon_{22} - \varepsilon_M) + (\varepsilon_{33} - \varepsilon_M) \tag{3.46}$$

$$= \varepsilon_{11} + \varepsilon_{22} + \varepsilon_{33} - 3\left[\frac{1}{3}(\varepsilon_{11} + \varepsilon_{22} + \varepsilon_{33})\right]$$

$$= 0.$$

This characterizes a change in *shape* of an element with no change in *volume*. The invariants of the strain deviator tensor (3.44) play an important role in the theory of plasticity since shape changes are of importance.

Corresponding to the deformations associated with the mean normal strain and strain deviator, we define the principal stresses associated with changes in the volume and the shape as

$$\boldsymbol{\sigma} = \boldsymbol{\sigma}_M + \boldsymbol{\sigma}_D \tag{3.47}$$

in which

$$\boldsymbol{\sigma}_M = \begin{bmatrix} \sigma_M & 0 & 0 \\ 0 & \sigma_M & 0 \\ 0 & 0 & \sigma_M \end{bmatrix} = \text{mean normal stress} \tag{3.48}$$

$$\boldsymbol{\sigma}_D = \begin{bmatrix} \sigma_{11} - \sigma_M & \sigma_{12} & \sigma_{13} \\ \sigma_{12} & \sigma_{22} - \sigma_M & \sigma_{23} \\ \sigma_{13} & \sigma_{23} & \sigma_{33} - \sigma_M \end{bmatrix} = \text{stress deviator} \tag{3.49}$$

and where

$$\sigma_M = \frac{1}{3}(\sigma_{11} + \sigma_{22} + \sigma_{33})$$
$$= \frac{1}{3}\left(\sigma^{(1)} + \sigma^{(2)} + \sigma^{(3)}\right). \tag{3.50}$$

In indicial form, the volumetric and deviatoric components of stress are, respectively,

$$\sigma_{M_{ij}} = \frac{1}{3}\sigma_{kk}\delta_{ij} \tag{3.51}$$

and

$$\sigma_{D_{ij}} = \sigma_{ij} - \sigma_{M_{ij}}$$
$$= \sigma_{ij} - \frac{1}{3}\sigma_{kk}\delta_{ij}. \tag{3.52}$$

For principal axes, the stress deviator becomes

$$\boldsymbol{\sigma}_D = \begin{bmatrix} \sigma^{(1)} - \sigma_M & 0 & 0 \\ 0 & \sigma^{(2)} - \sigma_M & 0 \\ 0 & 0 & \sigma^{(3)} - \sigma_M \end{bmatrix}$$

$$= \begin{bmatrix} \sigma_D^{(1)} & 0 & 0 \\ 0 & \sigma_D^{(2)} & 0 \\ 0 & 0 & \sigma_D^{(3)} \end{bmatrix}. \tag{3.53}$$

Note that the sum of the diagonals of $\boldsymbol{\sigma}_D$ is

$$\sigma_{D_{ii}} = \sigma^{(1)} + \sigma^{(2)} + \sigma^{(3)} - 3 \cdot \frac{1}{3}\left[\sigma^{(1)} + \sigma^{(2)} + \sigma^{(3)}\right] = 0. \tag{3.54}$$

The stress deviator tensor referred to the principal axes (3.53) is the basis of the renown von Mises yield condition, discussed in Sect. 12.3.3.

3.6 Compatibility

If we examine (3.14), $\varepsilon_{ij} = \frac{1}{2}\left(u_{i,j} + u_{j,i}\right)$, we see that the components of strain may be computed once the displacements are known, provided that they are once differentiable. However, the inverse problem of calculating the displacements from the strains is not so direct since there are six independent equations and only three unknowns. Conditions of *compatibility*, imposed on the components of strain ε_{ij}, may be shown to be necessary and sufficient [5] to insure a continuous single-valued displacement field **u**.

The usual procedure is to eliminate the displacements between the equations repeatedly to produce equations with only strains as unknowns. First we take two normal components $\varepsilon_{(ll)}$ and $\varepsilon_{(mm)}$ and the shearing component $\varepsilon_{(lm)}$ with the summation convention suspended.

Taking second partials, we have

$$\varepsilon_{(ll,mm)} = u_{(l,lmm)} \tag{3.55a}$$

$$\varepsilon_{(mm,ll)} = u_{(m,mll)} \tag{3.55b}$$

$$\varepsilon_{(lm,lm)} = \frac{1}{2}\left(u_{(l,mlm)} + u_{(m,llm)}\right)$$
$$= \frac{1}{2}\left(u_{(l,lmm)} + u_{(m,mll)}\right), \tag{3.55c}$$

with interchangeability of the order of differentiation assumed. Next, we equate (3.55a) plus (3.55b) to twice (3.55c) to get

$$\varepsilon_{(ll,mm)} + \varepsilon_{(mm,ll)} = 2\varepsilon_{(lm,lm)}. \tag{3.56}$$

Then we take one normal component $\varepsilon_{(ll)}$ and three shearing components ε_{lm}, ε_{ln}, and ε_{mn} and write the second partials

$$\varepsilon_{(ll,mm)} = u_{(l,lmn)} \tag{3.57a}$$

$$\varepsilon_{(lm,ln)} = \frac{1}{2}\left(u_{(l,mln)} + u_{(m,lln)}\right) \tag{3.57b}$$

$$\varepsilon_{(ln,lm)} = \frac{1}{2}\left(u_{(l,nlm)} + u_{(n,llm)}\right) \tag{3.57c}$$

$$\varepsilon_{(mn,ll)} = \frac{1}{2}\left(u_{(m,nll)} + u_{(n,mll)}\right). \tag{3.57d}$$

We find that (3.57a) can be equated to (3.57b) plus (3.57c) minus (3.57d) to give

$$\varepsilon_{(lm,ln)} + \varepsilon_{(ln,lm)} - \varepsilon_{(mn,ll)} = \varepsilon_{(ll,mn)}. \tag{3.58}$$

Equations (3.56) and (3.58) are six equations relating the components of strain that insure that they will integrate into a unique displacement field. They are of importance and may be written explicitly as

$$\varepsilon_{11,22} + \varepsilon_{22,11} = 2\varepsilon_{12,12} \tag{3.59a}$$

$$\varepsilon_{22,33} + \varepsilon_{33,22} = 2\varepsilon_{23,23} \tag{3.59b}$$

$$\varepsilon_{33,11} + \varepsilon_{11,33} = 2\varepsilon_{31,31} \tag{3.59c}$$

$$\varepsilon_{12,13} + \varepsilon_{13,12} - \varepsilon_{23,11} = \varepsilon_{11,23} \tag{3.59d}$$

$$\varepsilon_{23,21} + \varepsilon_{21,23} - \varepsilon_{31,22} = \varepsilon_{22,31} \tag{3.59e}$$

$$\varepsilon_{31,32} + \varepsilon_{32,31} - \varepsilon_{12,33} = \varepsilon_{33,12}. \tag{3.59f}$$

These are known as the $St.$ $Venant$ compatibility equations that may also be extracted from the tensor equation

$$\varepsilon_{ij,kl} + \varepsilon_{kl,ij} = \varepsilon_{ik,jl} + \varepsilon_{jl,ik} \tag{3.60}$$

which represents $3^4 = 81$ equations, only six of which are distinct; for example, for $i = 1, j = 1, k = 2$, and $l = 2$, we get

$$\varepsilon_{11,22} + \varepsilon_{22,11} = \varepsilon_{12,12} + \varepsilon_{12,12} = 2\varepsilon_{12,12} \tag{3.61}$$

which is also (3.59a). It has been shown that these equations are not strictly independent. A further differential combination of the six equations produces three identities connecting the equations which are known as the Bianchi formulas. However, one may not choose any three of the six, so the usual approach is to include all six in the formulation [7].

Even though we have the compatibility equations, the formulation is still incomplete in that there is no connection between the equilibrium equations (2.34), three equations in six unknowns σ_{ij}, and the kinematic equations (3.14), six equations in nine unknowns ε_{ij} and u_i. We will seek the connection between the equilibrium and the kinematic equations in the laws of physics governing material behavior, considered in the next chapter.

3.7 Incompatibility

The strain compatibility relationships given as (3.56) and (3.58) may not be satisfied by strain fields from nonmechanical loads such as thermal and plastic deformation [8]. This leads to residual stresses that are necessary to restore geometric continuity. The resulting *incompatible strain field* $\boldsymbol{\varepsilon}^*$ with components ε_{ij}^*, which are referred to as *eigenstrains* or *inherent strains*, can be described by a generalization of the $St.$ $Venant$ compatibility equations.

As suggested by Matos and Dodds [8], (3.56) and (3.58) can be combined and written in terms of the incompatibility strains as

$$R_{pq} = \varepsilon_{pki}\varepsilon_{qlj}\varepsilon_{ij,kl} \quad (i.j,k,l = 1,2,3). \tag{3.62}$$

In (3.62), ε_{pki} and ε_{qlj} are permutation tensors as defined in (1.27), and R_{pq} represents a residual incompatibility tensor. If R_{pq} vanishes, no residual stresses are required to restore geometric compatibility. In the more general case, the total compatibility strain tensor ε is decomposed as

$$\varepsilon = \varepsilon^e + \varepsilon^*, \tag{3.63}$$

where ε^* is the elastic strain tensor necessary to restore compatibility. The incompatibility was produced by the eigenstrain tensor ε^* and the separately applied nonmechanical loads.

The final linear elastic stresses are computed from (4.3) or simplifications thereof, with ε_{kl}^e evaluated from (3.63) by $\varepsilon_{kl}^e = \varepsilon_{kl} - \varepsilon_{kl}^*$. As an example, the treatment of thermal strains is given in Sect. 4.7.

3.8 Nonlinear Strain–Displacement Relations

The strain–displacement relations, stated in either the Lagrangian basis (3.8) or the Eulerian basis (3.12), are strictly geometric and do not deal with either the cause of the deformations or with the laws that describe the resistance of the deformations by the body. The l.h.s of the equations has two types of strains, extensional $\varepsilon_{(ii)}$ and shearing ε_{ij} ($i \neq j$). Also the rotations ω_{ij} are embedded as shown in (3.35). These equations include squares and products of the strain components in the last term $u_{k,i}u_{k,j}$ and are thus deemed nonlinear. The tensor formed by the products $u_{k,i}u_{k,j}$ plays an important role in nonlinear elasticity. Based on the physical nature of the specific problem, some of the product terms may be omitted leading to a variety of nonlinear strain–displacement relations. When all of the products are neglected, we come to the linear theory (3.14).

The "smallness" of strain and rotational components associated with the linear theory does not necessarily carry over to the magnitude of the eventual deformations. It has been argued by the eminent Russian mechanician V.V. Novozhilov [9] that associating the nonlinear theory with finite deformations, while commonly accepted, may not be strictly correct. There are cases in which finite deformations may occur with the so-called small strains and other cases where the nonlinear strain terms produce relatively small displacements. Some examples will be noted in later chapters.

To gain some insight into the participation of the product terms in the nonlinear equations, we again consider the specific term of (3.8) with $i = j = 2$ as expanded in Sect. 3.3 to produce (3.35):

$$\varepsilon_{22}^L = \varepsilon_{22} + \frac{1}{2}(\varepsilon_{22})^2 + \frac{1}{2}(\varepsilon_{12} + \omega_{12})^2 + \frac{1}{2}(\varepsilon_{23} - \omega_{23})^2. \tag{3.35}$$

Equation (3.35) clearly expresses the nonlinear Lagrangian strains in terms of the physically defined linear strains and rotations. We now look to these physical relationships to simplify the equations by eliminating some of the terms. The second term in (3.35) is the square of the first, and the direct strain is necessarily small compared to unity for most traditional elastic engineering materials, so this term is often summarily dropped since it appears with the like unsquared term. Likewise, the squared strain terms such as $(\varepsilon_{12})^2$ are of comparable small magnitude compared to unity but do not have an unsquared counterpart. Then we have the squared rotation terms such as $(\omega_{12})^2$, which represent rigid body motion that is not restricted by the elastic capacity of traditional materials and may be significant for some elastic bodies. Finally in $(\varepsilon_{23} - \omega_{23})^2$, we find products of strains and rotations such as $\varepsilon_{12}\omega_{12}$ that could possibly be significant. A similar argument may be made for each of the shearing strains ε_{ij}^L.

These considerations give rise to a variety of nonlinear theories based on the Lagrangian formulation that may be termed finite strain, small strain–moderate rotation, etc. Most of these theories lie beyond the scope of this introductory treatment, except perhaps for the case where all of the strain products other than those for rotations ω_{ij} are dropped on the basis of the preceding arguments. This is commonly called the theory of small strains and moderate rotations. For this simplification, (3.35) becomes

$$\varepsilon_{22}^L \approx \varepsilon_{22} + \frac{1}{2}\left[(\omega_{12})^2 + (\omega_{23})^2\right]. \tag{3.64}$$

In the general case of moderate rotations, the relationships take the form

$$\varepsilon_{ij}^L \cong \varepsilon_{ij} + \frac{1}{2}\omega_{ki}\omega_{kj}. \tag{3.65}$$

Bažant and Cedolin [10] noted that this form would describe the strains in bodies that are regarded as thin, such as plates and shells considered in Chap. 8. They argue that except for the case of very large deflections, these bodies (1) have strains ε_{ij} and in-plane rotations ω_{12} that are negligible compared to out-of-plane rotations ω_{13} and ω_{23} and (2) have out-of-plane shear and normal strains ε_{13}, ε_{23}, and ε_{33} that are negligible compared to the in-plane strains ε_{11}, ε_{22}, and ε_{12}. If the thin direction is defined by the normal $\mathbf{n}_3$, the sums in (3.65) are limited to $i = 1, 2$ and $j = 1, 2$.

When the deformations become greater than infinitesimal, it is common to find the extensional strain-type quantities represented as the ratio of the final length to the initial length of a fiber, e.g.,

$$\lambda_i = 1 + \varepsilon_{(ii)} \tag{3.66}$$

called the *stretch ratio* or just the *stretch* in the literature.

3.9 Strain Ellipsoid

In a similar manner to the stress ellipsoid described in Sect. 2.6.4, a 3-D strain ellipsoid and a contracted 2-D stain ellipse may be defined. The 2-D representation is particularly useful for visualizing the intensity of strain in a planar object or in slices of a solid object.

Fig. 3.3a Strain ellipse

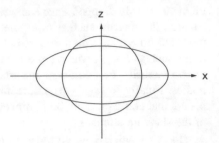

Basically the strain ellipse, Fig. 3.3a, is the product of a finite strain applied to a circle of unit radius. The circle becomes an ellipse with the major and minor principal axes equal to the principal stretch ratios, as defined in (3.66). The radius of the ellipse in any direction is proportional to the corresponding stretch ratio.

To understand the use of the strain ellipse, consider again the infinitesimal vectors dx_i and dX_i in Fig. 3.1. If we use a spatial transformation matrix F linking the undeformed and deformed vectors based on (3.6) and (3.10), we have

$$dx_i = x_{i,j}\,dX_j = F_{ij}\,dX_j \tag{3.67a}$$

$$dX_i = X_{i,j}\,dx_j = F_{ij}^{-1}\,dx_j. \tag{3.67b}$$

Substituting into (3.7) to get the Lagrangian strain,

$$
\begin{aligned}
(ds)^2 - (dS)^2 &= F_{ik}\,dX_k\,F_{il}\,dX_l - dX_m\,dX_m \\
&= (F_{ik}\,F_{il} - \delta_{km}\delta_{lm})dX_k dX_l \\
&= (F_{ik}\,F_{il} - \delta_{kl})dX_k\,dX_l \\
\varepsilon_{kl}^L &= \frac{1}{2}(F_{ik}\,F_{il} - \delta_{kl}).
\end{aligned}
\tag{3.68}
$$

Similarly we can get the Eulerian strain

$$\varepsilon_{kl}^E = \frac{1}{2}\left(\delta_{kl} - F_{ki}^{-1}F_{li}^{-1}\right). \tag{3.69}$$

The transformation matrix F is called the deformation gradient and is widely used in continuum mechanics [7] since its application is not constrained to small strain. The physical meaning is that F maps the undeformed vector into the deformed vector. Figure 3.3b illustrates the physical meaning of the deformation gradient with a strain ellipse. The blue circle is a unit circle with a radius of 1; it represents the undeformed configuration. After a certain deformation defined by a deformation gradient F, each point of the unit circle displaces to its deformed position. The green ellipse is the deformed state. If we use vector A to represent a certain point in the undeformed state, the deformed new position vector $a = FA$. The deformed shape of an ellipse can be determined by applying the same calculation for each point on the

Fig. 3.3b Strain ellipse for
2-D finite deformations

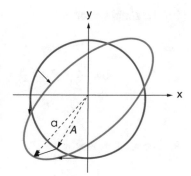

unit circle. For example, a deformation gradient $\mathbf{F} = \begin{bmatrix} 1 & 0.25 \\ 0.5 & 1 \end{bmatrix}$ is applied to a
unit circle to generate the strain ellipse. The Lagrangian tensor is $\boldsymbol{\varepsilon}^L = \begin{bmatrix} 0.0313 & 0.3750 \\ 0.3750 & 0.1250 \end{bmatrix}$ following (3.68). The component values of the deformation
gradient are not directly measurable from experiment. Usually the values are calcu-
lated based on displacements or strains measured [13]. An application of the strain
ellipse to illustrate finite deformations in brain tissue is presented in Ref. [11].

3.10   Computational Example

A displacement field is given by

$$\mathbf{u} = [(3x^2y + 6) \times 10^{-2}]\mathbf{e}_x + [(y^2 + 6xz) \times 10^{-2}]\mathbf{e}_y \\ + [(6z^2 + 2yz + 10) \times 10^{-2}]\mathbf{e}_z. \tag{3.70}$$

We seek to compute the components of the strain and rotation tensors ε_{ij} and ω_{ij}
and to verify the compatibility equations.

1. The partial derivatives $\times 10^{-2}$ with $x = 1$, $y = 2$, and $z = 3$ are

$$
\begin{array}{lll}
u_{1,1} = 6xy; & u_{1,2} = 3x^2; & u_{1,3} = 0; \\
u_{1,11} = 6y; & u_{1,12} = 6x; & u_{1,112} = 6. \\
u_{2,1} = 6z; & u_{2,2} = 2y; & u_{2,3} = 6x; \\
u_{2,13} = 6; & u_{2,22} = 2. & \\
u_{3,1} = 0; & u_{3,2} = 2z; & u_{3,3} = 12z + 2y; \\
u_{3,23} = 2; & u_{3,33} = 12. &
\end{array}
$$

2. The strain tensor is

$$
\varepsilon = \begin{bmatrix} 6xy & \frac{1}{2}(3x^2 + 6z) & 0 \\ \frac{1}{2}(3x^2 + 6z) & 2y & \frac{1}{2}(6x + 2z) \\ 0 & \frac{1}{2}(6x + 2z) & 12z + 2y \end{bmatrix} \times 10^{-2}.
$$

3. The rotation tensor is

$$
\omega = \begin{bmatrix} 0 & \frac{1}{2}(3x^2 - 6z) & 0 \\ -\frac{1}{2}(3x^2 - 6z) & 0 & \frac{1}{2}(6x - 2z) \\ 0 & -\frac{1}{2}(6x - 2z) & 0 \end{bmatrix} \times 10^{-2}.
$$

For any given points (x, y, z), the strains and average rotations may be computed as the elements of ε_{ij} and ω_{ij}, respectively, and the principal strains and directions determined.

4. Compatibility: since only nonzero second partial derivatives enter into (3.59a), (3.59b), (3.59c), (3.59d), (3.59e), and (3.59f), we consider $\varepsilon_{12,11} = 3$ and $\varepsilon_{21,11} = 3$. Neither of these appears in (3.59a), (3.59b), (3.59c), (3.59d), (3.59e), and (3.59f); thus, the latter equations are identically satisfied. Beyond this trivial example, we may observe that all strain fields which are *linear* will identically satisfy the compatibility relations. This is of some importance in generating numerical or approximate solutions to the equations of the theory of elasticity since linear approximations are otherwise convenient.

3.11 Computational Examples with MATLAB

The computational example in Sect. 3.10 can be easily solved by MuPAD. First, we write out each displacement components, by using ux, uy, and uz, to represent the three displacement components:

```
ux := 3*x^2*y + 6
3 y x² + 6

uy := y^2 + 6*x*z
y² + 6 x z

uz := 6*z^2 + 2*y*z + 10
6 z² + 2 y z + 10
```

The partial derivatives $\times 10^{-2}$ with $x = 1$, $y = 2$, and $z = 3$ can be derived. A lower bar is used to represent the comma derivative. For example, u1_1 represents $u_{1,1}$, and u1_11 represents $u_{1,11}$.

u1_1 := diff(ux, x)
6 x y

u2_3 := diff(uy, z)
6 x

u1_2 := diff(ux, y)
$3 x^2$

u2_13 := diff(u2_1, z)
6

u1_3 := diff(ux, z)
0

u2_22 := diff(u2_2, y)
2

u1_11 := diff(u1_1, x)
6 y

u3_1 := diff(uz, x)
0

u1_12 := diff(u1_1, y)
6 x

u3_2 := diff(uz, y)
2 z

u1_112 := diff(u1_11, y)
6

u3_3 := diff(uz, z)
2 y + 12 z

u2_1 := diff(uy, x)
6 z

u3_23 := diff(u3_2, z)
2

u2_2 := diff(uy, y)
2 y

u3_33 := diff(u3_3, z)
12

Finally, the strain tensor $(\times 10^{-2})$ can be derived.

e11 := u1_1
6 x y

e22 := u2_2
2 y

e12 := 1/2*(u1_2 + u2_1)
$\dfrac{3 x^2}{2} + 3 z$

e23 := 1/2*(u2_3 + u3_2)
3 x + z

e13 := 1/2*(u1_3 + u3_1)
0

e33 := u3_3
2 y + 12 z

The strain tensor $(\times 10^{-2})$ can be displayed using conventional matrix format.

Print out strain tensor ($\times 10^{-2}$)

```
ep := matrix([[e11, e12, e13], [e12, e22, e23], [e13, e23, e33]])
```

$$\begin{pmatrix} 6\,x\,y & \frac{3\,x^2}{2}+3\,z & 0 \\ \frac{3\,x^2}{2}+3\,z & 2\,y & 3\,x+z \\ 0 & 3\,x+z & 2\,y+12\,z \end{pmatrix}$$

Similarly, the rotation tensor ($\times 10^{-2}$) can be derived and displayed.

```
w12 := 1/2*(u1_2 - u2_1)
```

$$\frac{3\,x^2}{2}-3\,z$$

```
w13 := 1/2*(u1_3 - u3_1)
```

$$0$$

```
w23 := 1/2*(u2_3 - u3_2)
```

$$3\,x-z$$

```
rotW := matrix([[0, w12, w13], [w12, 0, w23], [w13, w23, 0]])
```

$$\begin{pmatrix} 0 & \frac{3\,x^2}{2}-3\,z & 0 \\ \frac{3\,x^2}{2}-3\,z & 0 & 3\,x-z \\ 0 & 3\,x-z & 0 \end{pmatrix}$$

3.12 Exercises

3.1 The components of a displacement field are (from [6])

$$u_x = (x^2 + 20) \times 10^{-4},$$
$$u_y = 2yz \times 10^{-3},$$
$$u_z = (z^2 - xy) \times 10^{-3}.$$

(a) Consider two points in the undeformed system, (2, 5, 7) and (3, 8, 9). Find the change in distance between these points.
(b) Compute the components of the strain tensor.
(c) Compute the components of the rotation tensor.
(d) Compute the strain at (2, −1, 3).
(e) Does the displacement field satisfy compatibility?

3.2 The strain field at a point $P(x, y, z)$ in an elastic body is given by

$$\varepsilon = \begin{bmatrix} 20 & 3 & 2 \\ 3 & -10 & 5 \\ 2 & 5 & -8 \end{bmatrix} \times 10^{-6}.$$

Determine the following values:

(a) The strain invariants
(b) The principal strains
(c) The octahedral shearing strain
(d) The absolute maximum shearing strain
(e) The mean normal strain and the strain deviator

3.3 With reference to Problem 3.2, consider a second coordinate system x_i' in which x_2' is parallel to x_2 and x_1' and x_3' are defined by a 60° counterclockwise rotation about x_2. Find the components of strain with reference to the x_i' system using the strain field from Problem 3.2.

3.4 The components of a strain tensor referred to the x_1-, x_2-, and x_3-axes in the figure are

$$\varepsilon = \begin{bmatrix} 0.02 & 0 & 0 \\ 0 & 0.01 & 0 \\ 0 & 0 & 0.03 \end{bmatrix}.$$

$OA = OB = OC$ and D is the midpoint of AC. The direction cosines of AC are $(1/\sqrt{2}, 0, \text{ and } -1/\sqrt{2})$, and those of BD are $(-1/\sqrt{6}, \sqrt{2}/\sqrt{3}, -1/\sqrt{6})$.
Find:

(a) The elongation of line AC
(b) The change of initial right angle BDA

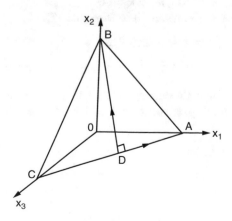

3.5 The components of a strain tensor referred to the x_1-, x_2-, and x_3-axes in the figure are given by

$$\varepsilon = \begin{bmatrix} 0.02 & -0.003 & 0 \\ -0.003 & 0.01 & 0.02 \\ 0 & 0.02 & 0.01 \end{bmatrix},$$

and they are constant in the region under consideration. The direction cosines of AC and DB are as given in Exercise 3.4:

(a) Find the relative elongations of the lines AC and DB.
(b) Find the change of the initially right angle ADB.
(c) Find the first, second, and third invariants of strain.
(d) Illustrate with sketches the deformation of a rectangular parallelepiped with its edges parallel to the x_1-, x_2-, and x_3-axes, respectively.

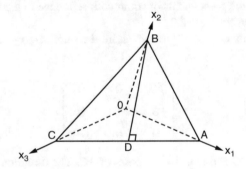

3.6 Compute the infinitesimal strains, and sketch the deformed configuration of an initially rectangular volume element subject to the following displacement fields (from [1]), where A, B, and $D_{ij} = $ constants:

(a) Simple extension: $u_1 = A_1 x_1$; $u_2 = u_3 = 0$.
(b) Simple shear: $u_1 = B x_2$; $u_2 = u_3 = 0$.
(c) Homogeneous deformation: $u_i = D_{ij} x_j$ [see 12].

3.7 Consider a (linear) strain field such that

$$\varepsilon_{11} = A x_2^2, \quad \varepsilon_{22} = A x_1^2, \quad \varepsilon_{12} = B x_1 x_2, \quad \varepsilon_{33} = \varepsilon_{32} = \varepsilon_{31} = 0.$$

Find the relationship between A and B such that it is possible to obtain a single-valued continuous displacement field which corresponds to the given strain field.

3.8 Consider the St. Venant compatibility equations (3.59a), (3.59b), (3.59c), (3.59d), (3.59e), and (3.59f)

(a) $R_3 = \varepsilon_{11,22} + \varepsilon_{22,11} - 2\varepsilon_{12,12} = 0$
$R_1 = \varepsilon_{22,33} + \varepsilon_{33,22} - 2\varepsilon_{23,23} = 0$
$R_2 = \varepsilon_{33,11} + \varepsilon_{11,33} - 2\varepsilon_{31,31} = 0$
$U_1 = -\varepsilon_{11,23} + (-\varepsilon_{23,1} + \varepsilon_{31,2} + \varepsilon_{12,3})_{,1} = 0$
$U_2 = -\varepsilon_{22,31} + (\varepsilon_{23,1} - \varepsilon_{31,2} + \varepsilon_{12,3})_{,2} = 0$
$U_3 = -\varepsilon_{33,12} + (\varepsilon_{23,1} + \varepsilon_{31,2} - \varepsilon_{12,3})_{,3} = 0$

(b) Show that these equations satisfy the Bianchi formulas [7]:

$$R_{1,1} + U_{3,2} + U_{2,3} = 0$$
$$U_{3,1} + R_{2,2} + U_{1,3} = 0$$
$$U_{2,1} + U_{1,2} + R_{3,3} = 0.$$

References

1. Tauchert TR (1974) Energy principles in structural mechanics. McGraw-Hill Book Company Inc., New York
2. Westergaard HM (1964) Theory of elasticity and plasticity. Dover Publications, Inc., New York
3. Filonenko-Borodich M (1965) Theory of elasticity, translated from Russian by M. Konayeon. Dover Publications Inc., New York
4. Ma Y, Desai CS (1990) Alternative definition of finite strains. J Eng Mech, ASCE 116 (4):901–919
5. Pearson CE (1959) Theoretical elasticity. Harvard University Press, Cambridge
6. Ugural AC, Fenster SK (1975) Advanced strength and applied elasticity. Elsevier North Holland, New York
7. Malvern LE (1969) Introduction to the mechanics of a continuous medium. Prentice-Hall, Englewood Cliffs
8. Matos CG, Dodds RH (2000) Modelling the effects of residual stresses on defects in welds of steel frame connections. Eng Struct 22(9):1103–1120
9. Novozhilov VV (1953) Foundations of the nonlinear theory of elasticity. Graylock Press, Rochester (Dover Publications, Inc., New York, 1999)
10. Bažant ZP, Cedolin L (1999) Stability of structures. Oxford University Press, New York
11. Feng Y et al (2010) Relative brain displacement and deformation during constrained mild frontal head impact. J R Soc Interface 7:1677–1688
12. Timoshenko SP, Goodier JN (1951) Theory of elasticity. McGraw-Hill Book Company Inc, New York
13. Feng Y et al (2016) On the accuracy and fitting of transversely isotropic material models. J Mech Beha Biomed Mat 61:554–566

Chapter 4
Material Behavior

Abstract The equilibrium equations derived in Chap. 2 must be connected with the kinematic relations developed in Chap. 3. This coupling is accomplished by considering the mechanical properties of the materials for which the theory of elasticity is to be applied and is expressed by *constitutive* or *material* laws. Beginning with the generalized Hooke's law, the relationships between stress and strain are progressively reduced to the familiar isotropic form. Several modern characterizations, including transverse isotropic and functionally graded materials, are also considered.

4.1 Introduction

The need to connect the equilibrium equations derived in Chap. 2 with the kinematic relations developed in Chap. 3 was pointed out earlier. This coupling is accomplished by considering the mechanical properties of the materials for which the theory of elasticity is to be applied and is expressed by *constitutive* or *material* laws.

4.2 Uniaxial Behavior

The most elementary description of material behavior is the well-known *Hooke's law* stated by Robert Hooke in the late seventeenth century [1] in terms of force and elongation, apparently before the more sophisticated concepts of stress and strain were introduced, leading to the present one-dimensional extensional test that produces

$$\sigma_{11} = E\varepsilon_{11}, \tag{4.1}$$

where E is called the *modulus of elasticity* or *Young's modulus* after Thomas Young who introduced the term in the early nineteenth century [1]. This test also reveals some additional bases of material characterization as shown in Fig. 4.1, which is an automatically recorded plot of two cycles of loading and unloading for mild steel.

© Springer International Publishing AG, part of Springer Nature 2018
P. L. Gould, Y. Feng, *Introduction to Linear Elasticity*,
https://doi.org/10.1007/978-3-319-73885-7_4

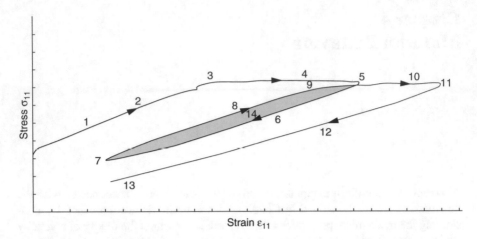

Fig. 4.1 Stress–strain curve

Branch (1) is the initial elastic response, from which E may be calculated as the slope. In the vicinity of (2), there is a decrease in the slope, commencing at the *proportional limit* and progressing until the *yield point* is reached at (3). Yielding progresses along branch (4) until unloading commences at (5) and continues along (6). The unloading ceases, and reloading begins at (7) to initiate another cycle that extends through (13) when the test is terminated. Region (14) is known as a *hysteresis loop* and is a measure of the energy dissipated through one excursion beyond the yield point.

A material that is loaded only to a level below the yield strain and that unloads along the same path is called *elastic*. If the path is linear as well, the material is said to be *linearly elastic*. It is also possible to have *nonlinear elastic* materials for which the coincident paths are curved. Both of these possibilities are addressed by the theory of elasticity, although only the former case is considered here in any detail. Past the yield point, we have the inelastic regime that requires additional considerations.

Beyond the uniaxial test, the characterization of material behavior is far more involved, from both a theoretical and an experimental standpoint, and is discussed in the following sections.

4.3 Generalized Hooke's Law

As a prerequisite to the postulation of a *linear* relationship between each component of stress and strain, it is necessary to establish the existence of a strain energy density W that is a homogeneous quadratic function of the strain components $W(\varepsilon_{ij})$ [2]. This concept is attributed to George Green [1]. For a body that is slightly strained by gradual application of the loading while the temperature remains constant, this will produce stress components derivable as

$$\sigma_{ij} = \frac{\partial W}{\partial \varepsilon_{ij}}. \tag{4.2a}$$

It is convenient initially to write W in terms of the engineering strains $\bar{\varepsilon}_{ij}$ as defined in Sect. 3.3. Thus,

$$\sigma_{ij} = \frac{\partial W(\bar{\varepsilon}_{ij})}{\partial \bar{\varepsilon}_{ij}}, \tag{4.2b}$$

where $\bar{\varepsilon}_{ij} = (2 - \delta_{ij})\varepsilon_{ij}$.

The function should have coefficients such that $W \geq 0$ in order to insure the stability of the body [2], with $W(0) = 0$ corresponding to a *natural* or *zero energy state* (unloaded). Energy principles are developed in Chap. 11.

Now, the generalized Hooke's law is written in the form of a fourth-order tensor:

$$\sigma_{ij} = E_{ijkl}\varepsilon_{kl} \tag{4.3}$$

in which the 81 coefficients E_{ijkl} are called the elastic constants. Since W is continuous, the order of differentiation in (4.2a) and (4.2b) is immaterial and E_{ijkl} is symmetric. Thus the number of independent equations reduces from nine to six, the elastic constants from 81 to 36, and then further to $[(36–6)/2] + 6 = 21$ when only half of the off-diagonal constants are counted. We represent this relationship in matrix form in terms of the engineering strains as

$$\boldsymbol{\sigma} = \mathbf{E}\bar{\boldsymbol{\varepsilon}}, \tag{4.4}$$

where

$$\begin{Bmatrix} \sigma_{11} \\ \sigma_{22} \\ \sigma_{33} \\ \sigma_{12} \\ \sigma_{23} \\ \sigma_{31} \end{Bmatrix} = \begin{bmatrix} E_{1111} & E_{1122} & E_{1133} & E_{1112} & E_{1123} & E_{1131} \\ & E_{2222} & E_{2233} & E_{2212} & E_{2223} & E_{2231} \\ & & E_{3333} & E_{3312} & E_{3323} & E_{3331} \\ & & & E_{1212} & E_{1223} & E_{1231} \\ & & & & E_{2323} & E_{2331} \\ & & & & & E_{3131} \end{bmatrix} \begin{Bmatrix} \bar{\varepsilon}_{11} \\ \bar{\varepsilon}_{22} \\ \bar{\varepsilon}_{33} \\ \bar{\varepsilon}_{12} \\ \bar{\varepsilon}_{23} \\ \bar{\varepsilon}_{31} \end{Bmatrix}. \tag{4.5}$$

Replacing the awkward quadruple subscripts, we have

$$
\left\{\begin{array}{c}
\sigma_{11} \\
\sigma_{22} \\
\sigma_{33} \\
\sigma_{12} \\
\sigma_{23} \\
\sigma_{31}
\end{array}\right\}
=
\left[\begin{array}{cccccc}
C_{11} & C_{12} & C_{13} & C_{14} & C_{15} & C_{16} \\
       & C_{22} & C_{23} & C_{24} & C_{25} & C_{26} \\
       &   & C_{33} & C_{34} & C_{35} & C_{36} \\
       &   &   & C_{44} & C_{45} & C_{46} \\
       &   &   &   & C_{55} & C_{56} \\
       &   &   &   &   & C_{66}
\end{array}\right]
\left\{\begin{array}{c}
\bar{\varepsilon}_{11} \\
\bar{\varepsilon}_{22} \\
\bar{\varepsilon}_{33} \\
\bar{\varepsilon}_{12} \\
\bar{\varepsilon}_{23} \\
\bar{\varepsilon}_{31}
\end{array}\right\}
\tag{4.6}
$$

or

$$
\boldsymbol{\sigma} = \mathbf{C}\bar{\boldsymbol{\varepsilon}}. \tag{4.7}
$$

The corresponding strain energy function is [2]

$$
\begin{aligned}
2W = {} & C_{11}\bar{\varepsilon}_{11}^{2} + 2C_{12}\bar{\varepsilon}_{11}\bar{\varepsilon}_{22} + 2C_{13}\bar{\varepsilon}_{11}\bar{\varepsilon}_{33} + 2C_{14}\bar{\varepsilon}_{11}\bar{\varepsilon}_{12} + 2C_{15}\bar{\varepsilon}_{11}\bar{\varepsilon}_{23} \\
& + 2C_{16}\bar{\varepsilon}_{11}\bar{\varepsilon}_{31} + C_{22}\bar{\varepsilon}_{22}^{2} + 2C_{23}\bar{\varepsilon}_{22}\bar{\varepsilon}_{33} + 2C_{24}\bar{\varepsilon}_{22}\bar{\varepsilon}_{12} + 2C_{25}\bar{\varepsilon}_{22}\bar{\varepsilon}_{23} \\
& + 2C_{26}\bar{\varepsilon}_{22}\bar{\varepsilon}_{31} + C_{33}\bar{\varepsilon}_{33}^{2} + 2C_{34}\bar{\varepsilon}_{33}\bar{\varepsilon}_{12} + 2C_{35}\bar{\varepsilon}_{33}\bar{\varepsilon}_{23} + 2C_{36}\bar{\varepsilon}_{33}\bar{\varepsilon}_{31} \\
& + C_{44}\bar{\varepsilon}_{12}^{2} + 2C_{45}\bar{\varepsilon}_{12}\bar{\varepsilon}_{23} + 2C_{46}\bar{\varepsilon}_{12}\bar{\varepsilon}_{31} \\
& + C_{55}\bar{\varepsilon}_{23}^{2} + 2C_{56}\bar{\varepsilon}_{23}\bar{\varepsilon}_{31} \\
& + C_{66}\bar{\varepsilon}_{31}^{2}.
\end{aligned}
\tag{4.8}
$$

from which (4.2a) and (4.2b) produce (4.6).

The preceding characterization is the most general with which we need to be concerned; such a material is termed *anisotropic* [3]. Fortunately, most engineering materials possess properties of symmetry about one or more of the planes or axes, which allow the number of independent constants to be reduced. Devising valid experiments to isolate each material constant can be tedious and difficult as described in Sect. 12.4.4.

The first reduction is for one plane of symmetry, for example, plane x_2x_3 in Fig. 4.2a. This implies that the x_1-axis can be reversed as shown in Fig. 4.2b which corresponds to a coordinate transformation with direction cosines as shown below [4].

	x_1	x_2	x_3
x_1'	-1		
x_2'		1	
x_3'			1

(4.9)

Fig. 4.2 (a) Reference
material coordinates;
(b) single-plane symmetry

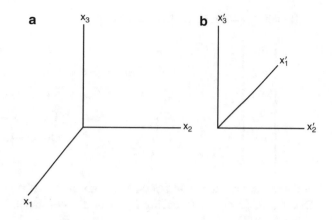

Using (2.13), $\sigma'_{ij} = \alpha_{ik}\alpha_{jl}\sigma_{kl}$, with the direction cosines given in (4.9), we find

$$\sigma'_{11} = \sigma_{11}; \quad \sigma'_{22} = \sigma_{22}; \quad \sigma'_{33} = \sigma_{33};$$
$$\sigma'_{12} = -\sigma_{12}; \quad \sigma'_{23} = \sigma_{23}; \quad \sigma'_{31} = -\sigma_{31}.$$

Similarly, using (3.16), $\varepsilon'_{ij} = \alpha_{ik}\alpha_{jl}\varepsilon_{kl}$, we find

$$\bar{\varepsilon}'_{11} = \bar{\varepsilon}_{11}; \quad \bar{\varepsilon}'_{22} = \bar{\varepsilon}_{22}; \quad \bar{\varepsilon}'_{33} = \bar{\varepsilon}_{33};$$
$$\bar{\varepsilon}'_{12} = -\bar{\varepsilon}_{12}; \quad \bar{\varepsilon}'_{23} = \bar{\varepsilon}_{23}; \quad \bar{\varepsilon}'_{31} = -\bar{\varepsilon}_{31}.$$

The preceding relations are reflected in the material law:

$$\left\{\begin{array}{c} \sigma'_{11} \\ \sigma'_{22} \\ \sigma'_{33} \\ \sigma'_{12} \\ \sigma'_{23} \\ \sigma'_{31} \end{array}\right\} = \left\{\begin{array}{c} \sigma_{11} \\ \sigma_{22} \\ \sigma_{33} \\ -\sigma_{12} \\ \sigma_{23} \\ -\sigma_{31} \end{array}\right\} = \left[\begin{array}{c} C_{ij} \end{array}\right] \left\{\begin{array}{c} \bar{\varepsilon}'_{11} \\ \bar{\varepsilon}'_{22} \\ \bar{\varepsilon}'_{33} \\ \bar{\varepsilon}'_{12} \\ \bar{\varepsilon}'_{23} \\ \bar{\varepsilon}'_{31} \end{array}\right\}$$

$$= \left[\begin{array}{c} C_{ij} \end{array}\right] \left\{\begin{array}{c} \bar{\varepsilon}_{11} \\ \bar{\varepsilon}_{22} \\ \bar{\varepsilon}_{33} \\ -\bar{\varepsilon}_{12} \\ \bar{\varepsilon}_{23} \\ -\bar{\varepsilon}_{31} \end{array}\right\}.$$

(4.10)

Rewriting (4.10) in the form of (4.6) gives

$$
\begin{Bmatrix} \sigma_{11} \\ \sigma_{22} \\ \sigma_{33} \\ \sigma_{12} \\ \sigma_{23} \\ \sigma_{31} \end{Bmatrix} =
\begin{bmatrix}
C_{11} & C_{12} & C_{13} & -C_{14} & C_{15} & -C_{16} \\
 & C_{22} & C_{23} & -C_{24} & C_{25} & -C_{26} \\
 & & C_{33} & -C_{34} & C_{35} & -C_{36} \\
 & & & C_{44} & -C_{45} & C_{46} \\
 & & & & C_{55} & -C_{56} \\
 & & & & & C_{66}
\end{bmatrix}
\begin{Bmatrix} \bar{\varepsilon}_{11} \\ \bar{\varepsilon}_{22} \\ \bar{\varepsilon}_{33} \\ \bar{\varepsilon}_{12} \\ \bar{\varepsilon}_{23} \\ \bar{\varepsilon}_{31} \end{Bmatrix}
\tag{4.11}
$$

but, since the constants do not change with the transformation, C_{14}, C_{16}, C_{24}, C_{26}, C_{34}, C_{36}, C_{45}, $C_{56} = 0$ leaving 21–8, or 13, constants. Such a material is called *monoclinic* [4].

Next, double symmetry is achieved by the transformation

	x_1	x_2	x_3
x_1'	-1		
x_2'		-1	
x_3'			1

$$\tag{4.12}$$

as shown in Fig. 4.3. Following a similar argument as before, the number of elastic constants is reduced from 13 to 9 producing an orthorhombic or *orthotropic* material [4]. No further reduction is achieved by taking $x_3' = -x_3$ since all of the axes could have been reversed initially in Fig. 4.2a.

The stress–strain law for an *orthotropic* material is written explicitly as

$$
\begin{Bmatrix} \sigma_{11} \\ \sigma_{22} \\ \sigma_{33} \\ \sigma_{12} \\ \sigma_{23} \\ \sigma_{31} \end{Bmatrix} =
\begin{bmatrix}
C_{11} & C_{12} & C_{13} & 0 & 0 & 0 \\
 & C_{22} & C_{23} & 0 & 0 & 0 \\
 & & C_{33} & 0 & 0 & 0 \\
 & & & C_{44} & 0 & 0 \\
 & & & & C_{55} & 0 \\
 & & & & & C_{66}
\end{bmatrix}
\begin{Bmatrix} \bar{\varepsilon}_{11} \\ \bar{\varepsilon}_{22} \\ \bar{\varepsilon}_{33} \\ \bar{\varepsilon}_{12} \\ \bar{\varepsilon}_{23} \\ \bar{\varepsilon}_{31} \end{Bmatrix}
\tag{4.13a}
$$

Fig. 4.3 Double-plane symmetry

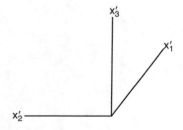

or, in terms of the elastic moduli, as

$$
\begin{Bmatrix} \sigma_{11} \\ \sigma_{22} \\ \sigma_{33} \\ \sigma_{12} \\ \sigma_{23} \\ \sigma_{31} \end{Bmatrix} = \begin{bmatrix} E_{1111} & E_{1122} & E_{1133} & 0 & 0 & 0 \\ & E_{2222} & E_{2233} & 0 & 0 & 0 \\ & & E_{3333} & 0 & 0 & 0 \\ & & & E_{1212} & 0 & 0 \\ & & & & E_{2323} & 0 \\ & & & & & E_{3131} \end{bmatrix} \begin{Bmatrix} \bar{\varepsilon}_{11} \\ \bar{\varepsilon}_{22} \\ \bar{\varepsilon}_{33} \\ \bar{\varepsilon}_{12} \\ \bar{\varepsilon}_{23} \\ \bar{\varepsilon}_{31} \end{Bmatrix}. \tag{4.13b}
$$

Additional simplifications are possible if directional independence of the material properties is present. The appropriate relationships are formally derived by interchanging axes. For example, we may interchange x_2 and x_3 by the transformation

	x_1	x_2	x_3
x_1'	1	0	0
x_2'	0	0	1
x_3'	0	1	0

$$\tag{4.14}$$

and x_1 and x_2 by

	x_1	x_2	x_3
x_1'	0	1	0
x_2'	1	0	0
x_3'	0	0	1

$$\tag{4.15}$$

The result is a *cubic* material with three independent constants.

A final simplification is to enforce rotational independence, in addition to the directional independence, by the transformation

	x_1	x_2	x_3
x_1'	1	0	0
x_2'	0	$\cos\theta$	$\sin\theta$
x_3'	0	$-\sin\theta$	$\cos\theta$

$$\tag{4.16}$$

as shown in Fig. 4.4. This reduces the number of constants from three to two, producing the familiar *isotropic* material. These moduli are known as the Lamé constants, since they were correctly established first by Gabriel Lamé in the middle

Fig. 4.4 Isotropic condition

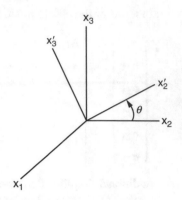

of the nineteenth century [1] and produce the stress–strain relationship written in terms of the tensor components ε_{ij} as

$$
\begin{Bmatrix} \sigma_{11} \\ \sigma_{22} \\ \sigma_{33} \\ \sigma_{12} \\ \sigma_{23} \\ \sigma_{31} \end{Bmatrix} = \begin{bmatrix} 2\mu+\lambda & \lambda & \lambda & 0 & 0 & 0 \\ \lambda & 2\mu+\lambda & \lambda & 0 & 0 & 0 \\ \lambda & \lambda & 2\mu+\lambda & 0 & 0 & 0 \\ & & & 2\mu & 0 & 0 \\ & & & & 2\mu & 0 \\ & & & & & 2\mu \end{bmatrix} \begin{Bmatrix} \varepsilon_{11} \\ \varepsilon_{22} \\ \varepsilon_{33} \\ \varepsilon_{12} \\ \varepsilon_{23} \\ \varepsilon_{31} \end{Bmatrix}
\tag{4.17}
$$

or, in indicial form,

$$
\sigma_{ij} = 2\mu\varepsilon_{ij} + \lambda\delta_{ij}\varepsilon_{kk}.
\tag{4.18}
$$

It is useful to invert (4.18), expressing the strains in terms of the stresses. This is best accomplished by first setting $i = j$ and expressing ε_{kk} in terms of σ_{kk} and then solving the resulting (4.18) for ε_{ij}. This gives

$$
\varepsilon_{ij} = \frac{1}{2\mu}\sigma_{ij} - \frac{\lambda}{2\mu(2\mu+3\lambda)}\delta_{ij}\sigma_{kk}.
\tag{4.19}
$$

Although the Lamé constants are perfectly suitable from a mathematical stand-point, it is common to use *engineering* material constants that are related to mea-surements from elementary mechanical tests.

For a uniaxial stress with σ_{11} constant and all other $\sigma_{ij} = 0$,

$$
\sigma_{11} = (2\mu+\lambda)\varepsilon_{11} + \lambda\varepsilon_{22} + \lambda\varepsilon_{33},
\tag{4.20a}
$$

$$
0 = \lambda\varepsilon_{11} + (2\mu+\lambda)\varepsilon_{22} + \lambda\varepsilon_{33},
\tag{4.20b}
$$

$$
0 = \lambda\varepsilon_{11} + \lambda\varepsilon_{22} + (2\mu+\lambda)\varepsilon_{33}.
\tag{4.20c}
$$

Solving (4.20b) and (4.20c) for $\varepsilon_{22} + \varepsilon_{33}$ and then substituting into (4.20a) gives the basic form of (4.1):

$$\sigma_{11} = \frac{\mu(2\mu + 3\lambda)}{\mu + \lambda}\varepsilon_{11} = E\varepsilon_{11}. \tag{4.21}$$

Young's modulus, E, was defined in Sect. 4.2.

From the same uniaxial stress state, the *fractional contraction* may be computed as

$$-\frac{\varepsilon_{22}}{\varepsilon_{11}} = -\frac{\varepsilon_{33}}{\varepsilon_{11}} = \frac{\lambda}{2(\mu + \lambda)} = \nu, \tag{4.22}$$

where ν is Poisson's ratio.

A third engineering constant is obtained from the state of *pure shear* in two dimensions, given by $\sigma_{12} = \sigma_{21} = $ constant, all other $\sigma_{ij} = 0$. From (4.18),

$$\sigma_{12} = 2\mu\varepsilon_{12} = 2G\varepsilon_{12}, \tag{4.23}$$

where G is the shear modulus.

Although the engineering constants E, ν, and G are convenient, we realize that only *two* of these constants are independent since $G = \mu$ and both E and ν are defined in terms of λ and μ. The relationships between the Lamé and engineering constants are collected as

$$\mu = \frac{E}{2(1 + \nu)}, \tag{4.24a}$$

$$\lambda = \frac{\nu E}{(1 + \nu)(1 - 2\nu)}, \tag{4.24b}$$

$$E = \frac{\mu(2\mu + 3\lambda)}{\mu + \lambda}, \tag{4.24c}$$

$$\nu = \frac{\lambda}{2(\mu + \lambda)}, \tag{4.24d}$$

$$G = \mu. \tag{4.24e}$$

Another relationship between the constants is defined as the bulk modulus

$$K = \frac{E}{3(1 - 2\nu)} \tag{4.24f}$$

which is used in the exercises, Sec. 4.9.

Note from (4.24b) that for λ to remain finite, Poisson's ratio must lie between

$$-1 < \nu < 0.5. \tag{4.25}$$

From (4.22) it is difficult to visualize a negative value of ν since it implies that an extension in the first direction would result in an expansion in the second direction as well; but see Chap. 13. Poisson's ratio is also related to the volume change or dilatation, introduced in Sect. 3.5. We write (3.41) for an extension of an isotropic material applied along the first principal axis as

$$\varepsilon^{(1)} = \bar{\varepsilon}, \tag{4.26}$$

whereupon (4.22) gives

$$\varepsilon^{(2)} = \varepsilon^{(3)} = -\nu\bar{\varepsilon}. \tag{4.27}$$

Then (3.41) becomes

$$\Delta = \bar{\varepsilon}(1 - 2\nu). \tag{4.28}$$

If $\nu = 0.5$, the upper positive limit, the dilatation $\Delta = 0$, and the material are said to be incompressible. Rubberlike and many biological materials exhibit this type of behavior.

The explicit form of the generalized Hooke's law in terms of the engineering constants and tensoral strains is best written in the inverse form from (4.17) as

$$\boldsymbol{\varepsilon} = \mathbf{C}^{-1}\boldsymbol{\sigma} \tag{4.29a}$$

or

$$
\begin{Bmatrix} \varepsilon_{11} \\ \varepsilon_{22} \\ \varepsilon_{33} \\ \varepsilon_{12} \\ \varepsilon_{23} \\ \varepsilon_{31} \end{Bmatrix} =
\begin{bmatrix}
\dfrac{1}{E} & -\dfrac{\nu}{E} & -\dfrac{\nu}{E} & 0 & 0 & 0 \\[8pt]
-\dfrac{\nu}{E} & \dfrac{1}{E} & -\dfrac{\nu}{E} & 0 & 0 & 0 \\[8pt]
-\dfrac{\nu}{E} & -\dfrac{\nu}{E} & \dfrac{1}{E} & 0 & 0 & 0 \\[8pt]
0 & 0 & 0 & \dfrac{1}{2G} & 0 & 0 \\[8pt]
0 & 0 & 0 & 0 & \dfrac{1}{2G} & 0 \\[8pt]
0 & 0 & 0 & 0 & 0 & \dfrac{1}{2G}
\end{bmatrix}
\begin{Bmatrix} \sigma_{11} \\ \sigma_{22} \\ \sigma_{33} \\ \sigma_{12} \\ \sigma_{23} \\ \sigma_{31} \end{Bmatrix}. \tag{4.29b}
$$

The elements of $\mathbf{C}^{-1}$ have been explicitly defined with reference to (4.21), (4.22), and (4.23). In general, these coefficients are called *compliances*. In indicial notation with G replaced by E using (4.24a) and (4.24e), (4.29b) becomes

$$\varepsilon_{ij} = \frac{1}{E}[(1 + \nu)\sigma_{ij} - \nu\delta_{ij}\sigma_{kk}]. \tag{4.30}$$

Recalling the discussion at the end of Sect. 3.6 and comparing the number of equations available (9) and the number of unknowns introduced (15), we see that the material law provides six additional equations yet involves no additional unknowns; thus, the set of equations for the linear theory of elasticity is consistent. In the next chapter, we will explore methods of reducing the equations as well as additional constraints on the theory.

4.4   Transverse Isotropic Material

In Fig. 4.5a, a configuration with the fibers oriented along the z-axis and perpendicular to the symmetric x–y plane is shown. This material model is appealing for biological applications because it is capable of describing the elastic properties of a fiber bundle aligned in one direction as illustrated in Fig. 4.5b. The stress–strain matrix $\mathbf{C_{TI}}$ is obtained by applying transformations (4.14) and (4.16) to (4.13a) with $1 = z$, $2 = x$, and $3 = y$.

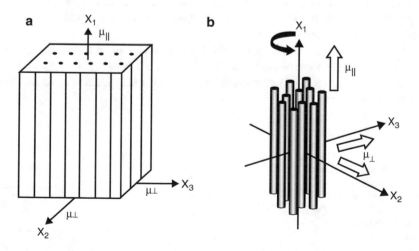

Fig. 4.5 Transverse isotropic material and bundled fiber configuration

This produces the transverse isotropic material with five elastic constants, $\lambda_\perp$, $\lambda_\parallel$, λ_M, $\mu_\perp$, and $\mu_\parallel$. The λ terms are related to the propagation of longitudinal waves in the different directions as described in Sect. 9.2, while the μ terms are shear moduli as discussed earlier. The expanded matrix for $\mathbf{C_{TI}}$ is given by [5]

$$
\mathbf{C_{TI}} =
\begin{bmatrix}
2\mu_\parallel + \lambda_\parallel & \lambda_M & \lambda_M & & & \\
\lambda_M & 2\mu_\perp + \lambda_\perp & \lambda_\perp & & 0 & \\
\lambda_M & \lambda_\perp & 2\mu_\perp + \lambda_\perp & & & \\
& & & \mu_\perp & & \\
& 0 & & & \mu_\parallel & \\
& & & & & \mu_\parallel
\end{bmatrix}
\tag{4.31}
$$

attributed to the general treatment in Ref. [6]. Applications are presented in Sect. 9.2.4, and it is also noted in [5] that $\mathbf{C_{TI}}$ and subsequent developments in Sect. 9.2.4 leading to (9.16) are valid only when the fiber axis of the material coincides with the z-axis of the coordinate system in Fig. 4.5. Tilting of the symmetry axis requires two rotational transformations similar to (4.16).

Note that (4.31) is expressed in terms of Lamé constants μ and λ, defined in (4.24a), (4.24b), (4.24c), (4.24d), and (4.24e). An alternate form in terms of the engineering constants has been suggested by Bower [6]. Referring to the $\mathbf{e_i}$ coordinate system shown in Fig. 4.6, for Young's modulus E, the subscript t represents the direction along the fibers $\mathbf{e_1}$; subscript p represents directions perpendicular to the fiber directions $\mathbf{e_2}$ and $\mathbf{e_3}$. The subscript order of t and p for Poisson's ratio ν represents the action from one direction to the other. Shear modulus μ_p is within the transverse symmetric plane, and shear modulus μ_t is within the plane that is perpendicular to the transverse symmetric plane. The physical meanings of these elastic constants are illustrated in Fig. 4.6.

Aligning the frame of reference with the unit vector $\mathbf{e_1}$, a linear elastic transversely isotropic constitutive model can be adopted for the gel [6] as described by the following compliance matrix relating the strain tensor components ε_{ij} to the stress tensor components σ_{ij}:

Fig. 4.6 Coordinates and elastic constants for a linear elastic, transversely isotropic model of fibrin gel. The unit vector $\mathbf{e_1}$ is parallel to the dominant fiber direction

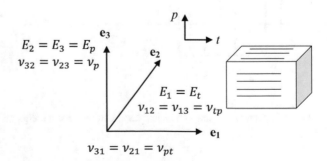

$$E_2 = E_3 = E_p$$
$$\nu_{32} = \nu_{23} = \nu_p$$

$$E_1 = E_t$$
$$\nu_{12} = \nu_{13} = \nu_{tp}$$

$$\nu_{31} = \nu_{21} = \nu_{pt}$$

$$
\begin{bmatrix} \varepsilon_{11} \\ \varepsilon_{22} \\ \varepsilon_{33} \\ 2\varepsilon_{23} \\ 2\varepsilon_{13} \\ 2\varepsilon_{12} \end{bmatrix} = \begin{bmatrix} \dfrac{1}{E_t} & -\dfrac{\nu_{pt}}{E_p} & -\dfrac{\nu_{pt}}{E_p} & 0 & 0 & 0 \\[2mm] -\dfrac{\nu_{tp}}{E_t} & \dfrac{1}{E_p} & -\dfrac{\nu_p}{E_p} & 0 & 0 & 0 \\[2mm] -\dfrac{\nu_{tp}}{E_t} & -\dfrac{\nu_p}{E_p} & \dfrac{1}{E_p} & 0 & 0 & 0 \\[2mm] 0 & 0 & 0 & \dfrac{1}{\mu_p} & 0 & 0 \\[2mm] 0 & 0 & 0 & 0 & \dfrac{1}{\mu_t} & 0 \\[2mm] 0 & 0 & 0 & 0 & 0 & \dfrac{1}{\mu_t} \end{bmatrix} \begin{bmatrix} \sigma_{11} \\ \sigma_{22} \\ \sigma_{33} \\ \sigma_{23} \\ \sigma_{13} \\ \sigma_{12} \end{bmatrix}. \tag{4.32}
$$

There are five independent scalar parameters for this material model; the seven variables in the compliance matrix in (4.32) are related by two additional equations:

$$
\frac{\nu_{pt}}{E_p} = \frac{\nu_{tp}}{E_t} \quad \text{or} \quad \frac{E_t}{E_p} = \frac{\nu_{tp}}{\nu_{pt}} \tag{4.33}
$$

$$
\mu_p = \frac{E_p}{2(1+\nu_p)}. \tag{4.34}
$$

Also, since the compliance matrix should be positive definite, the following five relationships for the Poisson's ratios and moduli exist for stability:

$$
\begin{cases} E_t, E_p, \mu_t, \mu_p > 0 \\[2mm] |\nu_p| < 1 \\[2mm] \nu_{pt} < \sqrt{\dfrac{E_p}{E_t}} \\[2mm] \nu_{tp} < \sqrt{\dfrac{E_t}{E_p}} \\[2mm] 1 - \nu_p^2 - 2\nu_{pt}\nu_{tp} - 2\nu_{pt}\nu_{tp}\nu_p > 0 \end{cases} \tag{4.35}
$$

4.5 Functionally Graded Materials

Functionally graded materials (FGMs) have mechanical properties that vary progressively within a body. Such materials may provide a special feature within the dominant material in a body such as a protective surface, a thermal barrier, or a piezoelectric layer.

Even though manufactured FGMs are a relatively recent development, such materials are common in nature. Graded structures are found in plant stems and tree trunks and in animal bones and teeth. The outer stronger surface provides structural support and protection from impact, while the less dense more flexible inner material allows for fluid circulation and cell growth.

Material property variations to provide locally different characteristics in a *discrete* manner, such as laminates and layered composites, are widely used in engineering applications. An FGM may be thought of as an extension of this concept so as to provide a *continuous* multipurpose-designed material cross section. In regenerative biomaterial applications, graded materials may provide superior transitions between components of greatly varying stiffness such as bones and tendons.

As illustrated in Fig. 4.7, the graded transition of material characteristics across an interface of two materials can reduce the thermal stresses and stress concentrations at the intersection with the free surface.

Our treatment is restricted to a two-dimensional or plane elasticity case which is developed extensively in Chap. 7. Here we introduce the coordinate system and basic material characterization. We refer to Fig. 4.2a and select the x_1–x_2 plane. This corresponds to the familiar *x*–*y* plane used extensively in two-dimensional elasticity as shown in Fig. 4.8.

The material is initially taken as orthotropic with the principal material directions *x* and *y*, corresponding to (4.13a) with the terms having a 3 subscript eliminated. The 2-D constitutive equation is

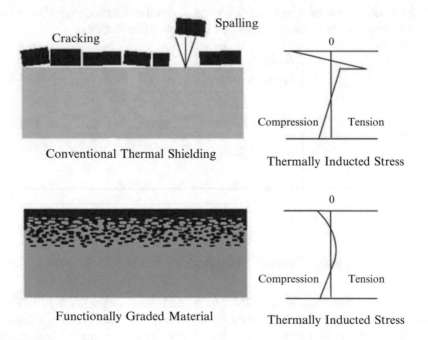

Fig. 4.7 Heat shield [7]

Fig. 4.8 Reference axes for
2-D functionally graded
materials

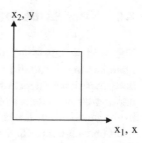

$$\sigma = C_{or}\varepsilon, \tag{4.36}$$

where $\boldsymbol{\sigma} = \{\sigma_{11}, \sigma_{22}, \sigma_{12}\} = \{\sigma_{xx}, \sigma_{yy}, \sigma_{xy}\}$, $\boldsymbol{\varepsilon} = \{\varepsilon_{11}\ \varepsilon_{22}\ \varepsilon_{12}\} = \{\varepsilon_{xx}\ \varepsilon_{yy}\ \varepsilon_{xy}\}$, and

$$C_{or} = \begin{bmatrix} C_{11} & C_{12} & 0 \\ C_{12} & C_{22} & 0 \\ 0 & 0 & C_{44} \end{bmatrix}. \tag{4.37}$$

It is also convenient to set the origin on one boundary surface of the member so that the variation through the thickness can be described without consideration of symmetry. Selecting the x_2-axis as the y-axis, the variation through the depth may be taken as exponential [8]:

$$\mathbf{C_{FG}}(y) = e^{\zeta y} \begin{pmatrix} C_{11}^0 & C_{12}^0 & 0 \\ C_{12}^0 & C_{22}^0 & 0 \\ 0 & 0 & C_{44}^0 \end{pmatrix}, \tag{4.38}$$

where $y = 0$ is the reference plane and ζ is a convenient parameter to represent a continuous variation of the material coefficients.

As a further simplification, the functionally graded cross section may be assumed to be isotropic at every point and Poisson's ratio as constant through the thickness. Substituting μ and λ from (4.24a) and (4.24b) into (4.17) allows (4.38) to be written as the final constitutive matrix:

$$\mathbf{C_{FGI}}(y) = \frac{E(y)}{(1-\nu)(1-2\nu)} \begin{pmatrix} 1-\nu & \nu & 0 \\ \nu & 1-\nu & 0 \\ 0 & 0 & 1-2\nu \end{pmatrix}. \tag{4.39}$$

Then the variation of Young's modulus is expressed as

$$E(y) = E(0)e^{\zeta y}. \tag{4.40}$$

The parameter ζ is set by specifying the value of Young's modulus at one point through the thickness in addition to the value at the reference plane $E(0)$. Applications are presented in Sect. 7.12.

4.6   Viscoelastic Materials

The stress–strain relationships derived in the preceding sections define elastic materials that respond immediately to the load applied, where the stress that the body experiences depends solely on strain. Viscous materials also resist loading and undergo strain, but the stress in a viscous body depends on the *strain rate* in addition to the strain itself. Since the material characterization, mathematical models, and solution techniques for viscoelastic materials are somewhat different than those for elastic materials, an independent treatment is presented in Chap. 10 based largely on Roylance [9].

4.7   Thermal Strains

Temperature changes may be a source of important stresses in elastic systems. It is reasonable to assume that a linear relationship exists between the temperature difference from the datum value and the corresponding strain:

$$\varepsilon_{(ii)}(T) = \alpha[T - T_0], \tag{4.41}$$

in which $\varepsilon_{(ii)}(T)$ is the linear strain due to the temperature difference $T - T_0$, T_0 is the datum temperature, and α is the *coefficient of thermal expansion*, assumed here to be a *constant*.

With this concept, it is easy to generalize (4.30) to

$$\varepsilon_{ij} = \frac{1}{E}[(1 + \nu)\sigma_{ij} - \nu\delta_{ij}\sigma_{kk}] + \alpha\delta_{ij}[T - T_0]. \tag{4.42}$$

We can also generalize (4.18) to

$$\sigma_{ij} = 2\mu\varepsilon_{ij} + \lambda\delta_{ij}\varepsilon_{kk} - \alpha\delta_{ij}(3\lambda + 2\mu)[T - T_0]. \tag{4.43}$$

When these expressions are used in subsequent applications, the thermal terms are carried through as constants and ultimately become grouped with the applied tractions and/or body forces. In this form, they have come to be known as *thermal loads* that can be treated along with mechanical loads as known quantities.

Obvious complications are produced when temperatures outside the normal environmental range are considered since both the material properties such as E and G and the thermal coefficient α may become possibly nonlinear functions of the temperature.

4.8   Physical Data

Since the implementation of the various material models is based on physical data, it is necessary either to obtain relevant data or to construct appropriate experiments. The literature is rich with data for a wide variety of materials, but, because they are oriented to specific applications, these data are somewhat scattered. Some procedures are discussed with respect to strength criteria in Chap. 11. For orthotropic materials, a detailed interpretation of the constants in (4.13a) and (4.13b) is provided by Vinson and Sierakowski [10]. A comprehensive guide to the techniques of constitutive modeling that treats metals, concrete, and soils from an elasticity basis may facilitate the collection and interpretation of physical data [11].

4.9   Computational Examples with MATLAB

In this example, we illustrate using MuPAD to derive compliance matrix S from elastic stiffness matrix C for transversely isotropic material. We first set up the matrix C, with elastic components $c11$, $c22$, $c33$, $c12$, $c23$, and $c55$:

```
C := matrix([ [c11, c12, c12, 0, 0, 0],
[c12, c22, c23, 0, 0, 0],
[c12, c23, c22, 0, 0, 0],
[0, 0, 0, (c22-c23)/2, 0, 0],
[0, 0, 0, 0, c55, 0],
[0, 0, 0, 0, 0, c55 ] ])
```

$$
\begin{pmatrix}
c11 & c12 & c12 & 0 & 0 & 0 \\
c12 & c22 & c23 & 0 & 0 & 0 \\
c12 & c23 & c22 & 0 & 0 & 0 \\
0 & 0 & 0 & \frac{c22}{2} - \frac{c23}{2} & 0 & 0 \\
0 & 0 & 0 & 0 & c55 & 0 \\
0 & 0 & 0 & 0 & 0 & c55
\end{pmatrix}
$$

Using the matrix inverse, it is very straightforward to calculate the compliance matrix S:

$$
\left[
\begin{array}{c}
\texttt{S := Simplify(1/C)} \\
\left(
\begin{array}{cccccc}
-\dfrac{c22+c23}{2\,c12^2-c11\,(c22+c23)} & \sigma_1\;\sigma_1 & 0 & 0 & 0 \\[2mm]
\sigma_1 & \sigma_3\;\sigma_2 & 0 & 0 & 0 \\[2mm]
\sigma_1 & \sigma_2\;\sigma_3 & 0 & 0 & 0 \\[2mm]
0 & 0 \quad 0 & \dfrac{2}{c22-c23} & 0 & 0 \\[2mm]
0 & 0 \quad 0 & 0 & \dfrac{1}{c55} & 0 \\[2mm]
0 & 0 \quad 0 & 0 & 0 & \dfrac{1}{c55}
\end{array}
\right)
\end{array}
\right.
$$

where

$$
\sigma_1 = \frac{c12}{2\,c12^2 - c11\,(c22+c23)}
$$

$$
\sigma_2 = \frac{c11}{2\left(-2\,c12^2 + c11\,c22 + c11\,c23\right)} - \frac{1}{2\,(c22-c23)}
$$

$$
\sigma_3 = \frac{1}{2\,(c22-c23)} + \frac{c11}{2\left(-2\,c12^2 + c11\,c22 + c11\,c23\right)}
$$

Derivation of constitutive equations using this approach can be useful for computational implementation in finite element method appplications.

4.10   Exercises

4.1 Derive the generalized Hooke's law for an orthotropic elastic solid, starting with the equations for an anisotropic material [3].

4.2 For an isotropic material, obtain the relationship

$$
\varepsilon_{ij} = \frac{1}{2\mu}\sigma_{ij} - \frac{\lambda}{2\mu(2\mu+3\lambda)}\delta_{ij}\sigma_{kk},
$$

from the expression

$$
\sigma_{ij} = 2\mu\varepsilon_{ij} + \lambda\delta_{ij}\varepsilon_{kk}.
$$

4.3 (a) Show that the principal axes of stress correspond with the principal axes of strain for an isotropic, linearly elastic material.

(b) Show that for an orthotropic material, principal axes of stress do not in general coincide with principal axes of strain.

(c) Under what conditions will these axes coincide with each other?

4.4 Extend the stress–strain laws to include thermal effects assuming that the coefficient of thermal expansion is equal to α. How does this complicate the formulation of the elasticity problem?

4.5 For steel, $E = 30 \times 10^6$ and $G = 12 \times 10^6$ (force/length^2). The components of strain at a point within this material are given by

$$\varepsilon = \begin{bmatrix} 0.004 & 0.001 & 0 \\ 0.001 & 0.006 & 0.004 \\ 0 & 0.004 & 0.001 \end{bmatrix}$$

Compute the corresponding components of the stress tensor.

4.6 Verify the relations (a) and (b) of (4.24a), (4.24b), (4.24c), (4.24d), and (4.24e) by using (c) and (d).

4.7 Determine the constitutive relations governing the material behavior of a point having the properties described below. Would the material be classified as anisotropic, orthotropic, or isotropic?

(a) A point in the material has a state of stress given by the following:

$$\sigma_{11} = 10.8; \quad \sigma_{22} = 3.4; \quad \sigma_{33} = 3.0$$
$$\sigma_{12} = \sigma_{13} = \sigma_{23} = 0$$

(b) The material develops the corresponding strain components:

$$\varepsilon_{11} = 10 \times 10^{-4}; \quad \varepsilon_{22} = 2 \times 10^{-4}; \quad \varepsilon_{33} = 2 \times 10^{-4};$$
$$\varepsilon_{12} = \varepsilon_{23} = \varepsilon_{31} = 0.$$

(c) When a state of stress of

$$\sigma_{11} = 10; \quad \sigma_{22} = 2; \quad \sigma_{33} = 2;$$
$$\sigma_{12} = \sigma_{23} = \sigma_{31} = 0$$

is applied, the strain components are

$$\varepsilon_{11} = 10 \times 10^{-4}; \quad \varepsilon_{22} = \varepsilon_{33} = \varepsilon_{12} = \varepsilon_{23} = \varepsilon_{31} = 0.$$

(d) When subjected to a shearing stress σ_{12}, σ_{13}, or σ_{23} of 10, the material develops no strains except the corresponding shearing strain, with tensor component ε_{12}, ε_{13}, or ε_{23} respectively of 20×10^{-4}.

4.8 Show that the stress–strain relationships for an isotropic material, (4.18), can be written in the general form of (4.3), where

(a) $E_{ijkl} = G(\delta_{ik}\delta_{jl} + \delta_{il}\delta_{jk}) + (K - \frac{2}{3}G)\delta_{ij}\delta_{kl}$,

where $K = \frac{E}{3(1-2\nu)}$ is the bulk modulus noted in (4.24f) [12].

(b) $E_{ijlm} = \lambda\delta_{ij}\delta_{lm} + \mu\delta_{il}\delta_{jm} + \mu\delta_{im}\delta_{jl}$ [13].

4.9 Consider an isotropic elastic medium subject to a hydrostatic state of stress:

$$\sigma^{(1)} = \sigma^{(2)} = \sigma^{(3)} = -p.$$

Show that for this state of stress

$$p = -KR_1,$$

where K is the bulk modulus and R_1 is the cubical strain (Sect. 3.4). Consider the implications of $\nu > 1/2$ and $\nu < -1$ [14].

4.10 For an elastic medium subject to a state of stress σ_{ij}, assume that the deformation is incompressible for $\sigma_{(ii)} \neq 0$.

(a) Verify that $\nu = 0.5$.
(b) Assume also that $\varepsilon_{33} = 0$ and determine ν [14].

4.11 The stress–strain relationship for a certain orthotropic material is given as

$$
\begin{Bmatrix} \sigma_1 \\ \sigma_2 \\ \sigma_3 \\ \sigma_4 \\ \sigma_5 \\ \sigma_6 \end{Bmatrix}
=
\begin{bmatrix}
C_{11} & C_{12} & C_{13} & 0 & 0 & 0 \\
C_{12} & C_{11} & C_{13} & 0 & 0 & 0 \\
C_{13} & C_{13} & C_{33} & 0 & 0 & 0 \\
0 & 0 & 0 & C_{44} & 0 & 0 \\
0 & 0 & 0 & 0 & C_{44} & 0 \\
0 & 0 & 0 & 0 & 0 & \dfrac{(C_{11}-C_{12})}{2}
\end{bmatrix}
\begin{Bmatrix} \varepsilon_1 \\ \varepsilon_2 \\ \varepsilon_3 \\ \varepsilon_4 \\ \varepsilon_5 \\ \varepsilon_6 \end{Bmatrix}.
$$

Show that the material exhibits rotational symmetry about the z-axis.

References

1. Westergaard HM (1964) Theory of elasticity and plasticity. Dover Publications, Inc., New York
2. Love AEH (1944) The mathematical theory of elasticity. Dover Publications, Inc., New York
3. Tauchert TR (1974) Energy principles in structural mechanics. McGraw–Hill Book Company, Inc, New York
4. Little RW (1973) Elasticity. Prentice–Hall, Englewood Cliffs

5. Sinkus R et al (2005) Imaging anisotropic and viscous properties of breast tissue by magnetic resonance-elastography. Magn Reson Med., Wiley Interscience 53:372–387
6. Namani R et al (2012) Elastic characterization of transversely isotropic soft materials by dynamic shear and asymmetric indentation. J Biomech Eng 136(6):061004
7. Cooley WG (2005) Application of functionally graded materials in aircraft structures. Masters thesis, AFIT/GAE/ENY/05-M04 Wright-Patterson AFB
8. Sankar BV (2001) An elasticity solution for functionally graded beams. Comp Sci Technol 61:689–696
9. Roylance D (1999) Engineering viscoelasticity, module for engineering/3–11. Massachusetts Institute of Technology, Cambridge, MA. 24 Oct 2001, 2 Dec 2009
10. Vinson JR, Sierakowski RL (1986) The behavior of structures composed of composite materials. Martinus Nijhoff, Dordrecht
11. Chen WF, Saleeb AF (1982) Constitutive equations for engineering materials, vol 1. Wiley, New York
12. Ohtani YC, Chen WF (1987) Hypoelasticity-perfectly plastic model for concrete materials. J Eng Mech, ASCE 113(12):1840–1860
13. Meek JL, Lin WJ (1990) Geometric and material nonlinear analysis of thin-walled beam columns. J Struct Eng 116(6):1473–1490
14. Boresi AP, Chong KP (1987) Elasticity in engineering mechanics. Elsevier, New York

Chapter 5
Formulation, Uniqueness, and Solution Strategies

Abstract The essential equations of the linear theory of elasticity have been derived. Restated concisely are the equilibrium conditions, the kinematic relations with the compatibility constraints, and the constitutive law for isotropic materials. These equations are combined in force, displacement, and mixed formulations. Also the necessary and sufficient conditions governing the uniqueness of a solution are established, and some fundamental approaches to the solution of an elasticity problem are introduced.

5.1 Introduction

In the previous chapters, we have derived the essential equations of the linear theory of elasticity. Repeated here in terse form, the equilibrium conditions (2.34) are

$$\sigma_{ij,i} + f_j = 0; \tag{5.1}$$

the kinematic relations (3.14) are

$$\varepsilon_{ij} = \frac{1}{2}\left(u_{i,j} + u_{j,i}\right), \tag{5.2}$$

with the compatibility constraints (3.60)

$$\varepsilon_{ij,kl} + \varepsilon_{kl,ij} = \varepsilon_{ik,jl} + \varepsilon_{jl,ik}; \tag{5.3}$$

and the constitutive law, (4.18) and (4.19), is given by

$$\sigma_{ij} = \lambda\delta_{ij}\varepsilon_{kk} + 2\mu\varepsilon_{ij} \tag{5.4}$$

or

$$\varepsilon_{ij} = \frac{1}{E}\left[(1+\nu)\sigma_{ij} - \nu\delta_{ij}\sigma_{kk}\right]. \tag{5.5}$$

© Springer International Publishing AG, part of Springer Nature 2018
P. L. Gould, Y. Feng, *Introduction to Linear Elasticity*,
https://doi.org/10.1007/978-3-319-73885-7_5

In this chapter, we combine and reduce these equations, establish the necessary and sufficient conditions governing the uniqueness of a solution, and introduce some approaches to the solution of an elasticity problem.

5.2 Displacement Formulation

With a view toward retaining only the displacements u_i, we substitute (5.2) into (5.4) to eliminate the strains, yielding

$$\sigma_{ij} = \lambda \delta_{ij} u_{k,k} + \mu \left(u_{i,j} + u_{j,i} \right), \tag{5.6a}$$

a relationship between stresses and displacements. Next, the stresses in (5.1) are replaced by (5.6a) to give

$$\lambda \delta_{ij} u_{k,ki} + \mu \left(u_{i,ji} + u_{j,ii} \right) + f_i = 0 \tag{5.6b}$$

which reduces to

$$\lambda u_{k,kj} + \mu (u_{i,ji} + u_{j,ii}) + f_i = 0 \tag{5.6c}$$

or, using i for the dummy index in all terms,

$$\mu u_{j,ii} + (\lambda + \mu) u_{i,ij} + f_j = 0. \tag{5.7}$$

Equations (5.7) are three equilibrium equations in terms of the displacements and constitute a classical *displacement* formulation. These equations are among the most famous in elasticity. Because they are written in terms of λ and μ, these are referred to as the *Lamé equations* [2] or the *Navier* equations [1] or the *Navier–Cauchy* equations. In the case where the displacement field u_j is continuous and differentiable, the corresponding strain field ε_{ij} always satisfies Eq. (5.3). It may be shown that according to the Helmholtz theorem, a vector field such as **u** may be resolved into the sum of a irrotational field **A** where $\nabla \times \mathbf{A} = 0$ and a divergence-free field **B** where $\nabla \cdot \mathbf{B} = 0$ [1]. This decomposition is cited here for completeness and developed further in Chap. 9.

The solution of (5.7) produces functions and constants of integration that are evaluated from boundary conditions expressed in terms of displacements. We may also encounter boundary conditions involving tractions. Fortunately, such conditions can often be incorporated into the loading terms, especially in numerical solutions.

In the case when the temperature of a body changes in the space, the thermal stress plays an role in the constitutive equation. We may also incorporate the thermal strain terms derived in Sect. 4.7 as (4.43) into (5.7). This leads to a thermal force f^{T} in each direction to be added to the mechanical and body forces.

$$f_j^{\mathrm{T}} = -\alpha(3\lambda + 2\mu)(T - T_0)_{,j} = -[\alpha E/(1 - 2\nu)](T - T_0)_{,j}. \tag{5.8}$$

5.3   Force Formulation

An alternate to the classical displacement formulation is to synthesize the equations in terms of the stresses σ_{ij}. If the stresses are known, the strains follow directly from (5.5), but the displacements remain to be evaluated from (5.2). In this case, these are six equations to find only three displacements. However, the strains must also satisfy the compatibility constraints (5.3) in order to insure that the integration of (5.2) provides a single-valued displacement field. With this in mind, it is helpful to rewrite (5.3) in terms of the stresses from the outset.

We start by substituting (5.5) into (5.3) and changing the dummy index k to t to produce

$$\sigma_{ij,kl} + \sigma_{kl,ij} - \sigma_{ik,jl} - \sigma_{jl,ik} = \frac{\nu}{1+\nu}\left(\delta_{ij}\sigma_{tt,kl} + \delta_{kl}\sigma_{tt,ij} - \delta_{ik}\sigma_{tt,jl} - \delta_{jl}\sigma_{tt,ik}\right) \quad (5.9a)$$

which are again 81 equations. We recall from the discussion in Sect. 3.6 that only six of (5.3), corresponding to the conditions $l = k$, are independent; so (5.9a) reduces to

$$\sigma_{ij,kk} + \sigma_{kk,ij} - \sigma_{ik,jk} - \sigma_{jk,ik} = \frac{\nu}{1+\nu}\left(\delta_{ij}\sigma_{tt,kk} + \delta_{kk}\sigma_{tt,ij} - \delta_{ik}\sigma_{tt,jk} - \delta_{jk}\sigma_{tt,ik}\right),$$
$$(5.9b)$$

which are nine equations with free indices i and j.

To simplify the equations somewhat, we recall (2.59a) where

$$Q_1 - \sigma_{ii} - \text{first stress invariant,} \quad (5.10a)$$

we use

$$(\ )_{,kk} = \nabla^2(\ ), \quad \text{from (1.10b)}, \quad (5.10b)$$

and we note that on the r.h.s.,

$$\delta_{kk}\sigma_{tt,ij} = 3\sigma_{tt,ij} = 3Q_{1,ij} \quad (5.11a)$$

$$\delta_{ik}\sigma_{tt,jk} = \sigma_{tt,ij} = Q_{1,ij} \quad (5.11b)$$

$$\delta_{jk}\sigma_{tt,ik} = \sigma_{tt,ij} = Q_{1,ij} \quad (5.11c)$$

so that (5.9b) becomes

$$\nabla^2\sigma_{ij} + Q_{1,ij} - \sigma_{ik,jk} - \sigma_{jk,ik} = \frac{\nu}{1+\nu}\left(\delta_{ij}\nabla^2 Q_1 + 3Q_{1,ij} - 2Q_{1,ij}\right). \quad (5.12)$$

Furthermore, we recall from (5.1) that

$$\sigma_{ik,jk} = -f_{i,j} \quad (5.13a)$$

and

$$\sigma_{jk,ik} = -f_{j,i}. \tag{5.13b}$$

Combining terms and considering the preceding relations, we have

$$\nabla^2 \sigma_{ij} + \frac{1}{1+\nu} Q_{1,ij} - \frac{\nu}{1+\nu} \delta_{ij} \nabla^2 Q_1 + f_{i,j} + f_{j,i} = 0. \tag{5.14}$$

Equations (5.14) are nine in number; however, only six are independent since the stress tensor is symmetric. It is possible to reduce the equations further by utilizing the additional (5.9a), when $k \neq l$. Letting $k = i$ and $l = j$, we have

$$\sigma_{ij,ij} + \sigma_{ij,ij} - \sigma_{ii,jj} - \sigma_{jj,ii} = \frac{\nu}{1+\nu} (\delta_{ij}\sigma_{tt,ij} + \delta_{ij}\sigma_{tt,ij} - \delta_{ii}\sigma_{tt,jj} - \delta_{jj}\sigma_{tt,ii}),$$

which is

$$2\sigma_{ij,ij} - 2\nabla^2 Q_1 = \frac{\nu}{1+\nu} (2\nabla^2 Q_1 - 6\nabla^2 Q_1),$$

so that

$$\sigma_{ij,ij} = \left(1 - \frac{2\nu}{1+\nu}\right) \nabla^2 Q_1$$

or

$$\nabla^2 Q_1 = \frac{1+\nu}{1-\nu} \sigma_{ij,ij} = -\frac{1+\nu}{1-\nu} f_{j,j} \tag{5.15}$$

from (5.1). Thus, additional compatibility conditions are introduced into (5.14) through the substitution of (5.15) for $\nabla^2 Q_1$:

$$\nabla^2 \sigma_{ij} + \frac{1}{1+\nu} Q_{1,ij} = \frac{-\nu}{1-\nu} \delta_{ij} f_{k,k} - (f_{i,j} + f_{j,i}). \tag{5.16}$$

If the body forces are constant, (5.16) reduces to

$$\nabla^2 \sigma_{ij} + \frac{1}{1+\nu} Q_{1,ij} = 0. \tag{5.17}$$

Furthermore, comparing (5.16) with (5.14), also with body forces constant, we deduce that

$$\nabla^2 Q_1 = 0, \tag{5.18}$$

that is, $\nabla^2 \sigma_{ii} = 0$ so that Q_1 is a harmonic function.

Equations (5.16) and (5.17) are a set of compatibility equations in terms of stresses (as opposed to the St. Venant equations that are in terms of strain) for a *body in equilibrium* and are known as the Beltrami–Michell equations [2] after the originator Eugenio Beltrami and John–Henry Michell who treated the case of nonconstant body forces. Together with (5.1), they constitute a set of six equations in six unknown components of stress, and they represent a classical *force* formulation. If a stress field is formed that satisfies these equations and the corresponding strains computed from (5.5) are integrated in (5.2), the displacement field is insured to be single-valued. However, this process may be quite tedious, as is demonstrated in for the prismatic bar solution in Sect. 6.2.

Consideration of boundary conditions involving tractions is straightforward; however, constraints on the displacements are generally deferred until the subsequent integration of the kinematic relations. It is sometimes rather difficult to express such constraints directly as equivalent statements on the stresses.

5.4 Other Formulations

Beyond the classical displacement and force formulations, there remains the possibility of using reduced equations that contain a mixture of stress and displacement and, possibly, strain terms. Such an approach is called a *mixed* formulation [2]. For example, the equilibrium Eq. (5.1) may be written in terms of strains by substituting the constitutive law (5.4) in terms of the engineering constants E and ν to get

$$\frac{E}{1+\nu}\left(\varepsilon_{ij,i} + \frac{\nu}{1+2\nu}\,\varepsilon_{kk,i}\right) = -f_j. \tag{5.19}$$

Additionally, there is the possibility of employing different formulations in various regions of the domain, for example, a stress formulation on the boundary and a displacement formulation in the interior. This is termed a *hybrid* approach and, obviously, covers many possibilities. Mixed and hybrid formulations are of considerable interest for numerical solutions in the theory of elasticity and may offer expedient alternatives to the more classical approaches.

5.5 Uniqueness

We seek to prove that a solution to the equations synthesized from (5.1), (5.2), (5.3), (5.4), and (5.5) is unique for a given region with reasonable boundary conditions.

A counterargument is followed whereby it is assumed that two distinct solutions, say $u_i^{(1)}$ and $u_i^{(2)}$, exist [2]. We designate the *differences* between the solutions by asterisk (*) superscripts, that is,

$$u_i^* = u_i^{(1)} - u_i^{(2)}, \tag{5.20}$$

and endeavor to show that $u_i^* = 0$.

The corresponding differences for the strains and stresses are

$$\begin{aligned} \varepsilon_{ij}^* &= \varepsilon_{ij}^{(1)} - \varepsilon_{ij}^{(2)} \\ &= \frac{1}{2}(u_{i,j}^* + u_{j,i}^*) \end{aligned} \tag{5.21a}$$

and

$$\sigma_{ij}^* = \sigma_{ij}^{(1)} - \sigma_{ij}^{(2)}. \tag{5.21b}$$

Since the body forces are identical, $f_j^* = 0$, and since both systems are in equilibrium,

$$\sigma_{ij,j}^* = 0. \tag{5.22}$$

To continue, we consider the energy difference

$$2U^* = \int_V \varepsilon_{ij}^* \sigma_{ij}^* \, dV, \tag{5.23}$$

where U is the strain energy. This expression is nonnegative, as discussed in Sect. 4.3 following 4.2(b). We replace ε_{ij}^* by (5.21a) to give

$$\begin{aligned} \int_V \varepsilon_{ij}^* \sigma_{ij}^* \, dV &= \int_V \frac{1}{2}\left(u_{i,j}^* + u_{j,i}^*\right) \sigma_{ij}^* \, dV \\ &= \int_V u_{i,j}^* \sigma_{ij}^* \, dV, \end{aligned} \tag{5.24}$$

since the i and j become repeated indices. Continuing with the r.h.s. of (5.24) and using the product differentiation rule $\left(u_i^* \sigma_{ij}^*\right)_{,j} = \sigma_{ij}^* u_{i,j}^* + u_i^* \sigma_{ij,j}^*$ with the r.h.s. $u_{i,j}^*$ σ_{ij}^* replaced by $\left(u_i^* \sigma_{ij}^*\right)_{,j} - u_i^* \sigma_{ij,j}^*$

$$\begin{aligned} \int_V \varepsilon_{ij}^* \sigma_{ij}^* \, dV &= \int_V \left(u_i^* \sigma_{ij}^*\right)_{,j} dV - \int_V u_i^* \sigma_{ij,j}^* \, dV \\ &= \int_A u_i^* \sigma_{ij}^* n_j dA - 0. \end{aligned} \tag{5.25}$$

The first term on the r.h.s. of (5.25) is converted to a surface integral using the divergence theorem in component form, (2.32), and the second term vanishes by (5.22). Noting that in the remaining term $\sigma_{ij}^* n_j = T_i^*$, we may finally write

$$\int_V \varepsilon_{ij}^* \sigma_{ij}^* \, dV = \int_A u_i^* T_i^* \, dA = \int_A \left(u_i^{(1)} - u_i^{(2)} \right) \left(T_i^{(1)} - T_i^{(2)} \right) dA. \qquad (5.26)$$

Now, if the r.h.s. of (5.26) vanishes, the *energy difference* between the two solutions also vanishes, and they are *identical*. Therefore, our proof of uniqueness covers the following conditions for which this occurs:

1. The displacement field u_i is prescribed on the entire surface, making $u_i^* = 0$.
2. The tractions T_i are prescribed over the entire surface, while overall equilibrium is satisfied, making $T_i^* = 0$.
3. Displacements u_i are prescribed on part of the surface and tractions T_i on the remainder so that either u_i^* or $T_i^* = 0$.

It is also of interest to note that the proof of uniqueness, while insuring the same set of stresses and strains in both solutions, does not automatically insure that the displacements are identical. In particular, for condition (2) where the tractions are prescribed over the entire surface, *rigid body displacements* could be added without affecting this condition [2].

The preceding constraints are limited to *regular*, *singly connected* regions where the basic form of the divergence theorem, used in (5.25), is applicable. Extensions to more complex domains are also possible [4].

5.6   Membrane Equation

Several problems in the theory of elasticity lend themselves to solution by analogy. One of the most popular is based on the deflection of a uniformly stretched membrane. In Fig. 5.1, we show such a membrane stretched over a cross section and inflated with a pressure q per unit area producing a uniform tension T per unit length of the boundary. We consider a rectangular element abcd and write the equilibrium equation in the z-direction, noting that the *slopes* of the membrane change by

$$\frac{\partial u_{z,x}}{\partial x} dx \quad \text{and} \quad \frac{\partial u_{z,y}}{\partial y} dy$$

Fig. 5.1 Pressurized
membrane. (From
Timoshenko and Goodier,
Theory of Elasticity, © 1951
(Reproduced with
permission of McGraw-
Hill))

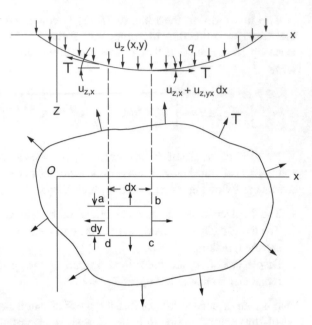

over the element in the respective directions. Also, the direction of the pressure q is
assumed to remain unchanged in the z-direction as opposed to following the
deformed shape of the membrane[1]. Thus, we have

$$qdxdy - Tdyu_{z,x} + Tdy[u_{z,x} + u_{z,xx}dx] - Tdxu_{z,y} + Tdx[u_{z,y} + u_{z,yy}dy] = 0$$

or

$$u_{z,xx} + u_{z,yy} = -\frac{q}{T}. \tag{5.27}$$

Also, at the boundary, the deflection $u_z = 0$.

Equation (5.27) is known as Poisson's equation and occurs repeatedly in the
theory of elasticity, thereby providing a basis for the so-called membrane analogies
which are used in Chap. 6.

5.7 Solution Strategies

The obvious course is to attempt to carry out a direct integration of the partial
differential equations and then to apply the appropriate boundary conditions. However,
the *direct method* is relatively difficult except for some fairly elementary problems.

---

[1]See comment on conservative force fields following (11.51).

A second approach is to select stress or strain fields that satisfy the governing differential equations and then to consider the boundary conditions that are satisfied. Finally, the boundary value problem that the solution represents is formulated. This is appropriately called the *inverse* method and, at first glance, seems to suggest that one may simply create a problem to fit the available solution, hardly a realistic approach.

The inverse method is meaningful when we look to the origin of stress or strain fields that may satisfy the governing equations. These may be borrowed in many cases from a less rigorous but hopefully approximately correct solution to a meaningful problem found from the strength of materials. It is soon realized, however, that such solutions are in some way deficient; so a third approach is to begin with part of the solution from this source and to develop the remainder by direct integration. This is known as the *semi-inverse* method and is one of many contributions of Barre'de A. J. C. St. Venant to the theory of elasticity. The principal historians of the subject, Isaac Todhunter and Carl Pearson, devote a major share of their first volume to his memory as the foremost of the modern elasticians [6].

In addition to analytical solutions of the governing differential equations, energy-based solutions play an important role in all of solid mechanics. Energy theorems also are the foundation for the powerful finite element method, which has become central to numerical techniques currently in use as discussed in Sect. 11.7.

Another notion attributed to St. Venant that aids immeasurably in obtaining practical solutions to elasticity problems is the concept of statically equivalent loads. *St. Venant's principle* [4] states that stresses and strains *reasonably distant* from the point of application of an external force are not significantly altered if the applied force is replaced by a *statically equivalent* load. An important consequence of this statement from an engineering standpoint is that satisfactory solutions may be obtained for the interior regions of a body without undue concern about the exact local distribution of the applied surface loading or the restraining boundary tractions. While there are some notable exceptions, such as cylindrical shells under concentrated axial edge load that tend to propagate far into the interior without much diffusion, St. Venant's principle is broadly applicable and is most useful in applied elasticity.

5.8 Computational Examples

Both the displacement and force formulations involve equations with potentially large amounts of computation and derivations, which are attractive for using computational tools. In this section, we take Eq. (5.16) as an example to check the compatibility of the following stress field:

$$\sigma_{xx} = 80x^2 + y, \sigma_{yy} = 100x^3 + 1600, \sigma_{zz} = 90y^2 + 100z^2$$
$$\sigma_{xy} = 1000x + 100y^2, \sigma_{xz} = 0, \sigma_{yz} = xz^3 + 100x^2y$$

- First, we define symbols and the stress field.

$$\begin{bmatrix} \text{nu} := \text{Symbol::nu} \\ v \end{bmatrix}$$

Given stress condition

$$\begin{bmatrix} \text{sigxx} := 80*x^2 + y \\ 80\,x^2 + y \end{bmatrix}$$

$$\begin{bmatrix} \text{sigyy} := 100*x^3 + y + 1600 \\ 100\,x^3 + y + 1600 \end{bmatrix}$$

$$\begin{bmatrix} \text{sigzz} := 90*y^2 + 100*z^2 \\ 90\,y^2 + 100\,z^2 \end{bmatrix}$$

$$\begin{bmatrix} \text{sigxy} := 1000*x + 100*y^2 \\ 100\,y^2 + 1000\,x \end{bmatrix}$$

$$\begin{bmatrix} \text{sigxz} := 0 \\ 0 \end{bmatrix}$$

$$\begin{bmatrix} \text{sigyz} := x*z^3 + 100*x^2*y \\ 100\,y\,x^2 + x\,z^3 \end{bmatrix}$$

- Then, we derive the body force using Eq. (5.1).

Solve body force

$$\begin{bmatrix} \text{f1} := - (\ \text{diff(sigxx, x)} + \text{diff(sigxy, y)} + \text{diff(sigxz, z)}\ ) \\ -160\,x - 200\,y \end{bmatrix}$$

$$\begin{bmatrix} \text{f2} := - (\ \text{diff(sigxy, x)} + \text{diff(sigyy, y)} + \text{diff(sigyz, z)}\ ) \\ -3\,x\,z^2 - 1001 \end{bmatrix}$$

$$\begin{bmatrix} \text{f3} := - (\ \text{diff(sigxz, x)} + \text{diff(sigyz, y)} + \text{diff(sigzz, z)}\ ) \\ -100\,x^2 - 200\,z \end{bmatrix}$$

- Finally, we use both the stress field and the derived body force to check compatibly. We take $i = 1, j = 1$ for a check.

Check compatibility

```
Q1 := sigxx + sigyy + sigzz
```
$$100\,x^3 + 80\,x^2 + 90\,y^2 + 2\,y + 100\,z^2 + 1600$$

```
LHS := linalg::laplacian(sigxx, [x,y,z]) + (1/(1+nu))*diff(Q1,x,x)
```
$$\frac{600\,x + 160}{v+1} + 160$$

```
RHS := (-nu/(1-nu)) * (diff(f1, x) + diff(f2, y) + diff(f3,z)) - 2*diff(f1,x)
```
$$320 - \frac{360\,v}{v-1}$$

- Therefore, the stress state does not satisfy the compatibility condition.

5.9 Exercises

5.1 The stress components at point $P(x, y, z)$ in an isotropic medium are (after [5])

$$\sigma_{xx} = y + 3z^2 \quad \sigma_{xy} = z^3,$$
$$\sigma_{yy} = x + 2z \quad \sigma_{xz} = y^2,$$
$$\sigma_{zz} = 2x + y \quad \sigma_{yz} = x^2.$$

No body forces are present.

(a) Does this stress field satisfy equilibrium?
(b) Does this stress field satisfy compatibility?

5.2 Write the Beltrami–Michell equations for the case where $\sigma_{zz} = 0$ (plane stress), and show that (after [3])

$$\sigma_{ii} = kz + f(x, y),$$

where k is a constant.

5.3 For the following stress field,

$$\sigma_{xx} = 80x^3 + y, \qquad \sigma_{xy} = 1,000 + 100y^2,$$
$$\sigma_{yy} = 100x^3 + 1,600, \quad \sigma_{xz} = 0,$$
$$\sigma_{zz} - 90y^2 + 100z^3, \qquad \sigma_{yz} = xz^3 + 100x^2y.$$

(a) What is the body force distribution required for equilibrium?
(b) What is the stress and body force at $(1, 1, 5)$?
(c) With $E = 30 \times 10^6$ (force/length^2) and $\nu - 0.3$, what is the strain at $(2, 2, 5)$?
(d) Does this stress distribution satisfy the equations of compatibility?

References

1. Chou PC, Pagano NJ (1992) Elasticity: tensor, dyadic and engineering approaches. Dover Publications, Inc., New York
2. Filonenko–Borodich M (1965) Theory of elasticity, trans. from Russian by M. Konayeon. Dover Publications, Inc., New York
3. Little RW (1973) Elasticity. Prentice–Hall, Inc., Englewood Cliffs
4. Timoshenko S, Goodier JN (1951) Theory of elasticity. McGraw–Hill Book Company, Inc., New York
5. Volterra E, Gaines JH (1971) Advanced strength of materials. Prentice–Hall Inc, Englewood Cliffs
6. Westergaard HM (1964) Theory of elasticity and plasticity. Dover Publications, Inc., New York

Chapter 6
Extension, Bending, and Torsion

Abstract Some classical problems that may be solved within the assumptions of the strength of materials are examined by applying the theory of elasticity. It is anticipated that the elementary solutions are approximately correct but deficient or incomplete in some way. In each case the isotropic material law is assumed to be applicable. The problems considered are a prismatic bar under axial load, an end-loaded cantilever beam, and torsion of circular shafts. Then a generalized theory of torsion applicable to members with noncircular prismatic cross, sections is developed.

6.1 Introduction

It is instructive to examine some familiar problems, readily solved by a strength-of-materials approach, using the theory of elasticity. We anticipate that the elementary solutions are approximately correct but deficient or incomplete in some way. In each case, the isotropic material law is assumed to be applicable.

6.2 Prismatic Bar Under Axial Loading

We consider the bar shown in Fig. 6.1a, suspended from a fixed support and loaded by self-weight γ (force/volume) [1].

First, we consider the equilibrium condition (2.46) with $1 = x$, $2 = y$, and $3 = z$; $f_x = f_y = 0$ and $f_z = -\gamma$; and $\sigma_{ij} = \sigma_{ji}$:

$$
\begin{aligned}
\sigma_{xx,x} + \sigma_{xy,y} + \sigma_{xz,z} &= 0, \\
\sigma_{xy,x} + \sigma_{yy,y} + \sigma_{yz,z} &= 0, \\
\sigma_{xz,x} + \sigma_{yz,y} + \sigma_{zz,z} &= \gamma.
\end{aligned}
\tag{6.1}
$$

© Springer International Publishing AG, part of Springer Nature 2018
P. L. Gould, Y. Feng, *Introduction to Linear Elasticity*,
https://doi.org/10.1007/978-3-319-73885-7_6

Fig. 6.1 (**a**) Prismatic bar in
extension; (**b**) deformed
cross section of prismatic
bar

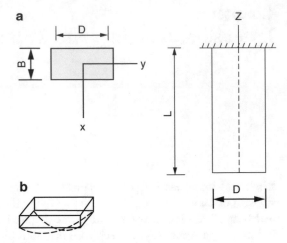

We also have an overall equilibrium relationship between the normal component
of the traction T_z at $z = L$, $\sigma_{zz}(x, y, L)$ and the self-weight given by

$$\int_{-B/2}^{B/2} \int_{-D/2}^{D/2} \sigma_{zz}(x, y, L)\, dy\, dx = \gamma BDL. \qquad (6.2)$$

Invoking St. Venant's principle and seeking a solution away from the support
region without concern for the exact stress distribution there, the semi-inverse
approach discussed in Sect. 5.7 is followed whereby the stresses are taken as

$$\sigma_{zz} = \gamma z,$$
$$\sigma_{xx} = \sigma_{yy} = \sigma_{xy} = \sigma_{xz} = \sigma_{yz} = 0, \qquad (6.3)$$

which satisfies (6.1) and (6.2) for equilibrium. All surfaces except $z = L$ are stress-
free, corresponding to condition (3) for uniqueness in Sect. 5.5; thus, the distribution
given by (6.3) is satisfactory.

Proceeding to the strain–stress relationships (4.29b), we have

$$\varepsilon_{xx} = \frac{1}{E}\big[\sigma_{xx} - \nu(\sigma_{yy} + \sigma_{zz})\big] = -\frac{\nu}{E}\gamma z$$

$$\varepsilon_{yy} = \frac{1}{E}\big[\sigma_{yy} - \nu(\sigma_{xx} + \sigma_{zz})\big] = -\frac{\nu}{E}\gamma z$$

$$\varepsilon_{zz} = \frac{1}{E}\big[\sigma_{zz} - \nu(\sigma_{xx} + \sigma_{yy})\big] = -\frac{1}{E}\gamma z$$

$$\varepsilon_{xy} = \frac{1}{2G}\sigma_{xy} = 0 \qquad (6.4)$$

$$\varepsilon_{yz} = \frac{1}{2G}\sigma_{yz} = 0$$

$$\varepsilon_{xz} = \frac{1}{2G}\sigma_{xz} = 0.$$

Since the strains are linear, the compatibility conditions (3.59) are automatically satisfied.

We should now be able to achieve a single-valued displacement field by the integration of (3.15) [1] with 1,2,3 = x,y,z, which take the form

$$\varepsilon_{xx} = u_{x,x} = -\frac{\nu}{E}\gamma z \tag{6.5a}$$

$$\varepsilon_{yy} = u_{y,y} = -\frac{\nu}{E}\gamma z \tag{6.5b}$$

$$\varepsilon_{zz} = u_{z,z} = \frac{1}{E}\gamma z \tag{6.5c}$$

$$\varepsilon_{xy} = \frac{1}{2}\left(u_{x,y} + u_{y,x}\right) = 0 \tag{6.5d}$$

$$\varepsilon_{yz} = \frac{1}{2}\left(u_{y,z} + u_{z,y}\right) = 0 \tag{6.5e}$$

$$\varepsilon_{xz} = \frac{1}{2}(u_{x,z} + u_{z,x}) = 0. \tag{6.5f}$$

Starting with (6.5c), we integrate to get

$$u_z = \frac{\gamma z^2}{2E} + f(x, y), \tag{6.6}$$

where $f(x, y)$ is an arbitrary function. Substitution of (6.6) into (6.5e) and (6.5f) yields

$$u_{y,z} = -f_{,y} \tag{6.7a}$$

$$u_{x,z} = -f_{,x}, \tag{6.7b}$$

which integrate into

$$u_y = -zf_{,y} + g(x, y) \tag{6.8a}$$

$$u_x = -zf_{,x} + h(x, y), \tag{6.8b}$$

where g and h are two more arbitrary functions. These expressions for u_y and u_x may now be substituted into (6.5a) and (6.5b), which become

$$-zf_{,xx} + h_{,x} = -\frac{\nu}{E}\gamma z \tag{6.9a}$$

$$-zf_{,yy} + g_{,y} = -\frac{\nu}{E}\gamma z. \tag{6.9b}$$

It has already been established that f, g, and h do not depend on z. Therefore, (6.9a) and (6.9b) may be separated into the following relationships:

$$h_{,x} = 0 \tag{6.10a}$$

$$g_{,y} = 0 \tag{6.10b}$$

$$f_{,xx} = \frac{\nu}{E}\gamma \tag{6.10c}$$

$$f_{,yy} = \frac{\nu}{E}\gamma. \tag{6.10d}$$

Thus far, only (6.5d) has not been considered. We substitute u_x and u_y from (6.8b) and (6.8a) into that equation producing

$$-2zf_{,xy} + h_{,y} + g_{,x} = 0 \tag{6.11}$$

that separates into

$$h_{,y} + g_{,x} = 0 \tag{6.12a}$$

$$f_{,xy} = 0 \tag{6.12b}$$

due to the independence of $h_{,y}$ and $g_{,x}$ from z. Equations (6.10a) and (6.10b) together with (6.12a) produce expressions for g and h of the form

$$g = C_1 r(x) + C_2 \tag{6.13a}$$

$$h = C_3 t(y) + C_4 \tag{6.13b}$$

$$C_3 t_{,y} = C_1 r_{,x} = 0, \tag{6.13c}$$

where r and t are again arbitrary functions. The foregoing are satisfied by choosing $t(y) = y$, $r(x) = x$, and $C_3 = -C_1$, whereupon

$$g = C_1 x + C_2 \tag{6.14a}$$

$$h = -C_1 y + C_4. \tag{6.14b}$$

If we take

$$f = \frac{\nu}{2E}\gamma(x^2 + y^2) + C_5 x + C_6 y + C_7, \tag{6.15}$$

we see that all of the constraints, (6.10c), (6.10d), and (6.12b), are satisfied, and sufficient rigid body terms are provided as shown below.

Collecting the evaluated integration functions, we are able to write the expressions for the displacements by substituting f, g, and h into (6.6), (6.8a), and (6.8b):

$$u_x = -\frac{\nu}{E}\gamma xz - C_1 y + C_4 - C_5 z \tag{6.16a}$$

$$u_y = -\frac{\nu}{E}\gamma yz + C_1 x + C_2 - C_6 z \tag{6.16b}$$

$$u_z = -\frac{\gamma}{2E}[z^2 + \nu(x^2 + y^2)] + C_5 x + C_6 y + C_7. \tag{6.16c}$$

The somewhat tedious calculations for the displacements contain six arbitrary constants, all of which are coefficients of rigid body terms. These constants may be evaluated by imposing constraints on the displacement and on the average rotation at the center of the fixed end, that is, at $(0, 0, L)$, where $u_x = u_y = u_z = 0$ and $\omega_{12} = \omega_{13} = \omega_{23} = 0$. These are written from (6.16a) to (6.16c) and (3.31b) to be

$$u_x(0, 0, L) = C_4 - C_5 L = 0 \tag{6.17a}$$

$$u_y(0, 0, L) = C_2 - C_6 L = 0 \tag{6.17b}$$

$$u_z(0, 0, L) = \frac{\gamma L^2}{2E} + C_7 = 0 \tag{6.17c}$$

$$\omega_{xy}(0, 0, L) = \frac{1}{2}(-C_1 - C_1) = 0 \tag{6.17d}$$

$$\omega_{xz}(0, 0, L) = \frac{1}{2}(-C_5 - C_5) = 0 \tag{6.17e}$$

$$\omega_{yz}(0, 0, L) = \frac{1}{2}(-C_6 - C_6) = 0 \tag{6.17f}$$

From the last three equations, $C_1 = C_5 = C_6 = 0$, while the first three give

$$C_4 = 0 \tag{6.18a}$$

$$C_2 = 0 \tag{6.18b}$$

$$C_7 = \frac{-\gamma L^2}{2E}. \tag{6.18c}$$

Thus, the final expressions for the displacements become

$$u_x = -\frac{\nu}{E}\gamma xz \tag{6.19a}$$

$$u_y = -\frac{\nu}{E}\gamma yz \tag{6.19b}$$

$$u_z = \frac{\gamma}{2E}[z^2 + \nu(x^2 + y^2) - L^2]. \tag{6.19c}$$

From these results, we make some observations not discernable from the strength-of-materials solution, a uniform axial extension along the z-axis corresponding to u_z $(0, 0, z)$ in (6.19c):

1. All points not on the center line have contractions in the x–y plane given by (6.19a) and (6.19b). This is sometimes called the "Poisson" effect, since it is evidently the motivation for the definition of Poisson's ratio as the fractional contraction (see Sect. 4.3).
2. The axial displacement is *not* uniform on the cross section but is a parabolic surface with a peak along the z-axis as shown in Fig. 6.1b.

Now in recalling the objectives of examining this problem as outlined in Sect. 6.1, we find that the elasticity solution has confirmed the adequacy of the strength-of-materials solution for defining the equilibrium state and has enhanced the description of the displacements.

6.3 Cantilever Beam Under End Loading

6.3.1 Elementary Beam Theory

Beams have been studied formally since the time of Galileo Galilei, who discussed the cantilever in the middle seventeenth century. Elastic beam theory evolved from the contributions of Edme Mariotte, James Bernoulli, Leonhard Euler, Charles A. Coulomb, and, finally L. M. H. Navier in the early nineteenth century. As an illustration, we consider the cantilever beam shown in Fig. 6.2a subjected to a concentrated end load P, neglecting the self-weight.

From elementary beam theory, we may write the solution for the stresses in terms of the shear V and the bending moment M as

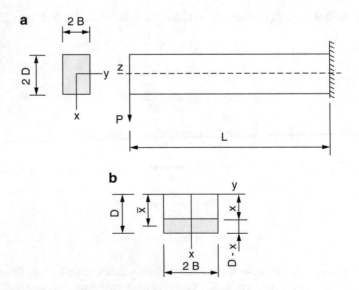

Fig. 6.2 (a) Cantilever beam in bending; (b) location of plane on which horizontal shear acts

$$\sigma_{xx} = \sigma_{yy} = 0 \tag{6.20a}$$

$$\sigma_{zz} = \frac{Mx}{I} = \frac{P}{I}(z - L)x \tag{6.20b}$$

$$\sigma_{xz} = \sigma_{zx} = \frac{VQ}{2IB} = \frac{P}{2I}(D^2 - x^2) \tag{6.20c}$$

$$\sigma_{xy} = \sigma_{yz} = 0, \tag{6.20d}$$

where I is the moment of inertia of the cross section equal to $\frac{1}{12} \times 2B \times (2D)^3 = \frac{4}{3}BD^3$; and Q is the first moment of the area between x and D, as shown on Fig. 6.2b. For a rectangular cross section,

$$Q = (D - x) \times 2B \times \bar{x} = (D - x) \times 2B \times \frac{(D + x)}{2} = (D^2 - x^2)B.$$

Implicit in these expressions is the assumption that the load P and the resulting stresses are uniform over the width of the beam, so that the solution is essentially two-dimensional. Later, we will test this solution against the theory of elasticity.

First, we consider equilibrium (2.46), with $1 = x$, $2 = y$, and $3 = z$:

$$\sigma_{xx,x} + \sigma_{xy,y} + \sigma_{xz,z} = 0 + 0 + 0 = 0 \tag{6.21a}$$

$$\sigma_{yx,x} + \sigma_{yy,y} + \sigma_{yz,z} = 0 + 0 + 0 = 0 \tag{6.21b}$$

$$\sigma_{zx,x} + \sigma_{zy,y} + \sigma_{zz,z} = -\frac{Px}{I} + 0 + \frac{Px}{I} = 0. \tag{6.21c}$$

Thus, the equilibrium is satisfied.

Next, we examine the compatibility conditions using the Beltrami–Michell form (5.16) since the solution being tested is in terms of stresses. From (5.10a),

$$Q_1 = \sigma_{tt} = \sigma_{zz} = \frac{P}{I}(z - L)x. \tag{6.22}$$

Equations (5.16) for $i, j = x, y, z$ become

$$\nabla^2 \sigma_{xx} + \frac{1}{1+\nu}Q_{1,xx} = 0 + 0 = 0 \tag{6.23a}$$

$$\nabla^2 \sigma_{xy} + \frac{1}{1+\nu}Q_{1,xy} = 0 + 0 = 0 \tag{6.23b}$$

$$\nabla^2 \sigma_{xz} + \frac{1}{1+\nu}Q_{1,xz} = -\frac{P}{I} + \frac{1}{1+\nu}\frac{P}{I} = \frac{-\nu}{1+\nu}\frac{P}{I} \neq 0 \tag{6.23c}$$

$$\nabla^2 \sigma_{yy} + \frac{1}{1+\nu}Q_{1,yy} = 0 + 0 = 0 \tag{6.23d}$$

$$\nabla^2 \sigma_{yz} + \frac{1}{1+\nu}Q_{1,yz} = 0 + 0 = 0 \tag{6.23e}$$

$$\nabla^2 \sigma_{zz} + \frac{1}{1+\nu} Q_{1,zz} = 0 + 0 = 0 \qquad (6.23f)$$

One of the compatibility equations, (6.23c), will be violated unless $\nu = 0$, which is implicitly assumed in elementary beam theory.

It is instructive also to consider the compatibility equations in terms of strain (3.59) and to identify specifically the violation. The relevant equation is (3.59e), which takes the form

$$\varepsilon_{yz,yx} + \varepsilon_{yx,yz} - \varepsilon_{zx,yy} = \varepsilon_{yy,zx}. \qquad (6.24)$$

The strains are calculated from (4.19) and (4.24) as

$$\varepsilon_{yz} = \varepsilon_{yx} = 0 \qquad (6.25a)$$

$$\varepsilon_{zx} = \frac{\sigma_{zx}}{2\mu} = \frac{1}{2G}\frac{P}{2I}\left(D^2 - x^2\right) \qquad (6.25b)$$

$$\varepsilon_{zx,yy} = 0 \qquad (6.25c)$$

$$\varepsilon_{yy} = \frac{-\lambda}{2\mu(2\mu + 3\lambda)}\sigma_{zz} = \frac{-\nu}{E}\frac{P}{I}(z - L)x \qquad (6.25d)$$

$$\varepsilon_{yy,zx} = \frac{-\nu P}{EI}. \qquad (6.25e)$$

Thus, (6.24) becomes

$$0 + 0 + 0 \neq \frac{-\nu P}{EI}. \qquad (6.26)$$

This compatibility equation can be satisfied only through a modification of the assumed stress distribution, considered in the next section. Also, we see that we have a strain ε_{yy}, normal to the plane of bending and proportional to Poisson's ratio, which does not enter into the elementary theory but which may be calculated from Hooke's law in terms of σ_{zz} (6.25d). This is shown on the deformed cross section in Fig. 6.3.

Fig. 6.3 Deformed cross section of a cantilever beam (after Ugural and Fenster [6] Reprinted with permission)

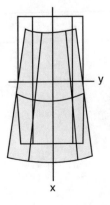

For $x < 0$ and $z < L$, ε_{yy} is negative, while for $x > 0$, it is positive. This distortion from the initially rectangular shape of the cross section is another example of a "Poisson" effect. Of course, so long as σ_{zz} and hence ε_{zz} remain independent of y and linear in x for a given value of z, the elementary postulate that "plane sections remain plane" is substantiated.

Next, we see if the elementary solution satisfies the boundary conditions. For the lateral boundaries, we have the components of the tractions given in terms of the stresses by (2.11), $T_i = \sigma_{ji} n_j$. For the top and bottom of the beam, we consider the normal traction at $x = \pm D$ for which $\mathbf{n} = \pm \mathbf{e}_x$ and

$$n_x = 1, \quad n_y = 0, \quad n_z = 0. \tag{6.27}$$

Then

$$T_x = \sigma_{jx}(\pm D, y, z)n_j = 0 \times 1 + 0 \times 0 + 0 \times 0 = 0. \tag{6.28}$$

It may be verified that the components T_y and T_z vanish as well. However, it is of interest to note that should the beam be loaded along the top or bottom surface, the solution would *not* automatically satisfy the boundary condition on the normal traction. This is an indication that the theory of elasticity solution must account for the variation of a distributed loading through the depth of the beam, in contrast to the elementary solution which is referred to as "neutral" axis, $x = 0$. We have avoided this problem for the present by considering a concentrated loading.

Evaluating the normal traction on the lateral boundaries $y = \pm B$ for which $\mathbf{n} = \pm \mathbf{e}_y$ and

$$n_x = 0, \quad n_y = 1, \quad n_z = 0 \tag{6.29}$$
$$T_y = \sigma_{jy}(x, \pm B, z)n_j = 0 \times 0 + 0 \times 1 + 0 \times 0 = 0. \tag{6.30}$$

The components T_x and T_z also vanish.

At the fixed end $z = 0$, our present solution is insufficient to check the displacement boundary conditions since it violates compatibility, while at $z = L$ we have

$$
\begin{aligned}
P &= \int_{-D}^{D} \int_{-B}^{B} \sigma_{xz} \, dy \, dx \\
&= \int_{-D}^{D} \int_{-B}^{B} \frac{P(D^2 - x^2)}{2I} \, dy \, dx \\
&= \frac{4}{3} \frac{P}{I} BD^3.
\end{aligned}
\tag{6.31}
$$

With $I = \frac{4}{3} BD^3$, the equation is satisfied. This implies that a concentrated load must be distributed over the depth in accordance with σ_{xz}, that is, parabolically, in order to satisfy the boundary condition exactly. Of course, other distributions are admissible through St. Venant's principle.

We have shown that the elementary beam solution, while satisfying equilibrium in general, fails to account for the lateral contraction or "Poisson" effect and does not recognize the distribution of the applied loading through the depth of the beam.

6.3.2 Elasticity Theory

Again drawing on the work of St. Venant, the semi-inverse approach is followed whereby it is assumed that the *normal* stresses are identical to those of the elementary theory as given by (6.20a) and (6.20b). Likewise, σ_{xy} is taken as zero as in (6.20d), but no stipulation is made on the remaining *shearing* stresses σ_{zx} and σ_{zy} that act in the plane of the cross-section x–y.

The equilibrium equations are generalized from (6.21a, 6.21b, 6.21c) to become

$$\sigma_{xz,z} = 0 \tag{6.32a}$$

$$\sigma_{yz,z} = 0 \tag{6.32b}$$

$$\sigma_{zx,x} + \sigma_{zy,y} = -\frac{Px}{I}. \tag{6.32c}$$

Equations (6.32a) and (6.32b) indicate that the shearing stresses do not depend on z and therefore are the same for all cross sections.

Recall the compatibility conditions (5.16) which were expanded as (6.23a), (6.23b), (6.23c), (6.23d), (6.23e), and (6.23f). For this case, two are relevant, (6.23c) and (6.23e). The stresses σ_{xz} and σ_{yz} should satisfy [1]

$$\nabla^2 \sigma_{xz} = -\frac{1}{1+\nu}\frac{P}{I} \tag{6.33a}$$

$$\nabla^2 \sigma_{yz} = 0. \tag{6.33b}$$

Also, we must verify that any modified solution continues to satisfy the boundary conditions. Particularly, we consider the components of traction T_z on each face of the beam, found on the top and bottom using (6.27) and (6.28) as

$$\sigma_{xz}(\pm D, y, z) = 0 \tag{6.34}$$

and on the sides using (6.29) and (6.30) as

$$\sigma_{yz}(x, \pm B, z) = 0. \tag{6.35}$$

To continue, we introduce a technique that is very effective in obtaining analytical solutions to a variety of problems in the theory of elasticity [1]. We define a *stress function* ϕ (x,y) such that

$$\sigma_{xz} = \phi_{,y} - \frac{Px^2}{2I} + f(y) \tag{6.36a}$$

and

$$\sigma_{yz} = -\phi_{,x}, \tag{6.36b}$$

which satisfies (6.32a), (6.32b), and (6.32c). Then (6.33a) and (6.33b) become, respectively,

$$(\phi_{,yxx} + \phi_{,yyy}) - \frac{P}{I} + f_{,yy} = \frac{-1}{1+\nu} \frac{P}{I},$$

or

$$(\phi_{,xx} + \phi_{,yy})_{,y} = \frac{\nu}{1+\nu} \frac{P}{I} - f_{,yy}. \tag{6.37a}$$

and

$$-\phi_{,xxx} - \phi_{,xyy} = 0,$$

or

$$\left(\phi_{,xx} + \phi_{,yy}\right)_{,x} = 0. \tag{6.37b}$$

Equation (6.37a) integrates to

$$\phi_{,xx} + \phi_{,yy} = \frac{\nu}{1+\nu} \frac{P}{I} y - f_{,y} + C, \tag{6.38}$$

where C is an arbitrary constant. Equation (6.38) also satisfies (6.37b).

To evaluate C, we consider the average rotation about the z-axis, ω_{xy}, as given by (3.32) with $i = x, j = y$:

$$\omega_{xy} = \frac{1}{2}\left(u_{x,y} - u_{y,x}\right). \tag{6.39}$$

The rate of change of ω_{xy} along z is given by

$$\begin{aligned}
\omega_{xy,z} &= \frac{1}{2}\left(u_{x,y} - u_{y,x}\right)_{,z} \\
&= \frac{1}{2}\left[\left(u_{x,z} + u_{z,x}\right)_{,y} - \left(u_{y,z} + u_{z,y}\right)_{,x}\right] \\
&= \varepsilon_{xz,y} - \varepsilon_{yz,z}
\end{aligned} \tag{6.40}$$

by (3.15e) and (3.15f). Converting to stresses through (4.29b), we have

$$\omega_{xy,z} = \frac{1}{2G}\left(\sigma_{xz,y} - \sigma_{yz,x}\right)$$

$$= \frac{1}{2G}\left(\phi_{,yy} + f_{,y} - \phi_{,xx}\right)$$

$$(6.41)$$

using (6.36a) and (6.36b). Finally, substituting (6.41) into the l.h.s. of (6.38), we find

$$2G\omega_{xy,z} = \frac{\nu}{1+\nu}\frac{Py}{I} + C. \tag{6.42}$$

We are considering a rectangular cross section [see Fig. 6.2a] in which the bending is symmetrical about the x-axis. So at $y = 0$, the average rotation ω_{xy} should be zero as well as its rate of change along z, $\omega_{xy,z}$; therefore $C = 0$.

Equation (6.38) now becomes

$$\phi_{,xx} + \phi_{,yy} = \frac{\nu}{1+\nu}\frac{P}{I}y - f_{,y}, \tag{6.43}$$

which is the governing equation for the stress function subject to satisfaction of the boundary conditions (6.34) and (6.35). If we introduce $f(y) = PD^2/(2I)$ into (6.36a), we may rewrite the boundary conditions directly in terms of the stress function as

$$\sigma_{xz}(\pm D, y, z) = \phi_{,y}(\pm D, y, z) = 0 \tag{6.44a}$$

$$\sigma_{yz}(x, \pm B, z) = -\phi_{,x}(x, \pm B, z) = 0. \tag{6.44b}$$

These are easily satisfied by taking ϕ in a form such that it *vanishes* on the boundary. In many cases, ϕ may be taken to be proportional to the equation of the perimeter of the cross section, which is for this case

$$(x^2 - D^2)(y^2 - B^2) = 0. \tag{6.45}$$

The choice of $f(y)$ to produce (6.44a) and (6.44b) then reduces (6.43) to

$$\phi_{,xx} + \phi_{,yy} = \frac{\nu}{1+\nu}\frac{Py}{I}. \tag{6.46}$$

Before proceeding with the solution, it is instructive to digress in order to draw an analogy between (6.46) and the equation of a uniform *membrane* stretched over the cross section of the beam in Fig. 6.4a. The equation for a membrane was derived in Sect. 5.6. Except near the boundaries, the height of the membrane $\phi(x, y)$ is proportional to the load intensity q, which is in turn analogous to the r.h.s. of (6.46) $[\nu/(1 + \nu)](Py/I)$ as shown in Fig. 6.4b [1]. In effect, the loading pattern shown in Fig. 6.4b would produce a deflected membrane as shown in Fig. 6.4c. At the edges and along the center line $y = 0$, the height of the membrane must be zero as indicated on the profile in Fig. 6.4c. Because of the constraint at $y = 0$, the profile is

Fig. 6.4 (**a**) Cross section of membrane analogous to beam; (**b**) loading and (**c**) profile of deformed membrane; (**d**) deep beam; (**e**) wide beam (From Ugural and Fenster [6] Reprinted with permission)

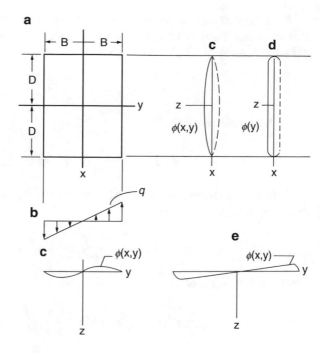

different than in Fig. 5.1. Now we restate the expressions for the stresses from (6.36a) and (6.36b), incorporating the chosen $f(y) = PD^2/(2I)$:

$$\sigma_{xz} = \phi_{,y} - \frac{Px^2}{2I} + \frac{PD^2}{2I}$$

$$= \frac{P(D^2 - x^2)}{2I} + \phi_{,y},$$

$$\sigma_{yz} = -\phi_{,x}.$$

(6.47)

From (6.47), we see that the *changes* in the shearing stresses from the expressions given by the elementary theory, (6.20c) and (6.20d), are proportional to the first derivative or *slope* of the membrane at any point. Examining the profile along y in Fig. 6.4c, we see that the correction to σ_{xz}, ϕ_y has a maximum positive value at the two sides and a maximum negative value in the center and is zero at the quarter-points. Thus, the *uniformity* of the stress state across the width of the beam inherent in the elementary theory is dispelled. Membrane analogies are often very useful in the theory of elasticity for visualizing and interpreting complex states of stress and strain.

The general solution to (6.46), as derived by Timoshenko and Goodier [1], is rather more complicated than our introductory objectives; however, it is possible to consider some approximate solutions reasoned from the membrane analogy [1].

If the depth of the beam is large compared with the width, $D \gg B$, we may assume that at points sufficiently distant from the short side $x = \pm D$, the surface is cylindrical. As shown in Fig. 6.4d, this implies that $\phi = \phi(y)$. Then (6.46) reduces to

$$\phi_{,yy} = \frac{\nu}{1+\nu} \frac{Py}{I}, \tag{6.48}$$

which has a solution of the form

$$\phi(y) = \frac{\nu}{1+\nu} \frac{P}{6I} (y^3 + C_1 y + C_2). \tag{6.49}$$

With the membrane condition $\phi(0) = 0$ (see Fig. 6.4c), $C_2 = 0$, and with $\phi(\pm B) = 0$, $C_1 = -B^3$, so that

$$\phi(y) = \frac{\nu}{1+\nu} \frac{P}{6I} (y^3 - B^3 y); \tag{6.50}$$

from which (6.47) gives

$$\sigma_{xz}(x, y) = \frac{P}{2I} \left[(D^2 - x^2) + \frac{\nu}{1+\nu} \left(y^2 - \frac{B^2}{3} \right) \right]. \tag{6.51}$$

Of course, the assumption on ϕ gives $\sigma_{yz} = 0$. At $x = 0$, where the stress computed by the elementary theory is largest,

$$\sigma_{xz}(0, y) = \frac{3}{2} \frac{P}{(2B \times 2D)},$$

the correction given by the second term of (6.51) is small.

The other extreme is when the width of the beam is large compared to the depth, $B \gg D$. Then, at points sufficiently distant from the short side $y = \pm B$, the function ϕ representing the deflection of the membrane is assumed to be linear in y as indicated in Fig. 6.4e. Thus, (6.46) reduces to

$$\phi_{,xx} = \frac{\nu}{1+\nu} \frac{Py}{I}, \tag{6.52}$$

which is solved as

$$\phi_{,x} = \frac{\nu}{1+\nu} \frac{Pyx}{I} + C_1$$

$$\phi(x, y) = \frac{\nu}{1+\nu} \frac{Py}{2I} x^2 + C_1 x + C_2. \tag{6.53}$$

With $\phi(\pm D, y) = 0$, we find

$$C_2 = \frac{-\nu}{1+\nu} \frac{Py}{2I} D^2,$$ (6.54)

and $C_1 = 0$ so that

$$\phi = \frac{\nu}{1+\nu} \frac{Py}{2I} (x^2 - D^2),$$ (6.55)

and, from (6.47),

$$\sigma_{xz} = \frac{P(D^2 - x^2)}{2I} \left[1 - \frac{\nu}{1+\nu} \right] = \frac{1}{1+\nu} \frac{P(D^2 - x^2)}{2I},$$ (6.56a)

$$\sigma_{yz} = \frac{-\nu}{1+\nu} \frac{P}{I} xy.$$ (6.56b)

In comparison to elementary theory, σ_{xz} is reduced by the factor $1/(1+\nu)$, and σ_{yz} is now available.

For very large values of B/D, both σ_{xz} and σ_{yz} may greatly exceed the $\frac{3}{2}P/(2B \times 2D)$ peak from elementary theory. At first, this does not seem possible from (6.56a) and (6.56b) with. $I = \frac{4}{3}BD^3$; however, we recall that the underlying assumption excluded regions near the short sides $y = \pm B$, where actual maximum stress occurs as depicted by the steep slope in Fig. 6.4e. Moreover, σ_{yz} near the corners, where (6.56b) would indicate the largest value occurs, may become numerically larger than σ_{xz}.

A final approximate solution is to take ϕ as a multiple of the boundary curve to satisfy the $\phi = 0$ condition. The form

$$\phi = (x^2 - D^2)(y^2 - B^2)(C_1 y + C_2 y^3)$$ (6.57)

is satisfactory, but the constants C_1 and C_2 depend on the D/B ratio and must be evaluated from a minimum energy criteria [1].

Although we have not derived an elasticity solution to cover all cases, we have shown the limitations of elementary beam theory and have demonstrated how a typical elasticity problem may be approached. Based on the general solution, the error in the maximum stress obtained by the elementary formula for a square beam cross section is about 10% [1].

6.4 Torsion

6.4.1 Torsion of Circular Shaft

The circular shaft with radius B shown in Fig. 6.5a is subjected to a twisting moment (torque) $M_z(L)$ at the free end and is restrained against both displacement and rotation at $z = 0$.

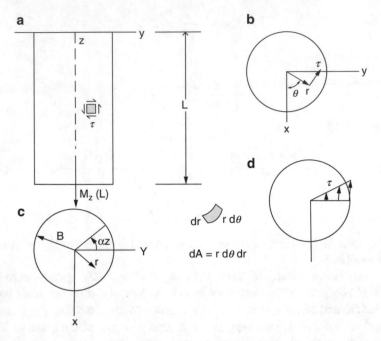

Fig. 6.5 (**a**) Circular shaft in torsion; (**b**) Coulomb torsion assumptions; (**c**) polar area element; (**d**) shearing stresses

The solution for this classical problem was presented by Charles A. Coulomb in the late eighteenth century and falls within the field of strength of materials. Referring to Fig. 6.5a, b, the Coulomb torsion solution is based on the following assumptions with respect to the shearing stress $\tau(x, y)$:

1. The shearing stress τ acts in a direction perpendicular to the radius vector **r**.
2. The shearing stress τ is proportional to $r = |\mathbf{r}|$.
3. The shearing stress τ is proportional to the angle of rotation per unit length, which is called the rate of twist α.

We see from the differential element on Fig. 6.5a that the twisting moment produces a state of pure shear stress, so that the proportionality constant in assumption (3) is the shear modulus G; therefore, assumptions (2) and (3) are expressed by

$$\tau = G\alpha r, \tag{6.58}$$

Equilibrium on the polar area element shown in Fig. 6.5c gives

$$M_z = \int_A \tau \times r \, dA, \tag{6.59}$$

or using (6.58) and taking $dA = r \, dr \, d\theta$,

$$M_z = G\alpha \int_0^B \int_{-\pi}^{\pi} r \times r \times r \, d\theta \, dr$$

$$= G\alpha \times 2\pi \int_0^B r^3 \, dr \qquad\qquad (6.60a)$$

$$= G\alpha J_c,$$

in which

$$J_c = \frac{\pi B^4}{2}, \qquad\qquad (6.60b)$$

where J_c is called the *polar moment of inertia*. Then, we may eliminate $G\alpha$ between (6.58) and (6.60a) to get

$$\tau = \frac{M_z r}{J_c}. \qquad\qquad (6.61)$$

Equation (6.61) permits the evaluation of the shearing stress for a known torque and cross section irrespective of the material and is obviously analogous to (6.20b) for beam bending. The linear distribution is shown in Fig. 6.5d.

Also, for a given twisting moment, the rate of twist may be computed from (6.60a) as

$$\alpha = \frac{M_z}{G J_c}. \qquad\qquad (6.62)$$

The quantity $G J_c$ is called the *torsional rigidity* and is a useful parameter for comparing the relative torsional stiffnesses of various cross sections.

With respect to compatibility, since the stresses are linear, the strains will also be linear, and the St. Venant equations are thus satisfied. Single-valued displacements may be evaluated from the kinematic relations, but this is not pursued since the angle of twist is of most interest.

Continuing to assess the admissibility of the strength-of-materials solution as an elasticity solution as well, we examine the boundary conditions. The fixed boundary is bypassed through St. Venant's principle, while the loaded boundary is treated by (6.59). The cylindrical lateral boundary is of primary interest. Assumption (1) guarantees that τ is tangent to the boundary circle and hence has no normal component, so that the boundary must be stress-free.

The satisfaction of the stress-free lateral boundary condition by a shearing stress distribution, that is, at the same time, perpendicular to the radius vector, is fortuitous, insofar as the solution for the torsional stresses on a *circular* cross section is concerned. The same can be said for an annular or hollow cylindrical section. On the other hand, it is easily shown that for *any* other cross section, it is not possible to have the shearing stress both *tangent* to the boundary and *perpendicular* to the radius vector. Simply visualize the stress distribution shown in Fig. 6.5d applied on a rectangular cross section. Except at the major and minor axis intersections with the perimeter, $x = 0$, $y = 0$, simultaneous satisfaction of these conditions cannot be attained. Moreover, the corner presents an additional complication. This condition perplexed early researchers in solid mechanics for some time and was ultimately resolved by a more fundamental approach based on the theory of elasticity, as developed in the following section.

6.4.2 Torsion of Solid Prismatic Shafts

The generalization of the uniform torsion problem from the circular cross-sectional solution is due to none other than St. Venant and was considered by Westergaard [2] to be the most important single contribution of St. Venant to the theory of elasticity. The solution for a noncircular cross section (see Fig. 6.6a) follows the now familiar semi-inverse approach whereby all stresses except the in-plane shearing stresses σ_{zx} and σ_{zy} are taken as zero, that is,

Fig. 6.6 (**a**) Prismatic shaft in torsion; (**b**) rotated axes; (**c**) stresses on lateral surface; (**d**) angle of rotation

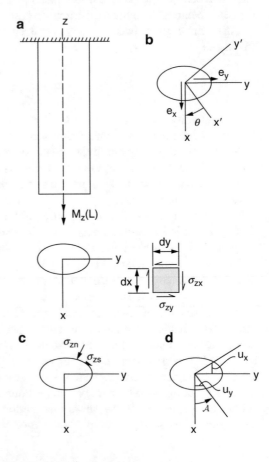

$$\sigma_{xx} = \sigma_{yy} = \sigma_{zz} = \sigma_{xy} = 0. \tag{6.63}$$

Considering the equilibrium equations (2.46) in view of (6.63) with $1 = x$, $2 = y$, $3 = z$, and no body forces,

$$\sigma_{xz,z} = 0, \tag{6.64a}$$

$$\sigma_{yz,z} = 0, \tag{6.64b}$$

$$\sigma_{xz,x} + \sigma_{yz,y} = 0. \tag{6.64c}$$

Equations (6.64a) and (6.64b) indicate that the shearing stresses are functions of x and y only.

The equations are solved by defining a stress function $\phi(x, y)$ such that

$$\sigma_{xz}(x, y) = \phi_{,y}, \tag{6.65a}$$

$$\sigma_{yz}(x, y) = -\phi_{,x}. \tag{6.65b}$$

The corresponding strains are computed from (4.29b) as

$$\varepsilon_{xz}(x, y) = \frac{1}{2G}\sigma_{xz}, \tag{6.66a}$$

$$\varepsilon_{yz}(x, y) = \frac{1}{2G}\sigma_{yz}. \tag{6.66b}$$

Then, we enter the compatibility equations (3.59), all of which are identically satisfied except (3.59d)

$$\varepsilon_{xy,xz} + \varepsilon_{xz,xy} - \varepsilon_{yz,xx} = \varepsilon_{xx,yz}.$$

After eliminating the zero terms $\varepsilon_{xy,xz}$ and $\varepsilon_{xx,yz}$, we have

$$\left(\varepsilon_{xz,y} - \varepsilon_{yz,x}\right)_{,x} = 0. \tag{6.67}$$

If $\varepsilon_{xz,y} - \varepsilon_{yz,x} = \text{constant}$, then (6.67) will be satisfied. We may express this requirement in terms of the stress function ϕ by considering (6.65) and (6.66),

$$\phi_{,xx} + \phi_{,yy} = H, \tag{6.68}$$

where H is the constant. Equation (6.68) is known as Poisson's equation and expresses the *compatibility* requirements for the torsion problem.

Since the *direction* of the shearing stress was a prime factor in motivating the more general study of torsion, we consider this aspect in detail. We first note that the components of the *gradient* of ϕ are

$$\nabla\phi = [\nabla\phi]_x\mathbf{e}_x + [\nabla\phi]_y\mathbf{e}_y$$
$$= \phi_{,x}\mathbf{e}_x + \phi_{,y}\mathbf{e}_y. \tag{6.69}$$

Then from (6.65a) and (6.65b),

$$\sigma_{xz} = [\nabla\phi]_y, \tag{6.70a}$$
$$\sigma_{yz} = -[\nabla\phi]_x. \tag{6.70b}$$

We next evaluate the shearing stresses for a rotated set of axes $x'\,y'$ (see Fig. 6.6b) using the transformation law (2.13)

$$\sigma_{x'z} = \alpha_{x'i}\alpha_{zj}\sigma_{ij}, \tag{6.71a}$$

$$\sigma_{y'z} = \alpha_{y'i}\alpha_{zj}\sigma_{ij}. \tag{6.71b}$$

Expanding these equations gives

$$\sigma_{x'z} = \alpha_{x'x}\alpha_{zz}\sigma_{xz} + \alpha_{x'y}\alpha_{zz}\sigma_{yz} + \alpha_{x'z}\alpha_{zx}\sigma_{zx} + \alpha_{x'z}\alpha_{zy}\sigma_{zy}, \tag{6.72a}$$

$$\sigma_{y'z} = \alpha_{y'x}\alpha_{zz}\sigma_{xz} + \alpha_{y'y}\alpha_{zz}\sigma_{yz} + \alpha_{y'z}\alpha_{zx}\sigma_{zx} + \alpha_{y'z}\alpha_{zy}\sigma_{zy}, \tag{6.72b}$$

where the direction cosines are

$$
\begin{array}{cccc}
 & x & y & z \\
x' & \cos\theta & \sin\theta & 0 \\
y' & -\sin\theta & \cos\theta & 0 \\
z' & 0 & 0 & 1.
\end{array}
$$

Therefore,

$$\sigma_{x'z} = \sigma_{xz}\cos\theta + \sigma_{yz}\sin\theta, \tag{6.73a}$$

$$\sigma_{y'z} = -\sigma_{xz}\sin\theta + \sigma_{yz}\cos\theta. \tag{6.73b}$$

Now, substituting (6.70a) and (6.70b) into (6.73a) and (6.73b), we get

$$\sigma_{x'z} = [\nabla\phi]_y\cos\theta - [\nabla\phi]_x\sin\theta, \tag{6.74a}$$
$$\sigma_{y'z} = -[\nabla\phi]_y\sin\theta - [\nabla\phi]_x\cos\theta. \tag{6.74b}$$

Also, we evaluate the unit vectors in the x' and y' vectors from Fig. 6.6b as

$$\mathbf{e}_{x'} = \mathbf{e}_x\cos\theta + \mathbf{e}_y\sin\theta, \tag{6.75a}$$

$$\mathbf{e}_{y'} = \mathbf{e}_y\cos\theta - \mathbf{e}_x\sin\theta. \tag{6.75b}$$

Then, we see that (6.74a) and (6.74b) can be written as

$$\sigma_{x'z} = [\nabla\phi]_y \cos\theta - [\nabla\phi]_x \sin\theta$$
$$= \nabla\phi \cdot \mathbf{e}_y \cos\theta - \nabla\phi \cdot \mathbf{e}_x \sin\theta$$
$$= \nabla\phi \cdot \mathbf{e}_{y'},$$
$$= [\nabla\phi]_{y'} \qquad (6.76a)$$

$$\sigma_{y'z} = -[\nabla\phi]_y \sin\theta - [\nabla\phi]_x \cos\theta$$
$$= -\nabla\phi \cdot \mathbf{e}_y \sin\theta - \nabla\phi \cdot \mathbf{e}_x \cos\theta$$
$$= -\nabla\phi \cdot \mathbf{e}_{x'}$$
$$= -[\nabla\phi]_{x'}. \qquad (6.76b)$$

From this exercise, we conclude that the shearing stress in *any* direction, say x' or y', at any point (x, y, z) is given by the component of $\nabla\phi$ at *right angles* to that direction, that is, y' or x'. This proves useful in the discussion that follows.

We are now in a position to examine the boundary conditions. At any point on the perimeter, the shearing stress perpendicular to the boundary σ_{zn} must vanish [see Fig. 6.6c]. Only stresses tangent to the boundary are permitted. From (6.76b) with $x' = s$, $y' = n$, and $\mathbf{e}_{x'} = \mathbf{s}$, we find

$$\sigma_{zn} = -\nabla\phi \cdot \mathbf{s} = -[\nabla\phi]_s = -\phi_{,s} = 0. \qquad (6.77)$$

Thus, the stress function ϕ must be *constant* on the boundary. In practice, it is usually sufficient to have ϕ vanish there, which is readily accomplished by selecting ϕ to be proportional to the equation of the perimeter of the cross section. This is similar to the bending stress function discussed in Sect. 6.3.

We now proceed to the evaluation of the stress function in terms of the applied torque M_z. First, we note that the resultant force on the face is zero. Referring to Fig. 6.6a and b,

$$\int_A \left(\sigma_{zx}\mathbf{e}_x + \sigma_{zy}\mathbf{e}_y\right) dA = 0. \qquad (6.78)$$

Substituting (6.65a) and (6.65b) for the stresses and $dx\,dy$ for dA, we have

$$\int_A \left(\phi_{,y}\mathbf{e}_x - \phi_{,x}\mathbf{e}_y\right)\, dx\,dy = 0, \qquad (6.79)$$

which separates into two scalar equations

$$\int dx \int \phi_{,y}\, dy = 0, \qquad (6.80a)$$

$$-\int dy \int \phi_{,x}\, dx = 0. \qquad (6.80b)$$

But $\int \phi_{,y} dy = \phi_2(x) - \phi_1(x)$ and $\int \phi_{,x} dx = \phi_4(y) - \phi_3(y)$, where $\phi_1 - \phi_4$ are values of $\phi(x, y)$ on the boundary. However, ϕ has been constrained to be *constant* on the boundary, so that the integrals in (6.80a) and (6.80b) are zero, and (6.78) is satisfied.

Next, we sum moments on the cross section as shown in Fig. 6.6a:

$$\int_A \left(-\sigma_{zx} y + \sigma_{zy} x \right) dx \, dy = M_z. \tag{6.81}$$

In terms of the stress function, the l.h.s. is

$$\int_A \left(-\phi_{,yy} - \phi_{,xx} \right) dx \, dy = -\int dx \int y\phi_{,y} dy - \int dy \int x\phi_{,x} dx. \tag{6.82}$$

We integrate by parts the inner terms on the r.h.s. of (6.82) to get

$$\int y\phi_{,y} \, dy = y\phi \big]_S - \int \phi \, dy \tag{6.83a}$$

$$\int x\phi_{,x} \, dx = x\phi \big]_S - \int \phi \, dx, \tag{6.83b}$$

where S indicates evaluation of the expression at end points that lie on the boundary. Since ϕ is constant there, the first term of each evaluated integral is zero. Therefore, returning to (6.81) in view of (6.82) and (6.83a, 6.83b),

$$M_z = \int dx \int \phi dy + \int dy \int \phi dx$$

$$= 2\int_A \phi dA. \tag{6.84}$$

To complete the solution, we consider the displacements. First, we have the rotation about the z-axis, given by (3.32) with $i = x$ and $j = y$:

$$\omega_{xy} = \frac{1}{2} \left(u_{x,y} - u_{y,x} \right). \tag{6.85}$$

The rate of change along z is

$$\omega_{xy,z} = \frac{1}{2} \left(u_{x,yz} - u_{y,xz} \right). \tag{6.86}$$

Considering the terms on the r.h.s. of (6.86), we have from (3.15) and (4.29b) with 1,2,3 = x, y, z:

$$2\varepsilon_{xz} = u_{x,z} + u_{z,x} = \frac{\sigma_{xz}}{G}, \tag{6.87a}$$

$$2\varepsilon_{yz} = u_{y,z} + u_{z,y} = \frac{\sigma_{yz}}{G}. \tag{6.87b}$$

Next, we take $\frac{\partial}{\partial y}$ (6.87a) and $\frac{\partial}{\partial x}$ (6.87b) to get

$$u_{x,yz} = -u_{z,xy} + \frac{1}{G}\sigma_{xz,y},$$ (6.88a)

$$u_{y,xz} = -u_{z,xy} + \frac{1}{G}\sigma_{yz,x}.$$ (6.88b)

Subtracting (6.88b) from (6.88a) and then substituting (6.65a) and (6.65b) gives

$$\begin{aligned}
\omega_{xy,z} &= \frac{1}{2G}(\sigma_{xz,y} - \sigma_{yz,x}) \\
&= \frac{1}{2G}(\phi_{,yy} + \phi_{,xx}) \\
&= \frac{H}{2G}
\end{aligned}$$ (6.89)

from (6.68). Then, the rate of twist α is defined in accordance with the sense of the stresses in Fig. 6.6a as

$$\alpha = -\omega_{xy,z} = -\frac{H}{2G},$$ (6.90)

expressed in terms of H which is determined in the course of the solution.

Finding that the angle of rotation per unit length is *constant*, we may compute the total angle of rotation at any cross section, designated as $\mathcal{A}$ on Fig. 6.6d, as

$$\mathcal{A} = \alpha z = -\frac{Hz}{2G},$$ (6.91)

and the in-plane displacements, shown in Fig. 6.6d, as

$$u_z(y,z) = -\mathcal{A}y = -\alpha yz,$$ (6.92a)
$$u_y(x,z) = \mathcal{A}x = -\alpha xz,$$ (6.92b)

The normal displacement u_z involves out-of-plane deformations of the cross section called *warping* and takes the form [1]

$$u_z = \alpha \psi(x,y),$$ (6.93)

where ψ is a so-called warping function. Rewriting (6.87a) and (6.87b), in view of (6.92a, 6.92b) and (6.93), we have

$$\alpha(-y + \psi_{,x}) = \frac{\sigma_{xz}}{G},$$ (6.94a)

$$\alpha(x + \psi_{,y}) = \frac{\sigma_{yz}}{G}.$$ (6.94b)

For elementary cases, it is sufficient to integrate the equations individually to eliminate α and produce

$$\psi = \frac{1}{2\alpha G}\left(\sigma_{xz}x + \sigma_{yz}y\right), \tag{6.95}$$

Anticipating that $u_z(x, y, 0) = 0$ eliminates the extraneous integration functions. Then,

$$u_z = \frac{1}{2G}\left(\sigma_{xz}x + \sigma_{yz}y\right). \tag{6.96}$$

We summarize the main steps to obtain this solution.

1. The stress function ϕ is selected in a form that is constant along the perimeter of the cross section to satisfy the boundary conditions.
2. The stress function ϕ is related to the constant H through (6.68), which expresses compatibility.
3. The stress function ϕ in terms of H is related to the twisting moment Mz by (6.84), which is an expression of overall equilibrium.
4. H is then found in terms of M_z so that ϕ becomes a function of M_z.
5. The stresses σ_{xz} and σ_{yz} are calculated from (6.65a) and (6.65b).
6. The rate of twist α is found from (6.90).
7. The total angle of rotation $\mathcal{A}$ is given by (6.91).
8. The displacements u_x, u_y, and u_z are computed from (6.92a), (6.92b), and (6.96).

6.4.3   Torsion of Elliptical Shaft

As an example of the application of the St. Venant torsion theory, we consider the cross section shown in Fig. 6.7a. The equation of the boundary is

Fig. 6.7 (a) Elliptical cross section; (b) shearing stresses; (c) contours for normal displacement (After Timoshenko and Goodier [1] Reprinted with permission)

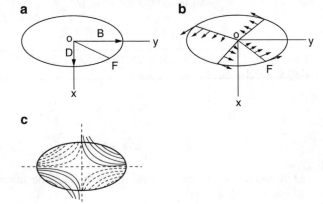

$$\frac{x^2}{D^2} + \frac{y^2}{B^2} - 1 = 0, \tag{6.97}$$

so we take

$$\phi = C\left(\frac{x^2}{D^2} + \frac{y^2}{B^2} - 1\right), \tag{6.98}$$

where C is a constant, to satisfy step (1) of the previous paragraph. Next, we calculate

$$H = 2C\left(\frac{1}{D^2} + \frac{1}{B^2}\right). \tag{6.99}$$

From (6.68) and (6.98) as suggested in step (2),

$$C = \frac{HD^2B^2}{2(D^2 + B^2)}, \tag{6.100}$$

whereupon

$$\phi = \frac{HD^2B^2}{2(D^2 + B^2)}\left(\frac{x^2}{D^2} + \frac{y^2}{B^2} - 1\right). \tag{6.101}$$

Proceeding to step (3), we substitute (6.101) into (6.84) which becomes

$$\begin{aligned}
M_z &= 2\int_A \frac{HD^2B^2}{2(D^2 + B^2)}\left(\frac{x^2}{D^2} + \frac{y^2}{B^2} - 1\right) dx\,dy \\
&= \frac{HD^2B^2}{D^2 + B^2}\left(\frac{1}{D^2}I_{yy} + \frac{1}{B^2}I_{xx} - A\right),
\end{aligned} \tag{6.102}$$

where the moments of inertia and the area for the ellipse are given by

$$I_{yy} = \frac{\pi BD^3}{4}, \tag{6.103a}$$

$$I_{xx} = \frac{\pi DB^3}{4}, \tag{6.103b}$$

$$A = \pi DB. \tag{6.103c}$$

Next, following step (4), we substitute (6.103a), (6.103b), and (6.103c) into (6.102) and simplify to find

$$M_z = -\frac{\pi D^3 B^3}{2(D^2 + B^2)}H, \tag{6.104}$$

so that

$$H = -\frac{2(D^2 + B^2)}{\pi D^3 B^3} M_z, \tag{6.105}$$

and from (6.101),

$$\phi = \frac{-1}{\pi DB} \left(\frac{x^2}{D^2} + \frac{y^2}{B^2} - 1\right) M_z. \tag{6.106}$$

Now, we evaluate the stresses as suggested in step (5). Equations (6.65a) and (6.65b) yield

$$\sigma_{xz} = \phi_{,y} = \frac{-2y}{\pi DB^3} M_z. \tag{6.107a}$$

$$\sigma_{yz} = \phi_{,x} = \frac{2x}{\pi D^3 B} M_z. \tag{6.107b}$$

Clearly, the maximum values occur at the boundaries and, if $B > D$, the absolute maximum is

$$\sigma_{yz}(D, 0) = \frac{2}{\pi D^2 B} M_z, \tag{6.108}$$

that is, at the extremity of the *minor* diameter. This is perhaps the most startling difference from the Coulomb approach where τ is proportional to r. It is also illustrative to consider a ray, say *of* in Fig. 6.7a, for which x/y is constant. Thus, the ratio of shearing stresses

$$\frac{\sigma_{yz}}{\sigma_{xz}} = -\frac{B^2 x}{D^2 y} = \text{constant}. \tag{6.109}$$

This indicates that the *direction* of the resultant shear must likewise be constant. From the boundary requirement, this direction is *tangent* to the perimeter at F. This is illustrated in Fig. 6.7b for several rays.

Then from step (6), the rate of twist is calculated from (6.90) and (6.105) as

$$\alpha = -\frac{H}{2G} = \frac{(D^2 + B^2)}{\pi G D^3 B^3} M_z = \frac{M_z}{GJ_e}, \tag{6.110a}$$

where

$$J_e = \frac{\pi D^3 B^3}{D^2 + B^2}. \tag{6.110b}$$

GJ_e is the torsional rigidity for the ellipse. The total angle of rotation $\mathcal{A}$, step (7), is evaluated at any point z along the shaft as $z\, \alpha$, (6.91).

Finally, the displacements may be found as indicated in step (8). From (6.92a), (6.92b) and (6.110a), (6.110b),

$$u_x = \frac{-M_z}{GJ_e} yz, \tag{6.111a}$$

$$u_y = \frac{M_z}{GJ_e} xz. \tag{6.111b}$$

Proceeding to the normal displacement and using (6.96), (6.107a), and (6.107b), we find

$$u_z = \frac{xy}{\pi G} \frac{(B^2 - D^2)}{D^3 B^3} M_z. \tag{6.112}$$

Of interest is the alternating algebraic sign of u_z in each quadrant as shown in Fig. 6.7c, illustrating the departure from the plane cross section (warping) and the hyperbolic contours of u_z [1]. Note that in the first and third quadrants where x and y are either positive or negative, the figure indicates a negative u_z for a positive M_z because $B < D$ in the illustration. Also, we observe that a classical "fixed" or "built-in" support which is intended to develop an applied M_z may not permit the u_z displacements to occur. If so, a self-equilibrating set of axial stresses would develop near the support, negating (6.63). This is of comparatively minor importance in solid shafts.

Torsion of rectangular bars is of practical interest but is considerably more complicated [3]. An approximate energy-based solution is presented in Sect. 11.7.3. We may infer some aspects from the ellipse solution but will defer until the membrane analogy for torsion is introduced in the next section.

6.4.4 Membrane Analogy

Since the governing equation for the stress function ϕ, (6.68), is the same as that describing the deflection u_z of a uniformly loaded membrane, (5.27), we may derive an analogy to assist our understanding and interpretation of the St. Venant torsion problem. This analogy is attributed to Ludwig Prandtl [1] and was extended into the inelastic range by Arpad Nádái [1].

We note from (6.91) that the r.h.s. of (6.68) is

$$H = -2G\alpha \tag{6.113}$$

so that replacing the quantity $-q/T$ in (5.27) by $-2G\alpha$ allows us to establish some important characteristics of the stress distribution on a solid bar under torsion.

1. From (6.84), the *total torque* is proportional to the *volume* between the membrane, which follows the stress function, and the cross section of the bar.

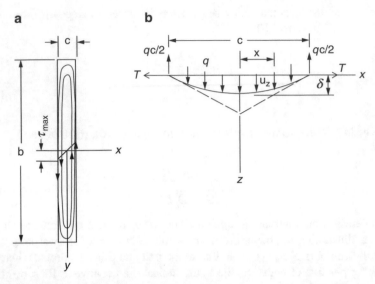

Fig. 6.8 Membrane stretched over narrow rectangular cross section (From Timoshenko and Goodier [6] Reproduced with permission of McGraw-Hill)

2. From (6.65a) and (6.65b), the *shearing stresses* are proportional to the *slope* of the membrane at any point.

In order to compare the *relative* torsional capacities of two cross sections, we need to only visualize a membrane with the same maximum slope (i.e., shearing stress) spanning both. It is apparent that a square or circular cross section has considerably more capacity than, say, a long narrow shape of the same area. Also, for a given cross-sectional area, St. Venant showed that the circular cross section is most efficient [1].

To illustrate a quantitative application of the membrane analogy, we consider a narrow rectangular bar as shown in Fig. 6.8a. For the moment we neglect the effects of the short side boundaries and assume that the surface of the membrane is a parabolic cylinder as shown in Fig. 6.8b. The deflection of the membrane u_z is written in terms of the maximum deflection at the center, $\delta = u_z(0, 0)$, by the well-known property of the offsets to parabolas,

$$\frac{\delta - u_z}{\delta} = \frac{x^2}{(c/2)^2} \tag{6.114}$$

so that

$$u_z = \frac{4\delta}{c^2}\left[\frac{c^2}{4} - x^2\right]. \tag{6.115}$$

The maximum slope is at the intersection of the *x*-axis with the boundary along the long side ($\pm c/2$) and is evaluated from

$$u_{z,x}(x,y) = 8\delta x/c^2 \qquad (6.116a)$$

as

$$u_{z,x}(\pm c/2, 0) = 4\delta/c \qquad (6.116b)$$

retaining only the positive value because of symmetry.

By considering the statical moment about $x = 0$ in Fig. 6.8b, we can write the equilibrium equation for a unit width of the parabolic cylinder, essentially the equation of a uniformly loaded string, as

$$-q\frac{c}{2}\left(\frac{c}{2}\right) + q\frac{c}{2}\left(\frac{c}{4}\right) + T\delta = 0$$

or

$$\delta = qc^2/(8T). \qquad (6.117)$$

Therefore, (6.116b) becomes

$$(u_{z,x})_{max} = qc/(2T). \qquad (6.118)$$

Now we compute the torsion in the bar by evaluating the volume under the membrane:

$$V = \frac{2}{3}cb\delta$$
$$= qbc^3/(12T). \qquad (6.119)$$

Using the analogy and replacing q/T by $2G\alpha$, we have

$$V = \frac{1}{6}bc^3 G\alpha. \qquad (6.120)$$

Inasmuch as M_z is *twice* the volume, as per (6.84),

$$M_z = \frac{1}{3}bc^3 G\alpha \qquad (6.121)$$

or the rate of twist α is given by

$$\alpha = \frac{M_z}{\left(\frac{1}{3}bc^3 G\right)}. \qquad (6.122)$$

Turning to the maximum stress along the long side of the bar, Fig. 6.8a, we have from (6.65a, 6.65b) and (6.68) with ϕ replaced by u_z in the analogy and $q/2T = G\alpha$

$$\tau_{\max} = |\sigma_{yz}|_{\max} = |\phi_{,x}|_{\max} = \frac{qc}{2T}$$

$$= cG\alpha = \frac{M_z}{\frac{1}{3}bc^2}. \tag{6.123}$$

The quantities $\left(\frac{1}{3}bc^3\right)$ and $\left(\frac{1}{3}bc^2\right)$ can be interpreted as section properties of rectangular bars related to torsional stiffnesses, much as moments of inertia relate to bending stiffness.

From (6.116a), the shearing stresses are distributed linearly across the width of the cross section. It may be shown that the shearing stresses are directed along the contours of the deflected membrane depicted in Fig. 6.8a [1]. Also, the shearing stresses acting near the short side σ_{xz} are not calculated by this simplification. They are smaller in magnitude but further separated than the σ_{yz} stresses and thus contribute equally to the total resistance provided by the cross section to M_z.

Equations (6.122) and (6.123) have great generality. If a thin-walled member, such as those shown in Fig. 6.9, is composed of several long rectangular pieces, each with $c = c_i$ and $b = b_i$, then we can imagine a parabolic cylindrical membrane spanning each segment independently so that

$$\alpha = \frac{M_z}{G \sum \frac{1}{3} b_i c_i^3} \tag{6.124}$$

and

$$(\tau_i)_{\max} = \frac{M_z c_i}{\sum \frac{1}{3} b_i c_i^3}. \tag{6.125}$$

For example, considering the flanges of the H-section shown in Fig. 6.9c,

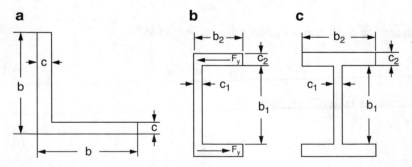

Fig. 6.9 Cross sections of thin-walled members (From Timoshenko and Goodier [6] Reproduced with permission of McGraw-Hill)

Fig. 6.10 Cross sections of
a triangular shaft

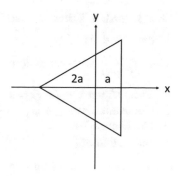

$$(\tau_2)_{\text{max}} = \frac{M_z c_2}{\frac{1}{3}b_1 c_1^3 + \frac{2}{3}b_2 c_2^3}. \tag{6.126}$$

The St. Venant torsional capacity of thin-walled members, comparatively small as previously argued with the membrane analogy, may be considerably enhanced by including the nonuniform torsion associated with the warping [4]. This discussion is beyond our present scope, but it is easily visualized for a section such as H-section or the channel in Fig. 6.9b, where torsion would be resisted far more efficiently by a couple consisting of opposing transverse forces $\mathbf{F_y}$ distributed along the top and bottom flanges with a nominal lever arm of $c_2 + b_1$. The transverse forces in turn cause so-called cross-bending of each flange about the x-axis, which must be accounted for in design.

Some details on a mathematical solution for torsion on general (non-narrow) rectangular cross sections are presented in Sect. 11.7.3. For now, we may discuss some qualitative aspects of such a solution based on visualizing an inflated membrane covering the cross section. This membrane would of course intersect the cross section along the perimeter and have the maximum amplitude in the center. The maximum slope and hence shearing stress is at the middle of a *long* side, which is *closer* to the center of the cross section than the middle of the short side. A second but lesser maximum will occur at the middle of a short side. Further both the deflection and the *slope* of the membrane are constrained to zero at the corners, removing the paradox of the shear stresses being tangent to the surface at each point along the perimeter.

The torsional resistance of members with noncircular cross sections is indeed quite different than might be inferred from the extension of Coulomb torsion theory.

6.5   Computational Examples with MuPAD

Example 6.1 Torsion of Elliptical Shaft (Sect. 6.4.3)
A formulation using a stress function is usually selected to derive analytical solutions of typical deformations such as extension, bending, and torsion. For an elliptical

shaft in torsion, a stress function was used to calculate stresses. If the stress function is expressed as $\phi = C\left(\frac{x^2}{D^2} + \frac{y^2}{M^2} - 1\right)$ where C, D, and B are constants, and $C = \frac{HD^2B^2}{2(D^2+B^2)}$, we want to derive or verify the following equations:

1. Shear stresses: $\sigma_{xz}(x, y) = \phi_{,y}$ and $\sigma_{yz}(x, y) = -\phi_{,x}$
2. Compatibility: $\phi_{,xx} + \phi_{,yy} = H$
3. Gradient of $\nabla\phi$
4. Integration of ϕ

We used the following steps:

- Define stress function—the stress function was assigned to ϕ.

```
phi := C*(x^2/DD^2 + y^2/B^2 - 1)
```
$$C\left(\frac{y^2}{B^2} + \frac{x^2}{DD^2} - 1\right)$$

- Note that we used DD rather than D in the expression. This is because identifier D is a protected variable.

```
D := 2
Error: The identifier 'D' is protected. [_assign]
```

- Similar to MATLAB, the symbolic equations can be assigned to the left-hand side symbol with ":=." The other arithmetic operations are the same as in MATLAB. In MuPAD, we do not have the dot product operation "*."
- The shear stresses are derived by differentiating the stress function.

```
sig_xz := diff(phi, y)
```
$$\frac{2\,C\,y}{B^2}$$

```
sig_yz := -diff(phi, x)
```
$$-\frac{2\,C\,x}{DD^2}$$

- The first-order differential operation is carried out by using function "diff." When the stress function has complicated forms, we can see that the symbolic calculation can be powerful and accurate.
- We can also use the Laplacian operator directly to check the compatibility.

```
linalg::laplacian(C*(x^2/DD^2 + y^2/B^2 - 1), [x,y])
```
$$\frac{2\,C}{B^2} + \frac{2\,C}{DD^2}$$

- In this command, we used the library "linalg," which contains a variety of linear algebra functions. At the same time, we can always use differentiation to check the compatibility.

```
comp := diff(diff(phi,x),x) + diff(diff(phi,y),y)
```
$$\frac{2\,C}{B^2} + \frac{2\,C}{DD^2}$$

- The gradient of the stress function can also be carried out with the gradient function.

```
linalg::gradient(phi, [x,y])
```
$$\begin{pmatrix} \dfrac{2\,C\,x}{DD^2} \\ \dfrac{2\,C\,y}{B^2} \end{pmatrix}$$

- To calculate the sum of moment on the cross section, we use

$M_z = \int_A(-\sigma_{zx}y + \sigma_{zy}x)dxdy$ to integrate the forces on the surface.

```
Mz := Simplify(-int(int(y*sig_xz, y), x) - int(int(x*sig_yz, x), y))
```
$$\frac{2\,C\,x^3\,y}{3\,DD^2} - \frac{2\,C\,x\,y^3}{3\,B^2}$$

Example 6.2 Torsion of Triangular Shaft (Sect. 6.6)
In this example, we illustrate solving a torsion problem using MuPAD. Consider the torsion of a cylinder with equilateral triangular section, as shown in the figure from Saad [7].
The following stress function satisfies the boundary equation:

$$\phi = K\left(x - \sqrt{3}y + 2a\right)\left(x + \sqrt{3}y + 2a\right)(x - a)$$

1. Calculate K.
2. Calculate torque T.
3. Derive the shear stresses σ_{xz} and σ_{yz}.

4. Derive the warping displacement w.

   We used the following steps:

- We first define the symbols in MuPAD.

```
mu  := Symbol::mu
μ
```
```
alpha := Symbol::alpha
α
```
```
phi := Symbol::phi
φ
```
```
sig := Symbol::sigma
σ
```
```
sig_xz := Symbol::subScript(sig, xz)
σ_xz
```
```
sig_xz := Symbol::subScript(sig, yz)
σ_yz
```

- Then define stress function ϕ.

```
phi := K*(x-sqrt(3)*y+2*a)*(x+sqrt(3)*y+2*a)*(x-a)
```
$$-K(a-x)(2a+x+\sqrt{3}\,y)(2a+x-\sqrt{3}\,y)$$

- Use Poisson's equation to solve K.

```
LapPhi := laplacian(phi, [x, y])
```
$$4K(a-x)+2K(2a+x+\sqrt{3}\,y)+2K(2a+x-\sqrt{3}\,y)$$
```
solve(LapPhi=-2*mu*alpha, K, IgnoreSpecialCases)
```
$$\left\{-\frac{\alpha\,\mu}{6\,a}\right\}$$

- Integrate the stress function over the cross-sectional area to get the torque.

```
T := 2*int(int(phi, y = -(x+2*a)/sqrt(3)..(x+2*a)/sqrt(3)), x = -2*a..a)
```
$$-\frac{54\sqrt{3}\,K\,a^5}{5}$$
```
subs(T, K=-alpha*mu/6/a)
```
$$\frac{9\sqrt{3}\,\alpha\,\mu\,a^4}{5}$$

- Derive shear stresses.

$$\left[\begin{array}{l} \texttt{sig\_xz := subs(Simplify(diff(phi, y)), K=-alpha*mu/6/a)} \\ - \dfrac{\alpha\,\mu\,y\,(a-x)}{a} \end{array} \right.$$

$$\left[\begin{array}{l} \texttt{sig\_yz := subs(Simplify(-diff(phi, x)), K=-alpha*mu/6/a)} \\ \dfrac{\alpha\,\mu\,\left(x^{2}+2\,a\,x-y^{2}\right)}{2\,a} \end{array} \right.$$

- Finally, integrate the stress function to get the warping displacement.

$$\left[\begin{array}{l} \texttt{int(sig\_xz/mu + alpha*y, x)} \\ \dfrac{\alpha\,x^{2}\,y}{2\,a} \end{array} \right.$$

$$\left[\begin{array}{l} \texttt{w := int(sig\_yz/mu - alpha*x, y)} \\ \dfrac{\alpha\,y\,\left(3\,x^{2}-y^{2}\right)}{6\,a} \end{array} \right.$$

6.6   Exercises

6.1 Consider the solution for the bending of a cantilever, described in Sect. 6.3, and sketch a lateral and a top view of the deflected beam using Fig. 6.2 as a reference.

6.2 Investigate the same cantilever beam considered in Problem 6.1 but with the load P uniformly distributed as $p = P/L$. Carry out the computations following the example in Sects. 6.3.1 and 6.3.2.

6.3 Consider the same cantilever beam but with an end moment M_y applied at $z = L$ and complete the solution.

6.4 Investigate the Coulomb (simple) torsion solution as developed for a circular cross section applied to a rectangular cross section. In what respect is it deficient insofar as the theory of elasticity is concerned?

6.5 Two bars made of the same material and having the same length are subjected to the same torque. One bar has a square cross section with sides equal to a, and the other bar has a circular cross section of diameter a. Which bar has the greater rate of twist as defined in (6.40)? (from [5])

6.6 The torsion solution for a cylinder of equilateral triangular cross section is derivable from the stress function [6]

$$\phi = C\left(x - \sqrt{3}y - \frac{2}{3}h\right)\left(x + \sqrt{3}y - \frac{2}{3}h\right)\left(x + \frac{1}{3}h\right).$$

(a) Verify the shear stress distribution shown along the x-axis.
(b) Develop expressions for the maximum and minimum shearing stresses and the rate of twist for a torque Mz.

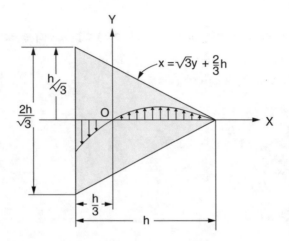

6.7 Consider an annular elliptical cross section with outside semiaxes B and D as shown in Fig. 6.7 and inner semiaxes kB and kD.

(a) Show that the stress function given by (6.98) is applicable for this cross section.
(b) Indicate the direction of resultant shear stress all along the interior and exterior boundaries.
(c) For a given M_t, what is the magnitude and location of the maximum shear stress?
(d) What is the twist per unit length?

6.8 The torsional rigidities of a circle, an ellipse, and an equilateral triangle are denoted by GJ_c, GJ_e, and GJ_t, respectively. If the cross-sectional areas of these sections are equal, demonstrate that the following relationships exist (from [6]):

$$J_e = \frac{2ab}{a^2 + b^2}J_c, \quad J_t = \frac{2\pi\sqrt{3}}{15}J_c,$$

where a and b are the semiaxes of the ellipse in the x and y directions.

6.9 A closed circular thin-walled tube has perimeter length L and uniform wall thickness t. An open tube is produced by making a fine longitudinal slit through the wall. Both tubes are subjected to torsion such that:

(a) The maximum shearing stresses are the same in each.

Show that $\dfrac{M_t^{\text{open}}}{M_t^{\text{closed}}} = \dfrac{Lt}{6A}$

(b) The twist per unit length is the same in each.

Show that $\dfrac{M_t^{\text{open}}}{M_t^{\text{closed}}} = \dfrac{L^2 t^2}{12A^2}$, where A is the area enclosed by the cross-sectional profile of the tube.

(c) Evaluate the results for the case of a hollow circular bar with $R = 30$ mm and $t = 3$ mm. Sketch the shear flow for the open and closed tube, respectively.

6.10 Consider the bending of a prismatic cantilever bar with a single axis of symmetry and loaded at the free end (i.e., $z = L$) by a force P along the axis of symmetry x as shown in Fig. 6.2.

(a) Show that a stress function may be proposed for the problem that satisfies following compatibility and boundary conditions, respectively,

$$\nabla^2 \phi = \frac{\nu}{1+\nu}\frac{Py}{I} - \frac{df}{dy}; \quad \frac{d\phi}{ds} = \left[\frac{Px^2}{2I} - f(y)\right]\frac{dy}{ds}$$

(b) Following the example of a bar with a rectangular cross section, determine ϕ and $f(y)$ for a bar with a circular and then an elliptical cross section, with its boundary given by (6.97).
(c) Find the distribution of shearing stresses over the cross section.

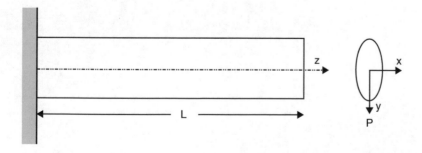

6.11 For the prismatic bar analyzed in Sect. 6.2, re-solve for the case of a concentrated load P applied at $z = 0$ and no self-weight [7].

References

1. Timoshenko S, Goodier JN (1951) Theory of elasticity, 2nd edn. McGraw-Hill Book Company Inc., New York
2. Westergaard HM (1964) Theory of elasticity and plasticity. Dover Publications, Inc., New York
3. Filonenko-Borodich M (1965) Theory of elasticity. Dover Publications Inc., New York
4. Timoshenko S, Gere JM (1951) Theory of elastic stability. McGraw-Hill Book Company Inc., New York
5. Volterra E, Gaines JH (1971) Advanced strength of materials. Prentice-Hall Inc., Englewood Cliffs, NJ
6. Ugural AC, Fenster SK (1975) Advanced strength and applied elasticity. Elsevier-North Holland, New York
7. Sadd MH (2014) Elasticity, 3rd edn. Elsevier, New York

Chapter 7
Two-Dimensional Elasticity

Abstract Many physical problems are reducible to two dimensions which facilitate their eventual solution. If there is no traction on *one* plane passing through the body, this state is known as *plane stress* since all nonzero stresses are confined to planes parallel to the traction-free plane. This is an obvious possibility for bodies with one dimension much *smaller* than the other two, such as a thin sheet or diaphragm loaded in the plane perpendicular to the small dimension. Another possibility is a body in which one dimension is much *greater* than the other two, such as a long pipe or a dam between massive end walls. This is known as *plane strain*.

The development is primarily confined to the realm of isotropic elasticity with some extensions to more complex material laws.

7.1 Introduction

We briefly mentioned in Sect. 2.6.2 that an elasticity problem may be reduced from three to two dimensions if there is no traction on *one* plane passing through the body. This state is known as *plane stress* since all nonzero stresses are confined to planes parallel to the traction-free plane.

Many physical problems are reducible to two dimensions which simplifies their eventual solution. In this regard, we should mention the important engineering theories of plate bending and thin shells. These theories may be regarded as extensions of the elementary theory of beam bending in that the concept of a reference datum (i.e., a middle plane or middle surface) is used to reduce the model to two dimensions. The description of stresses, strains, and displacements for points lying *away* from this datum is related to the corresponding quantities *on* the datum through maintenance of plane sections. Also, deformations in the direction normal to the reference datum are neglected. Such problems are addressed at an introductory level in the next chapter. Historically, the epic treatise of August E. H.

© Springer International Publishing AG, part of Springer Nature 2018   149
P. L. Gould, Y. Feng, *Introduction to Linear Elasticity*,
https://doi.org/10.1007/978-3-319-73885-7_7

Love [1] on the theory of elasticity includes plate and shell theories, as well as the theory of rods, beam bending, stability, and many other topics.[1]

Since we have identified problems *not* strictly within the two-dimensional theory of elasticity, a brief remark on the class of problems that *are* is in order. We have mentioned *plane stress* which is obviously a possibility for bodies with one dimension much *smaller* than the other two such as a thin sheet or diaphragm loaded in the plane perpendicular to the small dimension. A less obvious case is that of a body in which one dimension is much *greater* than the other two, such as a long pressurized pipe or perhaps a dam between massive end walls. This is known as *plane strain*.

Our development is primarily confined to the realm of isotropic elasticity with some extensions to more complex material laws.

7.2 Plane Stress Equations

For plane stress with the z-axis stress-free, we have

$$\sigma_{xz} = \sigma_{yz} = \sigma_{zz} = 0. \tag{7.1}$$

Also, we assume that the remaining stresses do not vary with z but are only functions of x and y. This reduces the equilibrium of Eq. (2.46), with $1 = x$, $2 = y$, $3 = z$, to

$$\sigma_{xx,x} + \sigma_{xy,y} + f_x = 0, \tag{7.2a}$$

$$\sigma_{xy,x} + \sigma_{yy,y} + f_y = 0, \tag{7.2b}$$

$$f_z = 0. \tag{7.2c}$$

For some linear problems, it is convenient to specify the body force vector in the form

$$\mathbf{f} = -\nabla V = -V_{,i}\mathbf{e}_i, \tag{7.3}$$

where V is a potential function. This corresponds to a body force field that is *conservative* (elaborated in Sect. 11.6.2). Introducing (7.3) into (7.2a, b, c) gives

$$\sigma_{xx,x} + \sigma_{xy,y} - V_{,x} = 0, \tag{7.4a}$$

---

[1]This prompted a favorite saying among graduate students and professors in an earlier era, that "All you really need is Love."

$$\sigma_{xy,x} + \sigma_{yy,y} - V_{,y} = 0, \tag{7.4b}$$

$$V_{,z} = 0. \tag{7.4c}$$

We now consolidate the equilibrium equations using a technique introduced in the study of bending in Sect. 6.3.2 and of torsion in Sect. 6.4.2. The stresses are written as

$$\sigma_{xx} = V + \phi_{,yy}, \tag{7.5a}$$

$$\sigma_{yy} = V + \phi_{,xx}, \tag{7.5b}$$

$$\sigma_{xy} = -\phi_{,xy}, \tag{7.5c}$$

in which ϕ is a stress function named after the astronomer George Airy. Substitution of (7.5a, b, c) into (7.4a) and (7.4b) satisfies these equations identically.

The constitutive equations are also expressed in terms of ϕ. First, the generalized Hooke's law (4.29b) for the plane stress case becomes

$$\varepsilon_{xx} = \frac{1}{E}(\sigma_{xx} - \nu\sigma_{yy}), \tag{7.6a}$$

$$\varepsilon_{yy} = \frac{\nu}{E}(\sigma_{yy} - \nu\sigma_{xx}), \tag{7.6b}$$

$$\varepsilon_{zz} = -\frac{\nu}{E}(\sigma_{xx} + \sigma_{yy}) \tag{7.6c}$$

$$\varepsilon_{xy} = \frac{1}{2G}\sigma_{xy}, \tag{7.6d}$$

$$\varepsilon_{xz} = \varepsilon_{yz} = 0. \tag{7.6e}$$

From (7.6a, b, c, d, e) we see that the strains do not depend on z, only on x and y. However, from (7.6c), we have a nonzero *strain* in the z-direction indicating that a state of plane stress does *not* imply a corresponding state of plane strain.

It is also convenient to write the stresses in terms of the strains by solving (7.6a) and (7.6b) for σ_{xx} and σ_{yy} and (7.6d) for σ_{xy}:

$$\sigma_{xx} = \frac{E}{(1-\nu^2)}(\varepsilon_{xx} + \nu\varepsilon_{yy}) \tag{7.7a}$$

$$\sigma_{yy} = \frac{E}{(1-\nu^2)}(\varepsilon_{yy} + \nu\varepsilon_{xx}) \tag{7.7b}$$

$$\sigma_{xy} = 2G\varepsilon_{xy}. \tag{7.7c}$$

Introducing (7.5a, b, c) into (7.6a, b, c, d, e) produces

$$\varepsilon_{xx} = \frac{1}{E}[(\phi_{,yy} - \nu\phi_{,xx}) + (1-\nu)V], \tag{7.8a}$$

$$\varepsilon_{yy} = \frac{1}{E}\left[(\phi_{,xx} - \nu\phi_{,yy}) + (1 - \nu)V\right], \tag{7.8b}$$

$$\varepsilon_{zz} = -\frac{\nu}{E}\left[(\phi_{,xx} + \phi_{,yy}) + 2V\right], \tag{7.8c}$$

$$\varepsilon_{xy} = -\frac{1}{2G}\phi_{,xy} \tag{7.8d}$$

$$\varepsilon_{xz} = \varepsilon_{yz} = 0. \tag{7.8e}$$

We may now establish the compatibility equations in terms of the stress function recalling (3.59). First (3.59a), with substitution of the strains from (7.8a, b, c, d, e), becomes

$$\frac{1}{E}[\phi_{,yyyy} - \nu\phi_{,xxyy} + (1 - \nu)V_{,yy} + \phi_{,xxxx} - \nu\phi_{,xxyy} + (1 - \nu)V_{,xx}]$$
$$= -\frac{1}{G}\phi_{,xxyy}. \tag{7.9a}$$

With $\frac{E}{G} = 2(1 + \nu)$ from (4.24a and e), (7.9a) reduces to

$$\nabla^4\phi = -(1 - \nu)\nabla^2 V, \tag{7.9b}$$

where the operator $\nabla^4(\ )$ is the two-dimensional contraction of the operator defined in (1.10c), i.e.,

$$\nabla^4(\ ) = (\ )_{,xxxx} + 2(\ )_{,xxyy} + (\ )_{,yyyy}. \tag{7.9c}$$

Equations (3.59d, e) are identically satisfied, while the others (3.59b, c, f) become

$$\varepsilon_{zz,yy} = \phi_{,xxyy} + \phi_{,yyyy} + 2V_{,yy} = 0 \tag{7.10a}$$

$$\varepsilon_{zz,xx} = \phi_{,xxxx} + \phi_{,yyxx} + 2V_{,xx} = 0 \tag{7.10b}$$

$$\varepsilon_{zz,xy} = \phi_{,xxxy} + \phi_{,yyyx} + 2V_{,xy} = 0. \tag{7.10c}$$

Elementary plane stress theory does not consider (7.10a, b, c) but only (7.9b). In a sense then this is an approximation, but (7.10a, b, c) only contain terms related to ε_{zz} which although not zero is but a "Poisson" effect as shown by (7.8c). Especially for thin members, the error should be negligible. Thus, (7.5a, b, c) and the single compatibility of Eq. (7.9b) constitute the governing equations for isotropic plane stress theory.

An extension of this theory, in which σ_{zz} is not equal to 0, as stated in (7.1), but is set to a known or assumed value, is known as generalized plane stress.

7.3   Plane Strain Equations

For plane strain with the z-axis strain-free, we have

$$\varepsilon_{xz} = \varepsilon_{yz} = \varepsilon_{zz} = 0. \tag{7.11}$$

As in the case of plane stress, all tractions and body forces are functions of x and y only.

To enforce these conditions, we use the generalized Hooke's law (4.29b), with $1 = x$, $2 = y$, and $3 = z$:

$$\varepsilon_{xx} = \frac{1}{E}[\sigma_{xx} - \nu(\sigma_{yy} + \sigma_{zz})], \tag{7.12a}$$

$$\varepsilon_{yy} = \frac{1}{E}[\sigma_{yy} - \nu(\sigma_{xx} + \sigma_{zz})], \tag{7.12b}$$

$$0 = \frac{1}{E}[\sigma_{zz} - \nu(\sigma_{xx} + \sigma_{yy})], \tag{7.12c}$$

$$\varepsilon_{xy} = \frac{1}{2G}\sigma_{xy}, \tag{7.12d}$$

$$\varepsilon_{xz} = \varepsilon_{yz} = 0. \tag{7.12e}$$

Equation (7.12c) indicates that

$$\sigma_{zz} = \nu(\sigma_{xx} + \sigma_{yy}), \tag{7.13}$$

so that the strain-free plane is not stress-free, in general. The equilibrium of Eq. (2.46) is almost identical to the plane stress case (7.2a, b, c), except for (7.2c) where $\sigma_{zz,z}$ remains while $f_z = 0$ since $\mathbf{f} = \mathbf{f}(x, y)$. With (7.3) in force, we have

$$\sigma_{xx,x} + \sigma_{xy,y} - V_{,x} = 0, \tag{7.14a}$$

$$\sigma_{xy,x} + \sigma_{yy,y} - V_{,y} = 0, \tag{7.14b}$$

$$\sigma_{zz} = \text{constant.} \tag{7.14c}$$

Turning to the compatibility of Eq. (3.59), all are identically satisfied except for (3.59a):

$$\varepsilon_{xx,yy} + \varepsilon_{yy,xx} = 2\varepsilon_{xy,xy}. \tag{7.15}$$

This is the same equation from which (7.9a) was derived. Thus, we again introduce the Airy stress function (7.5a, b, c) along with (7.13) into (7.12a, b, c) which become

$$\varepsilon_{xx} = \frac{1}{E}[(1 - \nu^2)\phi_{,yy} - \nu(1 + \nu)\phi_{,xx} + [1 - \nu(1 + 2\nu)]V]. \tag{7.16a}$$

$$\varepsilon_{yy} = \frac{1}{E}[(1 - \nu^2)\phi_{,xx} - \nu(1 + \nu)\phi_{,yy} + [1 - \nu(1 + 2\nu)]V]. \tag{7.16b}$$

$$\varepsilon_{xy} = -\frac{1}{2G}\phi_{,xy}. \tag{7.16c}$$

Then, we write (7.15) using (7.16a, b, c) as

$$\frac{1}{E}\Big[(1 - \nu^2)\phi_{,yyyy} - \nu(1 + \nu)\phi_{,xxyy} + [1 - \nu(1 + 2\nu)]V_{,yy}$$

$$+\ (1 - \nu^2)\phi_{,xxxx} - \nu(1 + \nu)\phi_{,xxyy} + [1 - \nu(1 + 2\nu)]V_{,xx} \tag{7.17a}$$

$$= -\frac{1}{G}\phi_{,xxyy}$$

which reduces to

$$\nabla^4\phi = -\frac{(1 - 2\nu)}{1 - \nu}\nabla^2 V, \tag{7.17b}$$

which is remarkably close to (7.9b). In fact, for *no body* forces, the two equations are identical. The resulting homogeneous compatibility equation

$$\nabla^4\phi = 0 \tag{7.18}$$

is known as the *biharmonic* equation.

We see that plane strain is exact insofar as satisfying the St. Venant compatibility equations, while plane stress violates some of the equations. For applications with no body forces, the solution for either case involves the same equations, (7.5a, b, c) and (7.18). While, if body forces are present, a slight alteration of the particular solution for the biharmonic equation is needed between the two theories. The solution to the biharmonic problem in Cartesian coordinates is most directly written in terms of polynomials having the general form

$$\phi = \sum_m \sum_n C_{mn}x^m y^n. \tag{7.19}$$

Obviously the lower-order terms, that is, $m + n \le 3$, *each* satisfy the equation identically. However, the linear terms $C_{00} + C_{10}x + C_{01}y$ do not contribute to stresses. The general strategy is to try and construct solutions from the lower-order terms that satisfy the biharmonic equation *individually* and the boundary conditions *collectively*. For more complicated problems, it is necessary to use higher-order terms, that is, $m + n > 3$, and achieve satisfaction of the equation by a combination of the individual terms. Finally, in the nature of general remarks, it is usually required to employ St. Venant's principle to satisfy some of the boundary conditions.

We present some specific applications of this approach in Sect. 7.9. However, it is instructive to consider the plane stress problem on a circular domain first, both from the standpoints of mathematical simplicity and practical importance.

7.4   Cylindrical Coordinates

7.4.1   *Geometric Relations*

There are many problems in two-dimensional elasticity that are most conveniently treated in cylindrical coordinates (which degenerate to polar coordinates). Therefore, we develop the appropriate transformations between Cartesian and polar coordinates.

Referring to Fig. 7.1, we have the relations [2] between x, y, and r, θ as

$$x = r \cos \theta, \tag{7.20a}$$

$$y = r \sin \theta, \tag{7.20b}$$

$$r = \left(x^2 + y^2\right)^{1/2}, \tag{7.20c}$$

$$\theta = \tan^{-1} \frac{y}{x}. \tag{7.20d}$$

Also, the derivatives of the polar coordinates with respect to the Cartesian coordinates are of interest:

$$r_{,x} = \frac{1}{2} \left(x^2 + y^2\right)^{-1/2} (2x) = \frac{x}{r} = \cos \theta, \tag{7.21a}$$

$$r_{,y} = \frac{y}{r} = \sin \theta, \tag{7.21b}$$

Fig. 7.1 Polar coordinates

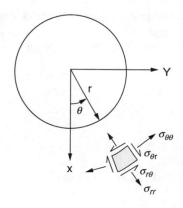

$$\theta_{,y} = \frac{1}{1 + (y/x)^2}\left(-\frac{y}{x^2}\right) = -\frac{y}{r^2} = -\frac{\sin\theta}{r}, \tag{7.21c}$$

$$\theta_{,y} = \frac{1}{1 + (y/x)^2}\left(\frac{1}{x}\right) = \frac{x}{r^2} = \frac{\cos\theta}{r}. \tag{7.21d}$$

Using (7.21a, b, c, d), we may write the differentiation formulas for functions specified in terms of r and θ as

$$\begin{aligned}
\frac{\partial()}{\partial x} &= \frac{\partial()}{\partial r}r_{,x} + \frac{\partial()}{\partial\theta}\theta_{,x} \\
&= \frac{\partial()}{\partial r}\cos\theta + \frac{\partial()}{\partial\theta}\left(\frac{-\sin\theta}{r}\right),
\end{aligned} \tag{7.22a}$$

$$\begin{aligned}
\frac{\partial()}{\partial y} &= \frac{\partial()}{\partial r}r_{,y} + \frac{\partial()}{\partial\theta}\theta_{,y} \\
&= \frac{\partial()}{\partial r}\sin\theta + \frac{\partial()}{\partial\theta}\left(\frac{\cos\theta}{r}\right).
\end{aligned} \tag{7.22b}$$

7.4.2 Transformation of Stress Tensor and Compatibility Equation

The stress components are transformed using (2.13) to become

$$\sigma_{rr} = \alpha_{ri}\alpha_{rj}\sigma_{ij}, \tag{7.23a}$$

$$\sigma_{\theta\theta} = \alpha_{\theta i}\alpha_{\theta j}\sigma_{ij}, \tag{7.23b}$$

$$\sigma_{r\theta} = \alpha_{ri}\alpha_{\theta j}\sigma_{ij}, \tag{7.23c}$$

where σ_{rr} is called the radial stress, $\sigma_{\theta\theta}$ the circumferential stress, and $\sigma_{r\theta}$ the shear stress.

The direction cosines are

$$\begin{array}{cccc}
 & x & y & z \\
r & \cos\theta & \sin\theta & 0 \\
\theta & -\sin\theta & \cos\theta & 0 \\
z & 0 & 0 & 1
\end{array} \tag{7.24}$$

and (7.23a, b, c) and (7.24) give

$$\sigma_{rr} = \sigma_{xx} \cos^2\theta + \sigma_{yy} \sin^2\theta + 2\sigma_{xy} \sin\theta \cos\theta,$$

$$\sigma_{\theta\theta} = \sigma_{xx} \sin^2\theta + \sigma_{yy} \cos^2\theta - 2\sigma_{xy} \sin\theta \cos\theta,$$

$$\sigma_{r\theta} = -\sigma_{xx} \sin\theta \cos\theta + \sigma_{yy} \sin\theta \cos\theta$$

$$+ \sigma_{xy}(\cos^2\theta - \sin^2\theta), \tag{7.25}$$

which should be familiar from the Mohr's circle concept of elementary strength of materials [3].

We now introduce the Airy stress function (7.5a, b, c) into (7.25), dropping the body force terms to produce

$$\sigma_{rr} = \phi_{,yy} \cos^2\theta + \phi_{,xx} \sin^2\theta - 2\phi_{,xy} \sin\theta \cos\theta,$$

$$\sigma_{\theta\theta} = \phi_{,yy} \sin^2\theta + \phi_{,xx} \cos^2\theta - 2\phi_{,xy} \sin\theta \cos\theta,$$

$$\sigma_{r\theta} = -\phi_{,yy} \sin\theta \cos\theta + \phi_{,xx} \sin\theta \cos\theta$$

$$- \phi_{,xy}(\cos^2\theta - \sin^2\theta). \tag{7.26}$$

The next step is to write (7.26) entirely in terms of cylindrical coordinates. This involves repeated application of (7.22a) and (7.22b), for example,

$$\phi_{,xx} = \frac{\partial}{\partial x}\left(\frac{\partial\phi}{\partial x}\right) = \frac{\partial}{\partial x}\left[\frac{\partial\phi}{\partial r}\cos\theta - \frac{\partial\phi}{\partial\theta}\frac{\sin\theta}{r}\right]$$

$$= \cos\theta\left[\frac{\partial^2\phi}{\partial r^2}\cos\theta - \frac{\partial^2\phi}{\partial r\partial\theta}\frac{\sin\theta}{r}\right] + \frac{\partial\phi}{\partial r}\left[0 + \frac{\sin^2\theta}{r}\right]$$

$$- \frac{\sin\theta}{r}\left[\frac{\partial^2\phi}{\partial r\partial\theta}\cos\theta - \frac{\partial^2\phi}{\partial\theta^2}\frac{\sin\theta}{r}\right]$$

$$- \frac{\partial\phi}{\partial\theta}\left[-\frac{\sin\theta}{r^2}\cos\theta - \frac{\cos\theta\sin\theta}{r^2}\right] \tag{7.27}$$

$$= \phi_{,rr}\cos^2\theta - 2\phi_{,r\theta}\frac{\sin\theta\cos\theta}{r} + 2\phi_{,\theta}\frac{\sin\theta\cos\theta}{r}$$

$$+ \phi_{,r}\frac{\sin^2\theta}{r} + \phi_{,\theta\theta}\frac{\sin^2\theta}{r^2}.$$

Continuing with $\phi_{,yy}$ and $\phi_{,xy}$ and back-substituting into (7.25) eventually produce the usable relations [4]:

$$\sigma_{rr} = \frac{1}{r}\phi_{,r} + \frac{1}{r^2}\phi_{,\theta\theta}, \tag{7.28a}$$

$$\sigma_{\theta\theta} = \phi_{,rr}, \tag{7.28b}$$

$$\sigma_{r\theta} = \frac{1}{r^2}\phi_{,\theta} - \frac{1}{r}\phi_{,r\theta}. \tag{7.28c}$$

The (invariant) biharmonic equation in cylindrical coordinates is written by using the appropriate operator. Thus, from (7.18), the expression

$$\nabla^4 \phi = \nabla^2 (\nabla^2 \phi) = 0 \tag{7.29}$$

holds, where

$$\nabla^2 \phi = \phi_{,rr} + \frac{1}{r} \phi_{,r} + \frac{1}{r^2} \phi_{,\theta\theta} \tag{7.30a}$$

and

$$\nabla^4 \phi = \left[\frac{\partial^2}{\partial r^2} + \frac{1}{r} \frac{\partial}{\partial r} + \frac{1}{r^2} \frac{\partial^2}{\partial \theta^2} \right] \left[\phi_{,rr} + \frac{1}{r} \phi_{,r} + \frac{1}{r^2} \phi_{,\theta\theta} \right], \tag{7.30b}$$

which constitute the transformation of (1–10 b and c) into cylindrical coordinates.

7.4.3   Axisymmetric Stresses and Displacements

Many problems in the circular domain are axisymmetric, depending only on r and not on θ. In these cases, we may use a simplified form of the preceding equations:

$$\sigma_{rr} = \frac{1}{r} \phi_{,r}, \tag{7.31a}$$

$$\sigma_{\theta\theta} = \phi_{,rr}, \tag{7.31b}$$

$$\sigma_{r\theta} = 0, \tag{7.31c}$$

$$\nabla^2 \phi = \phi_{,rr} + \frac{1}{r} \phi_{,r} + \frac{1}{r} \left[r(\phi)_{,r} \right]_{,r} \tag{7.31d}$$

$$\nabla^4 \phi = \left[\frac{d^2}{dr^2} + \frac{1}{r} \frac{d}{dr} \right] \left[\phi_{,rr} + \frac{1}{r} \phi_{,r} \right]$$

$$= \frac{1}{r} \left[r \left(\frac{1}{r} [r\phi_{,r}]_{,r} \right)_{,r} \right]_{,r} \tag{7.31e}$$

$$= 0.$$

The compact forms for $\nabla^2 \phi$ and $\nabla^4 \phi$ are easily verified by expansion of the terms and are readily integrated as shown symbolically in (7.32a) using the terms within the various brackets from the first equation:

$$\frac{1}{r}\left[r\left(\frac{1}{r}[r\phi,_r],_r\right),_r\right],_r = 0$$

$$\Big[\quad\Big],_r = 0$$

$$\Big[\quad\Big] = C_1$$

$$\Big(\quad\Big),_r = \frac{C_1}{r}$$

$$\Big(\quad\Big) = C_1\ln r + C_2$$ (7.32a)

$$\Big[\quad\Big],_r = C_1 r\ln r + C_2 r$$

$$\Big[\quad\Big] = C_1\left(\frac{r^2}{2}\ln r - \frac{r^2}{2}\right) + C_2\frac{r^2}{2} + C_3$$

$$= C_1' r^2\ln r + C_2' r^2 + C_3'$$

$$\phi,_r = C_1' r\ln r + C_2' r + \frac{C_3}{r}$$

$$\phi = C_1'\left(\frac{r^2}{2}\ln r - \frac{r^2}{2}\right) + C_2'\frac{r^2}{2} + C_3\ln r + C_4$$

or, redefining the constants,

$$\phi = C_1 r^2 \ln r + C_2 r^2 + C_3 \ln r + C_4.$$ (7.32b)

The corresponding stresses are found by substituting (7.32a) and (7.32b) into (7.31a, b, c):

$$\sigma_{rr} = C_1(1 + 2\ln r) + 2C_2 + C_3\frac{1}{r^2},$$ (7.33a)

$$\sigma_{\theta\theta} = C_1(3 + 2\ln r) + 2C_2 - C_3\frac{1}{r^2},$$ (7.33b)

$$\sigma_{r\theta} = 0.$$ (7.33c)

The transformation into cylindrical coordinates is sufficient to solve some representative stress analysis problems that are of great practical interest. However in order to derive the corresponding displacements, we would need to have the kinematic law (3.14) also transformed into cylindrical coordinates. This is rather involved but is carried out neatly by Little [4] using dyadic notation. Here, we simply record the results without derivation:

$$\varepsilon_{rr} = u_{r,r} \tag{7.34a}$$

$$\varepsilon_{\theta\theta} = \frac{u_r}{r} + \frac{1}{r} u_{\theta,\theta} \tag{7.34b}$$

$$u_r = \frac{1}{E} \Big\{ C_1 r[(1-v)(2\ln r - 1) - 2v] + 2C_2(1-v)r$$
$$\tag{7.34c}$$
$$- C_3 \frac{(1+v)}{r} + C_4 \sin\theta + C_5 \cos\theta$$

$$u_\theta = \frac{1}{E} [4C_1 r\theta + C_4 \cos\theta - C_5 \sin\theta + C_6 r] \tag{7.34d}$$

in which C_4–C_6 multiply rigid body terms. Notice that (7.34a, b, c, d) contain θ-dependent terms, although they are derived from the axisymmetric stress field (7.33a, b, c). This is a so-called quasi-axisymmetric case which occasionally proves to be useful, as we will see in Sects. 7.5 and 7.8.

7.5   Thick-Walled Cylinder or Disk

We examine a thick-walled cylinder constrained along the normal axis, plane strain, or its counterpart—a thin disk with a circular hole in the center, plane stress.

The cylinder is subjected to external and internal pressures shown in Fig. 7.2a, clearly an axisymmetrical loading case that should exhibit only an axisymmetric response. Thus, (7.33a, b, c) are applicable, and in (7.33a) and (7.33b), there are two obvious boundary conditions to evaluate the three constants:

$$\sigma_{rr}(R_0) = -p_0, \tag{7.35a}$$

$$\sigma_{rr}(R_i) = -p_i, \tag{7.35b}$$

which give

$$-p_i = C_1(1 + 2\ln R_i) + 2C_2 + \frac{C_3}{R_i^2}, \tag{7.35c}$$

$$-p_0 = C_1(1 + 2\ln R_0) + 2C_2 + \frac{C_3}{R_0^2}. \tag{7.35d}$$

To obtain a third condition, we must resort to the displacement expressions (7.34a, b, c, d). The expression for u_θ is multi-valued in θ and since $u_\theta(0) = u_\theta(2\pi) = u_\theta(4\pi)$, etc., we have $C_1 = 0$.

Fig. 7.2 (**a**) Thick-walled
cylinder under external and
internal pressure; (**b**) stress
distribution

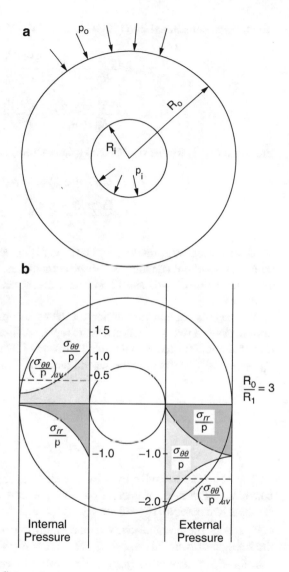

Then, we solve (7.35a, b, c, d) to get

$$C_2 = \frac{p_i R_i^2 - p_0 R_0^2}{2\left(R_0^2 - R_i^2\right)}, \tag{7.36a}$$

$$C_3 = \frac{R_i^2 R_0^2 (p_0 - p_i)}{\left(R_0^2 - R_i^2\right)}, \tag{7.36b}$$

which when substituted into (7.33a) and (7.33b) give the stress distributions:

$$\sigma_{rr} = \frac{p_i R_i^2 - p_0 R_0^2}{R_0^2 - R_i^2} + \frac{R_i^2 R_0^2 (p_0 - p_i)}{r^2 (R_0^2 - R_i^2)}, \tag{7.37a}$$

$$\sigma_{\theta\theta} = \frac{p_i R_i^2 - p_0 R_0^2}{R_0^2 - R_i^2} - \frac{R_i^2 R_0^2 (p_0 - p_i)}{r^2 (R_0^2 - R_i^2)}. \tag{7.37b}$$

and, from (7.13) for the constrained (plane strain) case,

$$\sigma_{zz} = \nu(\sigma_{xx} + \sigma_{yy}) = \nu(\sigma_{rr} + \sigma_{\theta\theta})$$

$$= \frac{2\nu(p_i R_i^2 - p_0 R_0^2)}{R_0^2 - R_i^2}. \tag{7.37c}$$

This solution was presented by G. Lamé [2] in the middle nineteenth century. The complete solution contains the displacements as well. These may be routinely evaluated from (7.34c) and (7.34d) and are given explicitly by Timoshenko and Goodier [2].

We may also obtain the solution for the case of pressure within a small hole in an infinite plane, a two-dimensional static representation of an explosion. We divide the numerator and the denominator of all terms in (7.37a, b, c) by R_0^2, take $p_0 = 0$, and let $R_0 \to \infty$. This leaves only the second terms of (7.37a) and (7.37b) simplified to

$$\sigma_{rr} = -p_i \frac{R_i^2}{r^2}, \tag{7.38a}$$

$$\sigma_{\theta\theta} = p_i \frac{R_i^2}{r^2}. \tag{7.38b}$$

The physical behavior of a thick-walled cylinder is easiest to appreciate by considering the cases of internal pressure and external pressure separately. Results for these two cases and $\frac{R_0}{R_i} = 3$ are shown through the wall thickness on Fig. 7.2b. For comparison, the nominal circumferential stress from the elementary formula from the strength of materials, $\sigma_{\theta\theta} = p\frac{R}{t}$, where $R = R_0$ or R_i for p_0 or p_i, respectively, and $t = R_0 - R_i$, is shown as $(\sigma_{\theta\theta})_{av}$. This value of $\sigma_{\theta\theta}$ may also be found from the cylindrical shell solution in Sect. 8.13. Obviously from Fig. 7.2b, the latter formula greatly underestimates the maximum circumferential stress that occurs at the inside wall and also does not recognize the radial stress.

The plane stress counterpart of this solution is a thin disk under radial loading p_0 and p_i. In this case, instead of (7.37c), we have from (7.6c)

$$\varepsilon_{zz} = -\frac{\nu}{E}(\sigma_{rr} + \sigma_{\theta\theta})$$

$$= \frac{2\nu}{E} \frac{(p_0 R_0^2 - p_i R_i^2)}{R_0^2 - R_i^2}. \tag{7.39}$$

Another variant to the solution is the case of a thick-walled pipe with closed ends. This is neither plane strain nor plane stress but can be viewed as a superposition of the two-dimensional solution in the cross section and an axial stress that is uniformly distributed and maintains overall equilibrium in the z-direction. The net force is

$$F_z = p_i \pi R_i^2 - p_0 \pi R_0^2 \tag{7.40a}$$

and

$$\sigma_{zz} = \frac{F_z}{\pi(R_0^2 - R_i^2)} = \frac{p_i R_i^2 - p_0 R_0^2}{(R_0^2 - R_i^2)}. \tag{7.40b}$$

The in-plane strains should account for σ_{zz} (7.12a) and (7.12b).

7.6   Sheet with Small Circular Hole

We consider a sheet of unit thickness under uniaxial uniform tension $\mathbf{T}_x = T_x \mathbf{e}_x$ with a small circular hole of radius A in the interior, as shown in Fig. 7.3a. Without the hole, the stress would be uniform:

$$\sigma_{xx} = T_x, \tag{7.41a}$$

$$\sigma_{yy} = \sigma_{xy} = 0. \tag{7.41b}$$

This distribution should be altered only in the vicinity of the hole, with the stresses maintaining the uniform values in the remainder of the domain by St. Venant's principle. Based on this reasoning, it is convenient to construct a hypothetical circle with a radius equal to the sheet width as shown on the figure, thus facilitating a solution in polar coordinates. Equation (7.25) is used to transform the uniform stress state at $r = B$.

$$\sigma_{rr}(B, \theta) = T_x \cos^2 \theta = \frac{T_x}{2}(1 + \cos 2\theta), \tag{7.42a}$$

$$\sigma_{\theta\theta}(B, \theta) = T_x \sin^2 \theta = \frac{T_x}{2}(1 - \cos 2\theta), \tag{7.42b}$$

$$\sigma_{r\theta}(B, \theta) = -T_x \sin \theta \cos \theta = -\frac{T_x}{2} \sin 2\theta. \tag{7.42c}$$

The stresses along the circular boundary at $r = B$ can be separated into two parts as shown in Fig. 7.3b, c. On Fig. 7.3b, we have the axisymmetric stress state:

$$\sigma_{rr}^{(1)}(B, \theta) = \frac{T_x}{2}, \tag{7.43a}$$

$$\sigma_{r\theta}^{(1)}(B, \theta) = 0. \tag{7.43b}$$

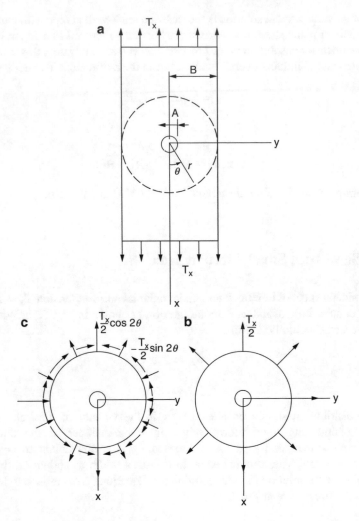

Fig. 7.3 (a) Sheet with small hole under tension; (b) axisymmetric state of stress; (c) θ-dependent state of stress

Of course, $\sigma_{\theta\theta}$ does not act on the boundary. On Fig. 7.3c, we have a stress state dependent on θ in the form

$$\sigma_{rr}^{(2)}(B,\theta) = \frac{T_x}{2}\cos 2\theta, \tag{7.44a}$$

$$\sigma_{r\theta}^{(2)}(B,\theta) = -\frac{T_x}{2}\sin 2\theta. \tag{7.44b}$$

The advantage to the separation is that we may use the solution for a thin disk under axisymmetrical radial load for the first part, (7.37a) and (7.37b). With $R_i = A$,

$R_0 = B$, $p_i = 0$, and $p_0 = -\frac{T_x}{2}$ [since it was originally shown as $(+)$ inward in Fig. 7.2], we have

$$\sigma_{rr}^{(1)} = \frac{T_x}{2}\frac{B^2}{B^2 - A^2} - \frac{T_x}{2}\frac{A^2 B^2}{B^2 - A^2}\frac{1}{r^2}, \tag{7.45a}$$

$$\sigma_{\theta\theta}^{(1)} = \frac{T_x}{2}\frac{B^2}{B^2 - A^2} + \frac{T_x}{2}\frac{A^2 B^2}{B^2 - A^2}\frac{1}{r^2}. \tag{7.45b}$$

Thus, we need only focus on the second part.

We return to the general form of the biharmonic Eq. (7.30b). Recognizing the periodicity of the boundary conditions (7.44a) and (7.44b), we assume a solution in the separated form:

$$\phi = F(r)\cos 2\theta. \tag{7.46}$$

If we substitute ϕ into (7.30b), we get an ordinary differential equation:

$$\left[\frac{\partial^2}{\partial r^2} + \frac{1}{r}\frac{\partial}{\partial r} + \frac{1}{r^2}\frac{\partial}{\partial \theta^2}\right]\left[\cos 2\theta\left(F_{,rr} + \frac{1}{r}F_{,r} - \frac{4}{r^2}F\right)\right] = 0, \tag{7.47a}$$

which expands to [2]

$$F_{,rrrr} + \frac{2}{r}F_{,rrr} - \frac{9}{r^2}F_{,rr} + \frac{9}{r^3}F_{,r} = 0. \tag{7.47b}$$

The solution to (7.47b) may be verified by back-substitution to be

$$F(r) = C_1 + C_2 r^2 + C_3 r^4 + C_4\frac{1}{r^2}. \tag{7.48}$$

Then,

$$\phi = \left(C_1 + C_2 r^2 + C_3 r^4 + C_4\frac{1}{r^2}\right)\cos 2\theta. \tag{7.49}$$

This expression for ϕ is inserted into (7.28a, b, c) to obtain

$$\sigma_{rr}^{(2)} = -\left(4C_1\frac{1}{r^2} + 2C_2 + 6C_4\frac{1}{r^4}\right)\cos 2\theta. \tag{7.50a}$$

$$\sigma_{\theta\theta}^{(2)} = \left(2C_2 + 12C_3 r^2 + 6C_4\frac{1}{r^4}\right)\cos 2\theta, \tag{7.50b}$$

$$\sigma_{r\theta}^{(2)} = \left(-2C_1\frac{1}{r^2} + 2C_2 + 6C_3 r^2 - 6C_4\frac{1}{r^4}\right)\sin 2\theta. \tag{7.50c}$$

The appropriate boundary conditions to evaluate the constants are (7.44a) and (7.44b) and the stress-free conditions at the hole:

$$\sigma_{rr}^{(2)}(A, \theta) = 0$$

$$\sigma_{r\theta}^{(2)}(A, \theta) = 0,$$

(7.51)

which leads to

$$
\begin{bmatrix}
-\dfrac{4}{A^2} & -2 & 0 & -\dfrac{6}{A^4} \\[2mm]
-\dfrac{2}{A^2} & 2 & 6A^2 & -\dfrac{6}{A^4} \\[2mm]
-\dfrac{4}{B^2} & -2 & 0 & -\dfrac{6}{B^4} \\[2mm]
-\dfrac{2}{B^2} & 2 & 6B^2 & -\dfrac{6}{B^4}
\end{bmatrix}
\begin{Bmatrix}
C_1 \\[2mm] C_2 \\[2mm] C_3 \\[2mm] C_4
\end{Bmatrix}
=
\begin{Bmatrix}
0 \\[2mm] 0 \\[2mm] \dfrac{T_x}{2} \\[2mm] -\dfrac{T_x}{2}
\end{Bmatrix}
$$

(7.52)

Equation (7.52) may be simplified for the case where the hole is small compared to the width of the sheet. We first multiply both sides of the fourth equation by $(A/B)^2$ and then let $A/B \to 0$ which gives $C_3 = 0$. Next, we multiply both sides of the third equation by A^2, let $A/B \to 0$, and get $C_2 = -T_x/4$. Then we have the first and second equations:

$$-\frac{4}{A^2}C_1 + \frac{T_x}{2} - \frac{6}{A^4}C_4 = 0$$

(7.53a)

$$-\frac{2}{A^2}C_1 - \frac{T_x}{2} - \frac{6}{A^4}C_4 = 0$$

(7.53b)

from which $C_1 = (A^2/2)T_x$ and $C_4 = (A^4/4)T_x$. Finally, the stresses are found from (7.50a)–(7.50c) to be

$$\sigma_{rr}^{(2)} = -\left(2\frac{A^2}{r^2} - \frac{1}{2} - \frac{3A^4}{2r^4}\right)T_x \cos 2\theta,$$

(7.54a)

$$\sigma_{\theta\theta}^{(2)} = -\left(\frac{1}{2} + \frac{3A^4}{2r^4}\right)T_x \cos 2\theta,$$

(7.54b)

$$\sigma_{r\theta}^{(2)} = \left(-\frac{A^2}{r^2} - \frac{1}{2} + \frac{3A^4}{2r^4}\right)T_x \sin 2\theta.$$

(7.54c)

We reevaluate (7.45a) and (7.45b) by dividing the numerator and denominator of each term by B^2, letting $A/B \to 0$, and then adding the results to (7.54a, b, c) to produce, finally, the total solution:

$$\sigma_{rr}(r,\theta) = \frac{T_x}{2}\left(1 - \frac{A^2}{r^2}\right) + \frac{T_x}{2}\left(1 - 4\frac{A^2}{r^2} + 3\frac{A^4}{r^4}\right)\cos 2\theta, \qquad (7.55a)$$

$$\sigma_{\theta\theta}(r,\theta) = \frac{T_x}{2}\left(1 + \frac{A^2}{r^2}\right) - \frac{T_x}{2}\left(1 + 3\frac{A^4}{r^4}\right)\cos 2\theta, \qquad (7.55b)$$

$$\sigma_{r\theta}(r,\theta) = -\frac{T_x}{2}\left(1 + 2\frac{A^2}{r^2} - 3\frac{A^4}{r^4}\right)\sin 2\theta. \qquad (7.55c)$$

Most interesting is to consider the maximum stress that occurs at $r = A$:

$$\sigma_{\theta\theta}(A,\theta) = T_x - 2T_x\cos 2\theta, \qquad (7.56)$$

which is largest when $\theta = \frac{\pi}{2}$ or $\frac{3\pi}{2}$, that is, at the sides of the hole where the tangent vector $\mathbf{s}$ is parallel to the direction of the applied $\mathbf{T}_x$,

$$(\sigma_{\theta\theta})_{\max} = \sigma_{\theta\theta}\left(A,\frac{\pi}{2}\right) = \sigma_{xx}(0,A) = 3T_x, \qquad (7.57a)$$

At $\theta = 0$ or π, the sides where the tangent vector $\mathbf{s}$ is perpendicular to $\mathbf{T}_x$, we find

$$(\sigma_{\theta\theta})_{\min} = \sigma_{\theta\theta}(A,0) = \sigma_{yy}(A,0) = -T_x. \qquad (7.57b)$$

The ratio of the maximum to the nominal stress is called the *stress concentration factor K*. In this case, $K = 3$. By superposition, it is easy to see that a uniform biaxial tension would produce $(\sigma_{\theta\theta})_{\max} = 3T_x - T_x = 2T_x$ and $K = 2$, while an equal tension in one direction and compression in the perpendicular direction would give $(\sigma_{\theta\theta})_{\max} = 3T_x + T_x = 4T_x$ and $K = 4$.

It is apparent from the $1/r^2$ and $1/r^4$ factors in (7.55a, b, c), which are indicative of the "strength" of the singularity, that the high stress concentration is quite localized and decays rapidly away from the hole. It is also obvious that K is *independent* of the actual size of the hole thus making the solution very general. Stress concentrations are of great importance in structural fatigue and crack propagation studies which are part of a relatively new and important branch of elasticity known as fracture mechanics.

7.7 Curved Beam

We now consider a beam that is curved in the plane of bending as shown in Fig. 7.4a. The outside and inside faces are cylindrical, the cross section is rectangular, and the loading is a constant bending moment M. With these restrictions, we may consider the stresses to be independent of θ and use the axisymmetric plane stress solution as given by (7.33a, b, c).

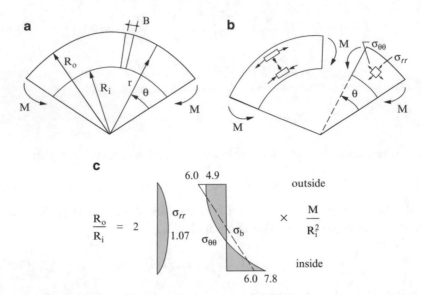

Fig. 7.4 (a) Curved beam in bending; (b) stresses in beam; (c) stress distribution through depth

We have the stress-free boundary conditions at the upper and lower faces:

$$\sigma_{rr}(R_0) = \sigma_{rr}(R_i) = 0 \tag{7.58a}$$

$$\sigma_{r\theta}(R_0) = \sigma_{r\theta}(R_i) = 0 \tag{7.58b}$$

along with the equilibrium conditions

$$\int_A \sigma_{\theta\theta}dA = 0 \tag{7.59a}$$

$$\int_A r\sigma_{\theta\theta}dA = M, \tag{7.59b}$$

where the stresses are shown in Fig. 7.4b. Obviously, (7.58b) is identically satisfied by the axisymmetric solution. However, it appears that the problem may be over-specified, that is, more conditions than unevaluated constants.

We evaluate (7.33a) for σ_{rr} on the boundaries using (7.58a) as

$$C_1(1 + 2\ln R_0) + 2C_2 + C_3 + \frac{1}{R_0^2} = 0, \tag{7.60a}$$

$$C_1(1 + 2\ln R_i) + 2C_2 + C_3\frac{1}{R_i^2} = 0. \tag{7.60b}$$

Next, we have (7.59a) that may be written in terms of the stress function, using (7.31b):

$$\int_A \sigma_{\theta\theta} dA = B \int_{R_i}^{R_0} \phi_{,rr} dr$$
$$= B \left[\phi_{,r}\right]_{R_i}^{R_0}. \tag{7.61}$$

where B is the width of the beam. Noting that $\phi_{,r}$ is proportional to σ_{rr} via (7.28a) and comparing the integrated function to (7.31a) for σ_{rr}, we see that the condition is identical to (7.58a) that produced (7.60a) and (7.60b). We conclude that $\phi_{,r} = 0$ on the boundaries. Then, the remaining equilibrium condition is (7.59b):

$$\int_A r \sigma_{\theta\theta} dA = B \int_{R_i}^{R_0} r \phi_{,rr} dr = M. \tag{7.62}$$

Considering the integral and using integration by parts,

$$\int_{R_i}^{R_0} r \phi_{,rr} dr = r \phi_{,r} \Big]_{R_i}^{R_0} - \int_{R_i}^{R_0} \phi_{,r} dr, \tag{7.63}$$

the first term on the r.h.s. is zero since $\phi_{,r}$ has been shown to be zero on the boundaries, while the second term, upon back-substitution into (7.62), leads to

$$\phi(R_0) - \phi(R_i) = -\frac{M}{B}, \tag{7.64}$$

or, by (7.32b),

$$C_1 \left(R_0^2 \ln R_0 - R_i^2 \ln R_i\right) + C_2 \left(R_0^2 - R_i^2\right) + C_3 \ln \frac{R_0}{R_i} - \frac{M}{B}. \tag{7.65}$$

Thus, we have produced three equations, (7.60a, b) and (7.65), for the constants C_1–C_3. These are solved [5] routinely as

$$C_1 = \frac{2M}{NB} \left(R_0^2 - R_i^2\right) \tag{7.66a}$$

$$C_2 = \frac{-M}{NB} \left[R_0^2 - R_i^2 + 2\left(R_0^2 \ln R_0 - R_i^2 \ln R_i\right)\right] \tag{7.66b}$$

$$C_3 = \frac{4M}{NB} R_0^2 R_i^2 \ln \frac{R_0}{R_i}, \tag{7.66c}$$

where

$$N = \left(R_0^2 - R_i^2\right)^2 - 4R_0^2 R_i^2 \left(\ln \frac{R_0}{R_i}\right)^2. $$

Finally, the stresses are written explicitly as [2]

$$\sigma_{rr} = \frac{4M}{NB}\left(R_0^2 R_i^2 \ln\frac{R_0}{R_i}\frac{1}{r^2} + R_0^2 \ln\frac{r}{R_0} + R_i^2 \ln\frac{R_i}{r}\right), \tag{7.67a}$$

$$\sigma_{\theta\theta} = \frac{4M}{NB}\left(R_0^2 R_i^2 \ln\frac{R_0}{R_i}\frac{1}{r^2} - R_0^2 \ln\frac{r}{R_0} - R_i^2 \ln\frac{R_i}{r} - R_0^2 + R_i^2\right), \tag{7.67b}$$

$$\sigma_{r\theta} = 0. \tag{7.67c}$$

To be applicable for the entire beam, the end moments M must be distributed in accordance with $\sigma_{\theta\theta}$; see (7.67b). Otherwise, the results are valid overall, except for the immediate region of moment application, through St. Venant's principle.

In order to study the stress distribution for a typical case, we choose $\frac{R_0}{R_i} = 2$. Then, in Fig. 7.4c, the stresses σ_{rr} and $\sigma_{\theta\theta}$ are plotted through the depth of the beam along with the straight line distribution obtained from elementary beam theory, $\sigma_b = \sigma_{zz}$ from (6.20b) [2]. Notice that the elementary beam solution underestimates the maximum stress that occurs along the inside boundary and, of course, does not indicate σ_{rr} at all. We may explain heuristically the increase in stress along the inside boundary and the decrease along the outside boundary as compared to elementary theory by recognizing that the inside fibers are shorter and the outside fibers are longer than those along the center line, which correspond to a straight beam of the same length. Therefore, the same extension or contraction would produce a correspondingly greater, or lesser, strain in these fibers. Also the sense of σ_{rr} may be related to the radial components of the stresses $\sigma_{\theta\theta}$ shown on the left portion of Fig. 7.4b. Near the outside face, this component is directed inward, while near the inside face, it also acts inward to produce the radial compression.

It should also be mentioned that there is a strength-of-materials solution for curved beams that produces results for $\sigma_{\theta\theta}$ practically identically to those obtained from the plane stress solution for beams in which the radius of the centroidal axis $1/2$ $(R_0 + R_i)$ is greater than the depth $R_0 - R_i$ but still no values for σ_{rr}. This solution is presented in detail in Ref. [6] and is adequate for many practical cases.

7.8   Rotational Dislocation

An interesting variant on the thick-walled cylinder problem treated in Sect. 7.5 is shown in Fig. 7.5, where the ring has been formed as open and separated by an angle α, assumed as the zero stress state, and then forced together and joined. The appropriate boundary conditions are

$$\sigma_{rr}(R_0) = \sigma_{rr}(R_i) = 0, \tag{7.68a}$$

$$u_\theta(r, 2\pi) = r\alpha. \tag{7.68b}$$

Fig. 7.5 Thick-walled
cylinder with rotational
dislocation

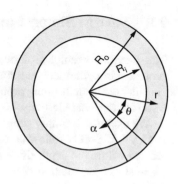

The utility of the "quasi-axisymmetric" displacement equations is now apparent,
since we may use (7.34d) to write (7.68b) as

$$r\alpha = \frac{1}{E} 8\pi C_1 r,$$

or

$$C_1 = \frac{E\alpha}{8\pi}. \tag{7.69}$$

We then evaluate (7.33a) for the boundary conditions (7.68a) with C_1 from
(7.69):

$$\frac{E\alpha}{8\pi}(1 + 2\ln R_0) + 2C_2 + \frac{C_3}{R_0^2} = 0, \tag{7.70a}$$

$$\frac{E\alpha}{8\pi}(1 + 2\ln R_i) + 2C_2 + \frac{C_3}{R_i^2} = 0, \tag{7.70b}$$

yielding

$$C_2 = -\frac{E\alpha}{16\pi} \frac{(1 + 2\ln R_0)R_0^2 - (1 + 2\ln R_i)R_i^2}{R_0^2 - R_i^2}, \tag{7.71a}$$

$$C_3 = \frac{E\alpha}{4\pi} \frac{R_0^2 R_i^2}{R_0^2 - R_i^2} \ln\frac{R_0}{R_i}, \tag{7.71b}$$

which can be back-substituted into (7.33a, b, c) along with C_1 to find σ_{rr} and $\sigma_{\theta\theta}$.
The total bending moment across the section is found from (7.66a) to be

$$M = \frac{C_1}{2} \frac{NB}{R_0^2 - R_i^2}$$

$$= \frac{E\alpha B}{16\pi} \frac{(R_0^2 - R_i^2)^2 - 4R_0^2 R_i^2 \ln(R_0/R_i)^2}{R_0^2 - R_i^2}. \tag{7.72}$$

7.9   Narrow Simply Supported Beam

The previous examples of two-dimensional elasticity solutions have focused on problems in cylindrical coordinates. We now turn to a Cartesian domain and again consider the beam bending problem which was analyzed in Sect. 6.3 in the context of three-dimensional elasticity.

We examine a simply supported beam of unit width under self-weight as shown in Fig. 7.6, closely following the treatment by Little [4]. The supports are precisely located within the depth since we are seeking an elasticity solution. We employ the plane stress formulation (7.9a, b, c) in terms of the Airy stress function. With the body forces given by

$$f = -\rho g e_y, \tag{7.73}$$

where ρ is the mass density and g the acceleration of gravity, the potential V as defined in (7.3) becomes

$$V = \rho g y. \tag{7.74}$$

Thus, (7.9b) takes the form

$$\nabla^4 \phi = -(1 - v)\nabla^2(\rho g y). \tag{7.75}$$

The boundary conditions on the top and bottom faces are

$$\sigma_{xy}(x, \pm D) = \sigma_{yy}(x, \pm D) = 0, \tag{7.76}$$

and on the ends,

$$\sigma_{xx}(\pm L, y) = 0, \tag{7.77a}$$

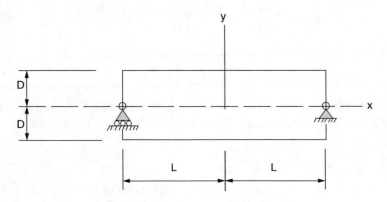

Fig. 7.6 Narrow beam in bending

$$\int_{-D}^{D} \sigma_{xy}(-L, y)dy = -2\rho gDL,$$ (7.77b)

$$\int_{-D}^{D} \sigma_{xy}(L, y)dy = 2\rho gDL.$$ (7.77c)

The latter two equations reflect the acceptance of St. Venant's principle for this case. Once the stress function is found, the stresses can be calculated from (7.5a, b, c) as

$$\sigma_{xx} = \rho gy + \phi_{,yy},$$ (7.78a)

$$\sigma_{yy} = \rho gy + \phi_{,xx},$$ (7.78b)

$$\sigma_{xy} = -\phi_{,xy}.$$ (7.78c)

Recalling the idea of expressing the solution ϕ in the form of selected polynomial terms, as suggested in (7.19), and recognizing that the stress would likely follow an *even* functional distribution in x and an *odd* distribution in y, we have

$$\phi = C_{21}x^2y + C_{23}x^2y^3 + C_{03}y^3 + C_{05}y^5$$ (7.79)

as proposed in Ref. [4]. From the discussion in Sect. 7.3, it is evident that the first and third terms satisfy the homogeneous part of (7.75) individually, but the second and fourth terms must satisfy the equation in combination. Substituting (7.79) into (7.78a, b, c) yields

$$\sigma_{xx} = \rho gy + 6C_{23}x^2y + 6C_{03}y + 20C_{50}y^3$$ (7.80a)

$$\sigma_{yy} = \rho gy + 2C_{21}y + 2C_{23}y^3$$ (7.80b)

$$\sigma_{xy} = -2C_{21}x - 6C_{23}xy^2$$ (7.80c)

for the stresses. Since we have chosen appropriate even or odd functions, we may select either the $(+)$ or $(-)$ conditions in (7.76). Using the condition at $(+) D$,

$$\sigma_{xy}(x, D) = -2C_{21}x - 6C_{23}xD^2 = 0$$ (7.81a)

$$\sigma_{yy}(x, D) = \rho gD + 2C_{21}D + 2C_{23}D^3 = 0,$$ (7.81b)

from which

$$C_{21} = -\frac{3}{4}\rho g,$$ (7.82a)

$$C_{23} = \frac{1}{4}\frac{\rho g}{D^2}.$$ (7.82b)

From (7.75) applied to (7.79) with the r.h.s $= 0$ since ρ is constant and y is linear and noting that only C_{23} and C_{05} remain,

$$C_{05} = -\frac{1}{5}C_{23}$$

$$= -\frac{1}{20}\frac{\rho g}{D^2}.$$

(7.83)

Finally, we turn to (7.77a). The form of σ_{xx} in (7.80a) makes it impossible to satisfy this equation identically in a pointwise fashion. Therefore, we invoke St. Venant's principle whereby

$$\int_{-D}^{D} \sigma_{xx}(L,y)dy = 0,$$

(7.84a)

$$\int_{-D}^{D} \sigma_{xx}(L,y)y\,dy = 0,$$

(7.84b)

Substituting (7.80a), along with (7.82a, b) and (7.83), into (7.84a), we find that it is satisfied identically since σ_{xx} is an odd function in y. Similarly, (7.84b) yields

$$\underbrace{\frac{2}{3}\rho g D^3}_{\text{from } \rho g y} + \underbrace{\rho g L^2 D}_{\text{from } c_{23}} + 4C_{03}D^3 - \underbrace{\frac{2}{5}\rho g D^3}_{\text{from } c_{05}} = 0,$$

(7.85)

from which

$$C_{03} = -\rho g\left(\frac{1}{15} + \frac{1}{4}\frac{L^2}{D^2}\right).$$

(7.86)

Inserting the evaluated constants into (7.80a, b, c) gives

$$\sigma_{xx} = 3\rho g\left[\frac{1}{5} + \frac{1}{2D^2}(x^2 - L^2)\right]y - \rho g\frac{y^3}{D^2}$$

$$= \frac{\rho g D}{1}\left[\frac{2}{5}D^2 y - \frac{2}{3}y^3 + \underline{(x^2 - L^2)y}\right],$$

(7.87a)

$$\sigma_{yy} = -\frac{1}{2}\rho g\left[1 - \frac{y^2}{D^2}\right]y$$

$$= \frac{\rho g D}{I}\left[-\frac{D^2 y}{3} + \frac{y^3}{3}\right],$$

(7.87b)

$$\sigma_{xy} = \frac{3}{2}\rho g\left[1 - \frac{y^2}{D^2}\right]x$$

$$= \frac{\rho g D}{I}[D^2 x - y^2 x],$$

(7.87c)

where I is the moment of inertia/unit width about the centroidal axis $= \frac{2}{3}D^3$.

Elementary beam theory, which was derived for a cantilever beam in Sect. 6.3.1, gives only the underlined term in (7.87a) for σ_{xx}, so that the other terms in (7.87a) and (7.87b) represent the modification due to the two-dimensional stress state produced by the presence of σ_{yy}. The value of σ_{xy} is not changed from the elementary theory result. The analysis may be carried on to produce strains and eventually displacements, but little that is new is revealed [4]. Of interest in the family of two-dimensional beam solutions are the various possibilities for applying the transverse loading other than the uniform distribution through the depth studied here. Loading on the top face or the bottom face produces alterations to (7.87a, b, c), especially in the σ_{yy} (transverse normal) stresses [5].

It is also notable that for both the curved beam in Sect. 7.7 and for the narrow beam in this section, one of the additional results provided by an elasticity solution over a strength-of-materials solution are the normal stresses in the transverse direction, σ_{rr} and σ_{yy}, respectively. For conventional homogeneous materials, such stresses have evidently not been a major concern since they are apparently far smaller than the longitudinal stresses, $\sigma_{\theta\theta}$ and σ_{xx}, respectively, as shown in the respective solutions. However, beams fabricated from anisotropic materials, such as laminated timber and modern composites, are composed of layers as briefly described in Sect. 4.5. Adequate resistance to transverse normal stresses, particularly tensile, is critical to the bonding of adjacent layers.

7.10 Semi-infinite Plate with a Concentrated Load

A semi-infinite plate of unit thickness under a concentrated vertical loading, shown in Fig. 7.7, may be solved in either polar or Cartesian coordinates. The general three-dimensional case is known as the Boussinesq problem and is renown in the theory of

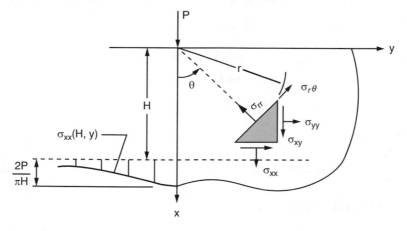

Fig. 7.7 Semi-infinite plate with concentrated load. (From Ugural and Fenster [10]. Reprinted with permission)

elasticity. Here, we present the two-dimensional version in polar coordinates and leave the Cartesian coordinate solution for the exercises.

It is postulated that any element located by (r, θ) is under a state of pure radial compression, i.e.,

$$\sigma_{rr}(r,\theta) = -2(P/\pi)\frac{\cos\theta}{r} \tag{7.88a}$$

$$\sigma_{\theta\theta} = 0 \tag{7.88b}$$

$$\sigma_{r\theta} = 0 \tag{7.88c}$$

which will be verified shortly. If the vertical component of σ_{rr}, $\sigma_{rr}\cos\theta$, is integrated over a semicircular cylindrical surface described by r, we have

$$2\int_0^{\pi/2} (\sigma_{rr}\cos\theta)(rd\theta) = -P \tag{7.89}$$

so that overall equilibrium is satisfied. Obviously the coefficient $2/\pi$ in (7.88a) was chosen to scale the integral to the r.h.s. of (7.89). The boundary conditions on the free surface are obviously satisfied since the stresses vanish at every point, except under the load where the infinite stress is given by (7.88a, b, c) with $\theta = 0$ and $r = 0$.

To complete the solution, it must be shown that local equilibrium and compatibility are also satisfied. It is convenient to use the stress function:

$$\phi = -(P/\pi)r\theta\sin\theta \tag{7.90}$$

which may be substituted into (7.28a, b, c) to confirm that (7.88a, b, c) produce equilibrium and then into (7.29) to verify compatibility.

We now consider a horizontal plane $x = H$ and compute the Cartesian components of stress using (2.13). From Fig. 7.7, the direction cosines are found as

	r	θ	z
x	$\cos\theta$	$-\sin\theta$	0
y	$\sin\theta$	$\cos\theta$	0
z	0	0	1

$$(7.91)$$

Since

$$\sigma_{rr}(r,\theta) = \sigma_{rr}\left(\frac{H}{\cos\theta}, \theta\right) = -2\left(\frac{P}{\pi H}\right)\cos^2\theta \tag{7.92}$$

we find from (2.13) $\sigma'_{ij} = a_{ik}a_{jl}\sigma_{kl}$:

$$\sigma_{xx}(H,\theta) = \sigma_{rr}\cos^2\theta = -2\left(\frac{P}{\pi H}\right)\cos^4\theta \tag{7.93a}$$

$$\sigma_{yy}(H, \theta) = \sigma_{rr} \sin^2\theta = -2\left(\frac{P}{\pi H}\right) \sin^2\theta \cos^2\theta \qquad (7.93b)$$

$$\sigma_{xy}(H, \theta) = \sigma_{rr} \sin\theta \cos\theta = -2\left(\frac{P}{\pi H}\right) \sin\theta \cos^3\theta. \qquad (7.93c)$$

For a point directly beneath the load, $\theta = 0$ and $\sigma_{rr} = \sigma_{xx} = -\frac{2P}{\pi H}$. As shown on the figure, the vertical compression attenuates away from the load.

A detailed study of the stress distribution is presented in Timoshenko and Goodier [2] along with a solution for a concentrated horizontal force at the origin. Also in Ref. [2], expressions for the displacements are derived and reveal singularities on the surface $x = 0$, even away from the point load. Solutions for distributed stress and constant displacement distributions of the vertical concentrated load that overcome the stress singularity at $r = 0$ in (7.88a) and the displacement singularities noted above are presented. These solutions are of great utility in determining the stresses and the corresponding displacements in soil deposits due to surface loading such as building columns or vehicular wheels and in numerous other problems involving concentrated surface loading or other pointwise excitation in many fields of engineering. A novel application is shown in the next section.

7.11   Tapered Thin-Wall Beam with Concentrated Loads

We have treated beams in several places, producing solutions for the bending stresses using the elementary theory of strength of materials (6.20) augmented by 3-D elasticity theory (6.56) and a 2-D theory allowing for the variation in stresses through the depth (7.87). Implicit in each of these cases is the assumption that the depth of the beam shown as 2-D on Fig.7.6 is constant along the span. Here we consider a tapered beam with a thin-walled cross section shown as Case c on Fig. 6.9. Such a beam is commonly known as an I-beam and is widely used in structural steel construction.

Here we consider a beam that has variable depth formed by a tapering web. A short section of such a beam is shown in Fig. 7.8a. Generally for mild tapers, the bending stresses can usually be calculated from the elementary theory, dependent on the moment of inertia and the area, but the shearing stresses are more complicated.

In uniform I-section beams, the normal stresses due to moments and axial forces are parallel to the centroidal axis, and the shear forces are assumed to be resisted by shear stresses in the web. However, in web-tapered beams, the flanges are inclined to the centroidal axis, as are the flange normal stresses, so the stress trajectories are inclined to the centroidal axis as well. In addition, the inclined flange forces have components transverse to the centroidal axis, which may participate in resisting the shear forces [7].

While this problem lies in the domain of the behavior of thin-walled steel beams, it is of interest here because of a connection to the radial normal stress solution in the

Fig. 7.8 (a) Web-tapered
I-beam (b) Wedge (Adapted
from Trahair and Ansourian
[7])

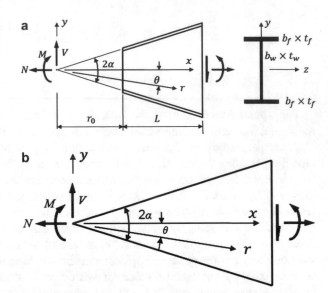

previous section, as noted by Trahair and Anssourian in [7, 8]. As a first step, we
present elasticity-based solutions to the wedge problem shown in Fig. 7.8b for an
axial load (N), shear (V), and bending moment (M) applied at the apex [2, 7].

Axial Force, N

$$\sigma_{rr} = -\frac{2N \cos \theta}{r(2\alpha + \sin 2\alpha)} \tag{7.94}$$

Shear, V

$$\sigma_{rr} = -\frac{2V \sin \theta}{r(2\alpha - \sin 2\alpha)} \tag{7.95}$$

Moment, M

$$\sigma_{rr} = -\frac{4M \sin 2\theta}{2r^2(\sin 2\alpha - 2\alpha \cos 2\alpha)} \tag{7.96a}$$

$$\sigma_{r\theta} = \frac{2M(\cos 2\theta - \cos 2\alpha)}{2r^2(\sin 2\alpha - 2\alpha \cos 2\alpha)} \tag{7.96b}$$

Now the connection between the radial normal stress problem and the wedge can
be established by setting $\alpha = \pi/2$ and $N = -P$ in Eq. (7.94) to match Eq. (7.88a). In
this sense, the half-space under concentrated loading is the limiting case of the
wedge. Moreover, the radial stress patterns in the wedge have similarities to the
corresponding web-tapered beam case. Our interest is on the application of the
elasticity solution for the wedge to the solution of the beam problem.

Fig. 7.9 Mid-span stresses in wedge (Adapted from Trahair and Ansourian [7])

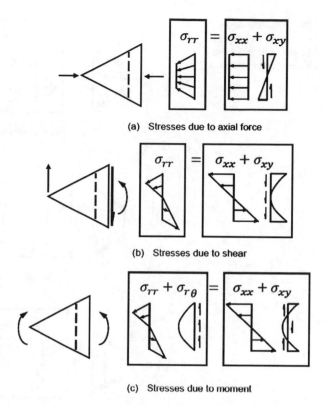

(a) Stresses due to axial force

(b) Stresses due to shear

(c) Stresses due to moment

To facilitate the comparison, the stress patterns for both the wedge in Fig. 7.9 and the web-tapered beam in Fig. 7.10 are shown for both polar and Cartesian coordinates. The qualitative comparisons are adapted (1) from the elasticity solutions for the wedge as given in Eqs. (7.94), (7.95), and (7.96) and (2) from an extension of uniform beam theory (CBA) to symmetric tapered I-beams (TBA). Briefly these solutions are facilitated by the assumption of inclined normal stress trajectories suggested by the wedge solution and additional modifications of the transverse shear stress distributions. Also, quantitative verification of some examples solved by the TBA equations is provided using a finite element program STRAND 7 [7, 8].

Axial Force

The tapered beam in Fig. 7.10a has equal and opposite axial forces $-N$ applied, which induce radial stresses σ_{rr} determined by TBA using the inclined stress trajectories. These stresses are constant across any cross section and are also accompanied by linear varying shear stresses $\sigma_{r\theta}$ not present in the wedge (Fig. 7.9a). The stresses are transformed to the Cartesian axes using the two-dimensional formulas (2.13) and produce almost constant normal stresses σ_{xx} and self-equilibrated tangential shear stresses σ_{xy}, as shown in Figs. 7.9a and 7.10a [7].

Fig. 7.10 Mid-span
stresses in web-tapered
beams (Adapted from
Trahair and Ansourian [7])

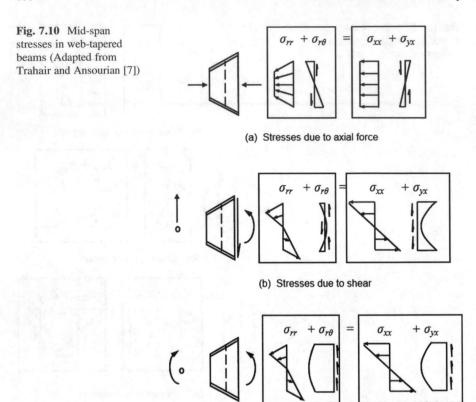

(a) Stresses due to axial force

(b) Stresses due to shear

(c) Stresses due to moment

Shear

In Fig. 7.10b, the shear force V is applied with the equilibrating moment $V(r_o + L)$ as shown in Fig.7.8a. Here, the stresses σ_{xx} vary linearly across any cross section for both the wedge and the beam, and the tangential shear stresses σ_{xy} vary parabolically. Since the total applied shear V needs to be resisted, a shear stress distribution following the CBA pattern (6.20c), which is similar to $\sigma_{r\theta}$ in Fig. 7.9c, might be anticipated. However, there is a reduction in the magnitude of the web shear and a reversal of the CBA pattern to one similar to the wedge, σ_{xx}, in Fig. 7.9b. These alterations are attributed to the reduction of the web shear by the net vertical components of the forces in the flanges V_f, which are proportional to σ_{rr} ($+\alpha$) and σ_{rr} ($-\alpha$). The reduced web shear $V - V_f$ is resisted primarily by the vertical component of the flange normal stress $\theta\sigma_{rr}$, where the approximation for sin θ following (3.29) is used, and by the circumferential shear stress $\sigma_{r\theta}$. As shown in Fig. 7.10b, the contribution of $\sigma_{r\theta}$ is greatly reduced from what might be expected from the CBA solution where the $\theta\sigma_{rr}$ contribution is not present [7].

Moment

The opposing constant moments M in Fig. 7.10c produce linear radial stresses σ_{rr} and nearly constant tangential shear stresses $\sigma_{r\theta}$ in the beam. The radial stresses are equivalent to linear normal stresses σ_{xx} and parabolic shear stresses $\theta\sigma_{rr}$. The parabolic transverse shear stresses σ_{xy} vary significantly from the zero values predicted by CBA. These stresses balance the vertical component of the flange forces V_f so that the net shear force on the section is zero [7]. In the wedge, the shear stresses $\sigma_{r\theta}$ are directly oriented with the applied moment M and are distributed like the shear stresses in the CBA analysis.

Comment

While the solution for a point-loaded half-space and the more general wedge has been available since the nineteenth century, the extension of the elementary theory beams to a configuration where the depth of the beam is varied by a tapered web is a relatively modern problem motivated perhaps by the development of technology to fabricate such beams. The radial normal stress patterns in point-loaded wedges, derived from the classical elasticity problem, were found to be similar to the stress patterns in web- tapered I-beams, solved by thin-walled beam theory. The similarity led to an improved solution for the normal stresses in the beam problem and provided some guidance in determining the shear stress distribution as well, demonstrating the continuing relevance of the theory of elasticity.

7.12 Functionally Graded Beam

7.12.1 Plane Stress Formulation for Orthotropic Material

In Sect. 4.5, a constitutive law was presented for two-dimensional functionally graded materials. Here we continue with an elasticity solution based in that law.

We begin with (7.2a) and (7.2b), substitute (4.36) for the stresses, and replace the strains by (3.15a, b, d) to produce two equilibrium equations in terms of $u_1 = u_x$ and $u_2 = u_y$:

$$\left(C_{11}u_{x,x} + C_{12}u_{y,y}\right)_{,x} + \frac{1}{2}\left(C_{44}u_{x,y} + C_{44}u_{y,x}\right)_{,y} + f_x = 0 \tag{7.97a}$$

$$\frac{1}{2}\left(C_{44}u_{x,y} + C_{44}u_{y,x}\right)_{,x} + \left(C_{12}u_{x,x} + C_{22}u_{y,y}\right)_{,y} + f_y = 0 \tag{7.97b}$$

7.12.2 Simply Supported FGB: Elasticity Analysis

Using this displacement formulation, Sankar [9] developed an elasticity solution for a simply supported beam along with a corresponding solution using conventional beam theory. The body forces f_x and f_y are set equal to zero for this case.

The beam shown in Fig. 7.11 with unit width is in a state of plane stress, with a stress-free top surface at $y = h$ and a symmetric transverse harmonic loading along the bottom surface ($y = 0$) equivalent to a normal stress:

$$\sigma_{yy}(x,0) = -p_y(x) = -p_{yj}\sin \xi x \qquad (7.98a)$$

and

$$\sigma_{xy}(x,0) = 0, \qquad (7.98b)$$

where

$$\xi = \frac{j\pi}{L}j = 1,3,5,\dots \qquad (7.98c)$$

Figure 7.11 shows the $j = 1$ case, but (7.98a) can treat a more fluctuating load distribution using Fourier series, described in detail in Sect. 8.4.1.

The stress–strain law is taken as (4.38) and the displacements are assumed in the form

$$u_x(x,y) = U(y) \cos \xi x \qquad (7.99a)$$

$$u_y(x,y) = V(y) \sin \xi x \qquad (7.99b)$$

corresponding to the boundary conditions:

$$u_y(0,y) = u_y(L,y) = 0 \qquad (7.100a)$$

$$\sigma_{xx}(0,y) = \sigma_{xx}(L,y) = 0 \qquad (7.100b)$$

Note that the boundary condition on u_y is a bit more general than shown on Fig. 7.11, where the simply supported condition requires that the displacement is constrained only at a single point through the depth on each vertical boundary. Here, this generalization is acceptable because the boundary conditions on u_y obviously match the displacement function selected in (7.99b). The satisfaction of the second boundary condition on σ_{xx} by (7.99a) is verified through the strain–displacement relations (3.15a, b) and (4.32), noting that $u_{y,y}$ and hence ε_{yy} at $x = 0$ and $x = L$ are equal to zero.

Fig. 7.11 FGB under symmetric transverse loading [9]

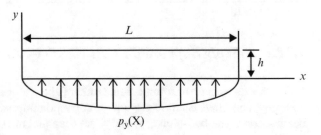

$p_y(X)$

Substituting (7.99a) and (7.99b) into (7.97a) and (7.97b) produces two ordinary coupled differential equations for the variables $U(y)$ and $V(y)$ with the common factors $e^{\zeta y} \cos \xi x$ and $e^{\zeta y} \sin \xi x$, respectively:

$$-C_{11}^0 \xi^2 U + C_{12}^0 \xi V_{,y} + \frac{1}{2} \left(C_{44}^0 U_{,yy} + C_{44}^0 \zeta U_{,y} + C_{44}^0 \xi V_{,y} + C_{44}^0 \zeta \xi V \right) = 0$$

(7.101a)

$$\frac{1}{2} \left(-C_{44}^0 \xi U_{,y} - C_{44}^0 \xi^2 V \right) - C_{12}^0 \xi U_{,y} - C_{12}^0 \zeta \xi U + C_{22}^0 V_{,yy} + C_{22}^0 \zeta V_{,y} = 0$$

(7.101b)

Recall that ζ was defined in (4.38) to provide a continuous variation of the material coefficients. To verify (7.101a) and (7.101b) from (7.97a) and (7.97b), it should be noted that terms such as $(C_{44}u_{y,x})_{,y}$ are differentiated by the product rule as $(C_{44,y}u_{y,x} + C_{44}u_{y,xy})$ and that $\left(C_{44}^0 \right)_{,y} = \zeta C_{44}^0$.

Proceeding, Sankar suggests further assumptions whereby the FGM is taken as isotropic at each point and Poisson's ratio is constant throughout the thickness [9]. This leads to the variation of Young's modulus given in (4.40) and the stress–strain law in (4.39). The solution of (7.101a) and (7.101b) is taken in the form

$$U(y) = \sum_{i-1}^{4} a_i e^{\alpha_i y}$$

(7.102a)

$$V(y) = \sum_{i=1}^{4} b_i e^{\alpha_i y}$$

(7.102b)

As shown below, four terms are taken in each series to correspond to the available traction boundary conditions.

Next, a single term of each series, $U(y) = a\, e^{\alpha y}$ and $V(y) = b\, e^{\alpha y}$, is substituted into (7.101a) and (7.101b), and the α_i are roots of the characteristic of equation [6]:

$$|A| = 0$$

(7.103)

where

$A_{11} = 1/2(1-2\nu)\alpha^2 + 1/2\,(1-2\nu)\zeta\alpha - (1-\nu)\,\xi^2$
$A_{12} = 1/2\xi\alpha + 1/2(1-2\nu)\zeta\xi$
$A_{21} = -\,1/2\xi\alpha - \nu\,\zeta\xi$
$A_{22} = (1-\nu)\alpha^2 + (1-\nu)\zeta\alpha - 1/2(1-2\nu)\,\xi^2$

The constants a_i and b_i in (7.102a) and (7.103b) are related by

$$r_i = \frac{b_i}{a_i} = \frac{-(1-2\nu)\alpha_i(\zeta + \alpha_i) - 2(1-\nu)\xi^2}{\xi\alpha_i + (1-2\nu)\zeta\xi}$$

(7.104)

and are to be determined from the traction boundary conditions on the top and bottom surfaces of the beam as shown in Fig. 7.11.

The boundary conditions are $\sigma_{xy}(x,0) = 0$, $\sigma_{xy}(x,h) = 0$, $\sigma_{yy}(x,0) = -p_{yj} \sin \xi x$, and $\sigma_{yy}(x,h) = 0$ and must be expressed in terms of the displacements u_x and u_y. We return to (4.37) with C_{or} replaced by $C_{FGI}(y)$, Eq. (4.39) recalled and the strains written in terms of the displacements by (3.15a, b, d) with $1 = x$ and $2 = y$. Then we have

$$\sigma_{xy}(x, 0) = G(0)\left(u_{x,y} + u_{y,x}\right)\big|_{y=0} = 0 \tag{7.105a}$$

$$\sigma_{xy}(x, h) = G(h)\left(u_{x,y} + u_{y,x}\right)\big|_{y=h} = 0 \tag{7.105b}$$

$$\sigma_{yy}(x, 0) = C_{FGI12}(0)u_{x,x}\big|_{y=0} + C_{FGI22}(0)u_{y,y}\big|_{y=0} = -p_{yj} \sin \xi x \tag{7.105c}$$

$$\sigma_{yy}(x, h) = C_{FGI12}(h)u_{x,x}\big|_{y=h} + C_{FGI22}(h)u_{y,y}\big|_{y=h} = 0, \tag{7.105d}$$

where the elements of $C_{FGI}, C_{FGIij}(y)$, are given by (4.39) partitioned as shown on the r.h.s. of (4.36). In (7.105a) and (7.105b), element C_{FGI44} has been directly replaced by its value G(y).

Substituting for u_x and u_y as given in (7.99a) and (7.99b) into (7.105a, b, c, d), we obtain the boundary conditions in terms of U and V:

$$U_{,y}(0) + \xi V(0) = 0 \tag{7.106a}$$

$$U_{,y}(h) + \xi V(h) = 0 \tag{7.106b}$$

$$-C_{FGI\,12}(0)\xi U(0) + C_{FGI\,22}(0)V_{,y}(0) = -p_{yj} \tag{7.106c}$$

$$-C_{FGI\,12}(h)\xi U(h) + C_{FGI\,22}(h)V_{,y}(h) = 0 \tag{7.106d}$$

Finally substituting for U and V from (7.102a) and (7.102b) produces four equations for a_i:

$$\sum_{i=1}^{4}(\alpha_i + \xi r_i)a_i = 0 \tag{7.107a}$$

$$\sum_{i=1}^{4}\exp(\alpha_i h)(\alpha_i + \xi r_i)a_i = 0 \tag{7.107b}$$

$$\sum_{i=1}^{4}[-C_{FGI\,12}(0)\xi + C_{FGI\,22}(0)r_i\alpha_i]a_i = -p_{yj} \tag{7.107c}$$

$$\sum_{i=1}^{4}\exp(\alpha_i h)[-C_{FGI\,12}(h)\xi + C_{FGI\,22}(h)r_i\alpha_i]a_i = 0 \tag{7.107d}$$

Then b_i would be calculated using (7.104). The solutions for the displacements and the stresses are found in series form similar to (7.102a) and (7.102b).

We will proceed to derive a comparable solution using elementary beam theory as introduced in Sect. 6.3.1 and then compare some results in order to understand the special characteristics of beams composed of FGM.

7.12.3   Simply Supported FGB: Elementary Beam Theory Analysis

In Sect. 6.3.1, some limited results from an analysis of a cantilever beam using elementary beam theory were presented in order to show the refinements obtained using elasticity theory. The complete beam theory analysis, including the computation of the displacements, was not continued there since it is readily available in introductory texts and was not particularly germane to our focus. However for the topic of FGB, the situation is somewhat different since the accompanying beam theory is not widely disseminated at elementary levels. Therefore the essentials of the theory will be presented in order to facilitate the comparison with results obtained using the elasticity solution described in the preceding section. We will use the geometry and the definition of the axes as introduced in Fig. 7.11.

With the assumptions that plane sections remain plane and no thickness changes and the choice of coordinates as shown in Fig. 7.11, the displacements are

$$u_y(x, y) = u_{yb}(x) \tag{7.108a}$$

$$u_{xb}(x, y) = u_{xb}(x, 0) - y u_{y\,b,x}, \tag{7.108b}$$

where the subscript b indicates beam theory displacements measured from the neutral axis and $u(x,0)$ denotes displacements on the bottom surface of the beam, $y = 0$.

With the normal stresses σ_{yy} neglected, the stress–strain relations take the form

$$\sigma_{xx}(x, y) = \bar{E}(y)\varepsilon_{xx}(x, y) \tag{7.109a}$$

$$\sigma_{xy}(x, y) = 2G(y)\varepsilon_{xy}(x, y), \tag{7.109b}$$

where

$$\bar{E}(y) = E(y)/(1 - v^2) \tag{7.109c}$$

as found from the plane stress relations (7.6a) and (7.6b) with $\sigma_{yy} = 0$. Since v is constant through the depth,

$$\bar{E}(y) = \bar{E}(0)e^{\zeta y} \tag{7.109d}$$

from (4.40). The axial strain and stress are found from (3.15a) and (7.109a) as

$$\varepsilon_{xx}(x,y) = u_{xb,x}(x,y) = u_{xb}(x,0)_{,x} - y u_{yb,xx} = \varepsilon_{xx}(x,0) + y\kappa \qquad (7.110a)$$

$$\sigma_{xx}(x,y) = \bar{E}(y)\varepsilon_{xx}(x,y) = \bar{E}(y)[\varepsilon_{xx}(x,0) + y\kappa], \qquad (7.110b)$$

where ε_{xx} $(x,0)$ is the strain on the reference plane and κ is the beam curvature.
The resultant axial force and bending moment resultants are

$$N(x) = \int_0^h \sigma_{xx}\, dy \qquad (7.111a)$$

and

$$M(x) = \int_0^h \sigma_{xx} y\, dy \qquad (7.111b)$$

whereupon substituting (7.110b) for σ_{xx} into (7.111a) and (7.111b) gives a relationship between the force and moment resultants and the reference strain and curvature:

$$\mathbf{F} = \mathbf{K}\boldsymbol{\kappa}, \qquad (7.112a)$$

where $\mathbf{F} = \{N\ M\}$, $\boldsymbol{\kappa} = \{\varepsilon_{xx}\ (x,0)\ \kappa\}$ and

$$\mathbf{K} = \begin{bmatrix} A & B \\ B & D \end{bmatrix} \qquad (7.112b)$$

In Ref. [9], elements of A, B, and D are defined as

$$\{A, B, D\} = \int_0^h \bar{E}\{1, y, y^2\}\, dy \qquad (7.113)$$

and, using (7.109d), they are evaluated as [9]

$$A = \frac{\bar{E}(h) - \bar{E}(0)}{\zeta}$$

$$B = \frac{h\bar{E}(h) - A}{\zeta}$$

$$D = \frac{h^2\bar{E}(h) - 2B}{\zeta}.$$

Equation (7.112a) is inverted to give

$$\boldsymbol{\kappa} = \mathbf{K}^{-1}\mathbf{F}, \qquad (7.114a)$$

where

$$\mathbf{K}^{-1} = \begin{bmatrix} A^* & B^* \\ B^* & D^* \end{bmatrix} \qquad (7.114b)$$

Since the axial force $N = 0$ in (7.112a), (7.114a) and (7.114b) relate the deformations to the bending moment resultant

$$\varepsilon_{xx}(x,0) = B^* M(x) \tag{7.115a}$$

and

$$\kappa = D^* M(x). \tag{7.115b}$$

Substituting (7.115a) and (7.115b) into (7.110b) gives the axial stress in the FGB in the basic form of (6.20b), $\sigma = My/I$, i.e., proportional to the moment M, the distance from the reference axis y and inversely proportional to the moment of inertial

$$\sigma_{xx}(x,y) = \bar{E}(y)M(x)D^*(B^*/D^* + y). \tag{7.116}$$

The initially unanticipated term $\bar{E}(y)$ is countered by the inverse of (7.112b) given as (7.114b) and is a reflection of the material parameter ζ. D^* is nominally proportional to $1/h^2$ and augmented by the $1/h$ in ζ introduced in the following (7.122) to produce the moment of inertial type term.

From (7.116), the neutral axis is established where $\sigma_{xx}(x,y) = 0$.

$$y = y_{NA} = -\frac{B^*}{D^*}. \tag{7.117}$$

The transverse shear stress σ_{xy} is found by integrating (7.2a) noting that $\sigma_{xy}(x,0) = 0$. At a distance y^* from the bottom of the beam,

$$\sigma_{xy}(x, y^*) = -\int_0^{y*} \sigma_{xx,x} dy. \tag{7.118}$$

With σ_{xx} given by (7.119) and the shear force $V_y(x) = M(x)_{,x}$

$$\sigma_{xy}(x, y^*) = -V_y \int_0^{y*} \bar{E}(y)(B^* + D^* y) dy \tag{7.119}$$

Finally, substituting (7.109d) for $\bar{E}(y)$, the transverse shear stress for an FGB is found in [9]:

$$\sigma_{xy}(x,y) = -V_y \left[(B^*/\zeta)[\bar{E}(y) - \bar{E}(0)] + (D^*/\zeta^2)(\zeta y - 1)\bar{E}(y) + \bar{E}(0) \right], \tag{7.120}$$

where the shear force V_y is related to the applied transverse loading by

$$p_y(x)| = -V_{y,x} \tag{7.121}$$

and the maximum shear stress occurs at the neutral axis y_{NA}, found from (7.117).

7.12.4 Comparative Results

There are two main objectives in the comparative study: (1) to determine the conditions for which elementary beam theory is adequate for an FGB and (2) to study the differences in the elasticity solution between a homogeneous beam and an FGB. It is noted that for a given value of the parameter ζ, which describes the variation of the elastic moduli through the depth, the elasticity solution depends only on the non-dimensional parameter $\xi h = j\pi h/L$ from (7.98c). Small values of $\xi h/L$ represent long or shallow beams with more uniform loading, as indicated by small values of j. On the other hand, larger values of $\xi h/L$ correspond to short or deep beams and possibly more rapidly varying loading with larger values of j.

In the numerical solutions presented in Ref. [9], $E(0)$ is taken as 1 GPa and ν as 0.25. Also, the parameter ζ is set implicitly by the ratio of the elastic moduli, $E(h)/E(0)$, i.e.,

$$\zeta = \frac{1}{h}[\ln E(h) - \ln E(0)]. \tag{7.122}$$

By plotting normalized values of the stresses and displacements, the dependence on the axial coordinate x is removed because of the harmonic form of the elasticity solution: (7.99a) and (7.99b).

First we study the axial displacements $u_x(x,y)$. On Fig. 7.12, the displacements, normalized by the top unloaded surface value, i.e., at $y = 0$, are shown for varying values of ξh through a highly graded beam. $\xi h \leq 1$ corresponds to elementary beam theory behavior where plane sections remain plane and the variation of the axial

Fig. 7.12 Normalized axial displacements U_x through the thickness of the FGB for various values of ξh ($E_h = 10E_0$)

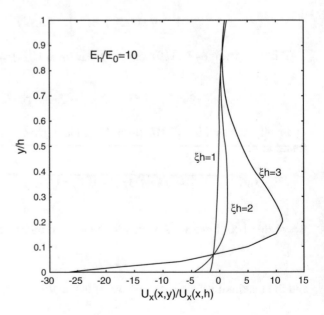

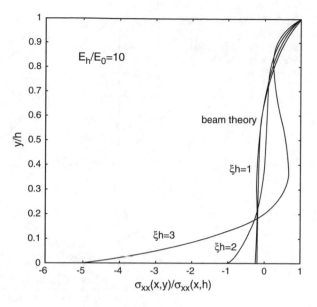

Fig. 7.13 Normalized axial stresses σ_{xx} through the thickness of the FGB for various values of ξh ($E_h = 10E_0$)

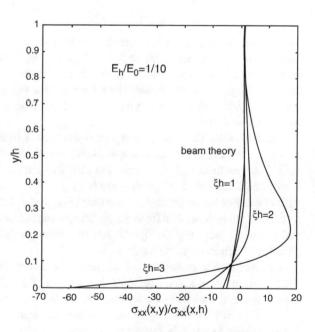

Fig. 7.14 Normalized axial stresses σ_{xx} through the thickness of the FGB for various values of ξh ($E_h = 0.1E_0$)

displacements through the depth is confirmed as linear, while larger values of ξh reveal warping of the cross section toward the loaded bottom surface.

The normal stresses σ_{xx} (x,y) are shown on Figs. 7.13 and 7.14 normalized by the top unloaded surface value σ_{xx} (x,h). The two cases chosen are for the softer loaded face $E(h)/E(0) = 10$ and the stiffer loaded face $E(h)/E(0) = 1/10$, respectively. Again $\xi h \leq 1$ corresponds to elementary beam behavior, but while the axial displacements

Fig. 7.15 Transverse shear
stresses through the
thickness of the FGB for
various values of ξh
($E_h = 10E_0$)

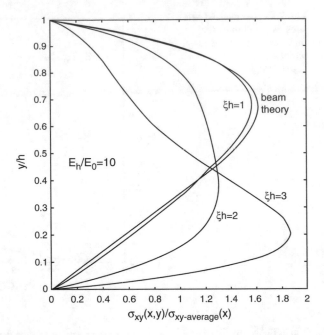

were linear for this case as in elementary beam theory, the stresses become nonlinear
near the stiffer faces reflecting the gradation. For the higher values of ξh, the stress
concentration is higher for the case where the loaded face has the larger Young's
modulus (Fig. 7.14), while the softer loaded face represented by Fig. 7.13 has a
stress-mitigating effect. Of course, the material would likely be stronger for the
stiffer loaded face case.

The transverse shear stresses σ_{xy} (x,y) are shown in Figs. 7.15 and 7.16 for the
softer and stiffer loaded faces, respectively, normalized by the average shear stress
V_y/h. The results for $\xi h \le 1$ show the classical parabolic variation close to beam
theory amplification of the average stress by 1.5 but somewhat displaced toward the
stiffer face. For the softer loaded face case (Fig. 7.15), the stresses hardly exceed the
maximum value computed from beam theory even for larger ξh values, while for the
stiffer loaded face (Fig. 7.16), the maximum beam theory value is exceeded for
$\xi h > 1$ and almost doubled for $\xi h = 3$.

Transverse displacements $u_y(x,y)$ normalized by the maximum beam theory
deflections are plotted over one-half wavelength of loading on Figs. 7.17 and 7.18.
To clarify the abscissa values, consider the loading case shown in Fig. 7.11 for which
$j = 1$. Then from (7.98c), $\xi = \pi/L$ and the abscissa $2\xi x/\pi = 2x/L$, so a value of
1 corresponds to $x = L/2$, the center of the beam.

For the $\xi h = 2$ case in Fig. 7.18, we find from (7.98a) that the maximum load
intensity along the beam occurs when $\xi x = \pi/2$ and $\sin \xi x = 1$. The maximum
deflection likewise occurs at this point where $2\xi x/\pi = 1$, as shown in the figure.
Again for $j = 1$ as shown in Fig. 7.11, this is the center of the beam. Recall from
Fig. 7.11 that the loading is applied to the bottom face $y = 0$ acting upward in the $(+)$

Fig. 7.16 Transverse shear
stresses through the
thickness of the FGB for
various values of ξh
$(E_h = 0.1E_o)$

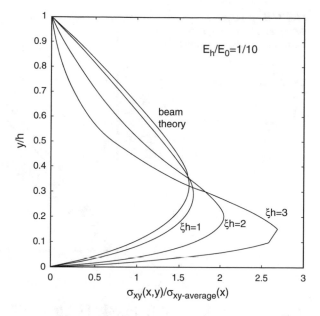

Fig. 7.17 Transverse
displacements $U_y(x,z)$ in the
FGB for $\xi h = 1$ $(E_h = 10E_o)$

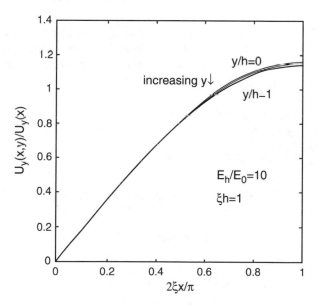

y direction. The curves progress *downward* with *increasing y* and represent the total
deflections from the softer bottom level to the stiffer top level of the beam. These
values are the sum of the through-thickness compression and the beam theory
deflection. This compression was found to be minimal for $\xi h = 1$ as shown in
Fig. 7.17, but the maximum deflections were about 15% greater than computed by
beam theory [9].

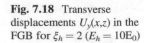

Fig. 7.18 Transverse
displacements $U_y(x,z)$ in the
FGB for $\xi_h = 2$ ($E_h = 10E_0$)

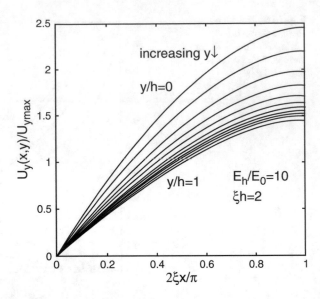

7.13   Computational Example Using MATLAB

In Sect. 7.4, the transformation of the 2-D formulation to cylindrical coordinates
required a lot of tedious operations. However, using MuPAD, the process becomes
very convenient. First, we set up the direction cosines, and then we derive (7.25).

- Define symbols

```
th := Symbol::theta
  θ
sig := Symbol::sigma
  σ
sigxx := Symbol::subScript(sig,xx)
  σxx
sigyy := Symbol::subScript(sig,yy)
  σyy
sigxy := Symbol::subScript(sig,xy)
  σxy
sigrr := Symbol::subScript(sig,rr)
  σrr
sigtt := Symbol::subScript(sig,th)
  σθ
sigrt := Symbol::subScript(sig,rth)
  σrth
```

- Stress tensor

```
stress := matrix([[sigxx, sigxy, 0],[sigxy, sigyy, 0],[0, 0, 1]])
```
$$\begin{bmatrix} \begin{pmatrix} \sigma_{xx} & \sigma_{xy} & 0 \\ \sigma_{xy} & \sigma_{yy} & 0 \\ 0 & 0 & 1 \end{pmatrix} \end{bmatrix}$$

- Direction cosine matrix

```
rotMat := matrix([[cos(th), sin(th), 0],[-sin(th), cos(th), 0],[0, 0, 1]])
```
$$\begin{bmatrix} \begin{pmatrix} \cos(\theta) & \sin(\theta) & 0 \\ -\sin(\theta) & \cos(\theta) & 0 \\ 0 & 0 & 1 \end{pmatrix} \end{bmatrix}$$

```
sig_cylindrical := Simplify(rotMat * stress * transpose(rotMat))
```
$$\begin{pmatrix} \sigma_{xx}\cos(\theta)^2 + 2\,\sigma_{xy}\cos(\theta)\sin(\theta) + \sigma_{yy}\sin(\theta)^2 & \sigma_1 & 0 \\ \sigma_1 & \sigma_{yy}\cos(\theta)^2 - 2\,\sigma_{xy}\cos(\theta)\sin(\theta) + \sigma_{xx}\sin(\theta)^2 & 0 \\ 0 & 0 & 1 \end{pmatrix}$$

where

$$\sigma_1 = \sigma_{xy}\cos(2\,\theta) - \frac{\sigma_{xx}\sin(2\,\theta)}{2} + \frac{\sigma_{yy}\sin(2\,\theta)}{2}$$

7.14 Exercises

7.1 Using the element shown in the figure below, solve the following problems.

  (a) Derive the equilibrium equations in polar coordinates, neglecting body forces.

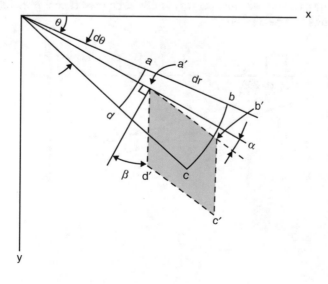

Answer:

$$\sigma_{rr,r} + \frac{1}{r}\sigma_{r\theta,\theta} + \frac{\sigma_{rr} - \sigma_{\theta\theta}}{r} = 0$$
$$\frac{1}{r}\sigma_{\theta\theta,\theta} + \sigma_{r\theta,r} + \frac{2}{r}\sigma_{r\theta} = 0 \qquad (7.123)$$

(b) Verify that the above equations are satisfied by (7.28a, b, c).

7.2 Consider a thick-walled cylinder with an outside diameter of 5. The internal pressure is 0.5 (force/length^2).

(a) For an inside diameter of 4, compute the maximum normal and circumferential stresses.
(b) How much error would be involved if part (a) were carried out using thin-walled cylinder theory.
(c) Determine the radial displacement of a point on the inside surface. Assume that $\nu = 0.3$ and $E = 30 \times 10^6$.

7.3 A cylinder with inside radius a and outside radius $b = 1.20a$ is subjected to:

1. Internal pressure p_i only
2. External pressure p_0 only
   Determine the ratio of maximum to minimum circumferential stress for each case (from [10]).

7.4 Determine the value of the constant C in the stress function

$$\phi = C\left[r^2(\alpha - \theta) + r^2 \sin\theta \cos\theta - r^2 \cos^2\theta \tan\alpha\right] \qquad (7.124)$$

required to satisfy the conditions on the upper and lower edges of the triangular plate shown in the figure. Evaluate the stress components σ_{xx}, σ_{yy}, and σ_{xy} at M for the case $\alpha = 30°$. In what way is this solution different from that found by the elementary beam theory?

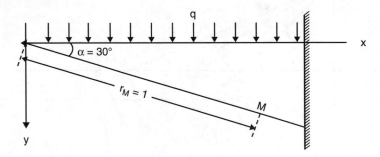

7.5 Determine the stress concentration factor for the plate as shown in the figure.

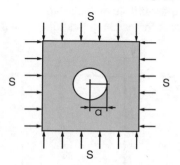

7.6 Consider a curved beam as shown in the figure below.

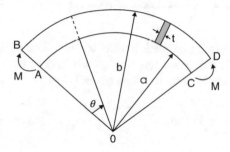

The values of the dimensions and moments are

$$a - 10 \ b - 13; \ M - 11,000 \text{ (length-force)}, \ t - 0.8 \qquad (7.125)$$

Calculate the maximum stress arising from M, and compare this with $\sigma_{\theta\theta} = M$ $(b-a)/2I$, computed from elementary beam theory.

7.7 For problem 5.2, show that the corresponding Airy stress function may be written as

$$\phi = -\frac{1}{2}\frac{\nu}{1+\nu}f(x,y)z^2 + A + Bx + Cy + \phi_1(x,y)z + \phi_0(x,y), \qquad (7.126)$$

where ϕ, A, B, and C may be functions of z and ϕ_0 is a specialized Airy stress function for problems symmetrical about the plane $z = 0$ (from [4]).

7.8 A stress distribution is given by

$$\sigma_{xx} = C_1yx^3 - 2C_2xy + C_3y$$
$$\sigma_{yy} = C_1xy^3 - 2C_1x^3y$$
$$\sigma_{xy} = -\frac{3}{2}C_1x^2y^2 + C_2y^2 + \frac{1}{2}C_1x^4 + C_4, \qquad (7.127)$$

where C_1–C_4 are constants (from [10]).

(a) Show that this field represents a solution for a thin plate as shown below.
(b) Derive the corresponding stress function.
(c) Evaluate the surface forces along the edges $y = 0$ and $y = b$ of the plate.

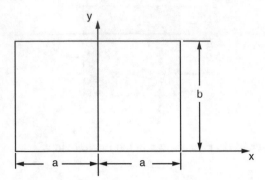

7.9 Consider the semi-infinite plate treated in Sect. 7.10 using Cartesian coordinates. Show that the stress function $\phi = -(P/\pi)y \tan^{-1}(y/x)$ results in the following stress field within the plate (from [10]):

$$\sigma_{xx} = -\frac{2P}{\pi}\frac{x^3}{(x^2+y^2)^2}; \quad \sigma_{yy} = -\frac{2P}{\pi}\frac{xy^2}{(x^2+y^2)^2}; \quad \sigma_{xy} = -\frac{2P}{\pi}\frac{yx^2}{(x^2+y^2)^2}.$$

(7.128)

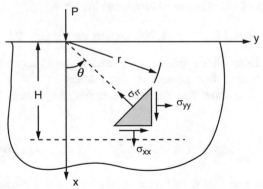

Also, plot the resulting stress distribution for σ_{xx} and σ_{xy} at a constant depth H below the boundary (from [10]).

7.10 The thin cantilever beam shown is subjected to a uniform shearing stress τ_0 along the upper surface, $y = +D$, while the surfaces $y = -D$ and $x = L$ are stress-free. Determine if the Airy stress function

$$\phi = \frac{1}{4}\tau_0\left(xy - \frac{xy^2}{D} - \frac{xy^3}{D^2} + \frac{Ly^2}{D} + \frac{Ly^3}{D^2}\right)$$

(7.129)

satisfies the required conditions for this problem (from [10]).

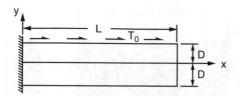

7.11 Show that for the case of plane stress with no body forces present:

(a) The equations of equilibrium may be expressed in terms of displacements
u_x and u_y as follows (from [10]):

$$\nabla^2 u_x + \left(\frac{1+\nu}{1-\nu}\right)(u_{x,x} + u_{y,y})_{,x} = 0$$
$$\nabla^2 u_y + \left(\frac{1+\nu}{1-\nu}\right)(u_{y,y} + u_{x,x})_{,y} = 0$$
(7.130)

(b) Verify that the indicial form of these equations, following steps analogous
to those in Sect. 5.2, is

$$\nabla^2 u_i + \left(\frac{1+\nu}{1-\nu}\right)u_{j,ji} = 0 \quad i,j = 1,2$$
(7.131)

7.12 For the stress function in (7.93),

$$\phi = -(P/\pi)r\theta \sin\theta,$$
(7.132)

verify that equilibrium and compatibility are satisfied.

7.13 Refer to the solution in Sect. 7.10.

(a) Compute the Cartesian components of stress as a function of r and θ.
(b) Show that the integral of σ_{rr} along any semicircle about the origin is equal
to P (from [10]).

7.14 A diaphragm or plate loaded in its plane is called a "deep beam" when it spans
between supports and has a depth/span ratio outside the limitations of simple
beam theory. For the simply supported case shown [11], based on the assumption that the column reactions are uniformly distributed over the width c,
formulate the problem within the theory of elasticity:

(a) State the governing equation.
(b) State the boundary conditions on all edges for a uniform load applied along
the top edge.
(c) Suppose the load is applied along the lower edge instead. How will the
boundary conditions be altered? What limitation of elementary beam
theory does this illustrate?

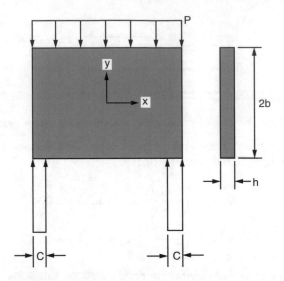

7.15 Derive (7.28a, b, c).

7.16 A pipe is composed of two concentric cylinders. The outer radius is $3a$, the radius along the interface is $2a$, and the inner radius is a. For the outer cylinder, $E = E_0$, while for the inner cylinder, $E = E_i = \frac{1}{2}E_0$:

(a) For an internal pressure p, compute the circumferential and radial stress distributions in the pipes.

(b) Evaluate the shear stress along the interface.

(c) Compute the radial displacement at both surfaces.

7.17 Consider the material law stated in (4.43):

(a) Rewrite these equations in terms of the engineering constants E and ν.

(b) Specialize and expand the equations for the cylindrical coordinates r, θ, and z introduced in Sect. 7.4.

(c) Consider the concentric cylinders described in Problem 7.16. When the outer cylinder is much stiffer than the inner, $E_o \ll E_i$, the inner cylinder may be considered to be in a state of *generalized plane stress*, with σ_{rr} taken as equal to $-p$, or an assumed $\bar{\sigma}_{rr}(p, z)$, throughout. Show that the remaining stress–strain equations are

$$\sigma_{\theta\theta} = \frac{E_i}{1 - \nu^2}[\varepsilon_{\theta\theta} + \mu\varepsilon_{zz} - (1 + \nu)\alpha\Delta T] + \frac{\nu}{1 - \nu}\bar{\sigma}_{rr}$$

$$\sigma_{ZZ} = \frac{E_i}{1 - \nu^2}[\mu\varepsilon_{\theta\theta} + \varepsilon_{zz} - (1 + \nu)\alpha\Delta T] + \frac{\nu}{1 - \nu}\bar{\sigma}_{rr}.$$

(7.133)

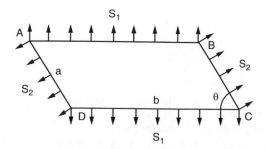

7.18 A skewed plate of unit thickness is loaded by uniformly distributed stresses, S_1 and S_2, applied perpendicular to the sides of the plate as shown in the figure [12].

(a) Solve the equilibrium equations for the plate in terms of S_1, S_2, a, b, and θ.
(b) Determine the elongation of the diagonal AC under the action of the stresses, S_1 and S_2, assuming that the material is linearly elastic and isotropic and that $\theta = 90°$.

7.19 Now, consider the stresses, S_1 and S_2, to be applied so that they are directed parallel to the edges AB and AD, respectively, of the skewed plate. Derive expressions for the principal stresses and the principal strains in terms of S_1, S_2, a, b, θ, E, and ν, where E and ν denote Young's modulus and Poisson's ratio, respectively [12].

7.20 Consider the stress–strain relationships in terms of the engineering constants (7.7a, b, c) and (4.13b) simplified for the plane stress case in the x–y plane.

Show that (7.7a, b, c) may be generalized for a homogeneous orthotropic material defined by

$$E_x\nu_{yx} = E_y\nu_{xy} \tag{7.134}$$

to the form [13]

$$\sigma_{xx} = \frac{E_x}{(1 - \nu_{xy}\nu_{yx})}(\varepsilon_x + \nu_{yx}\varepsilon_y)$$

$$\sigma_{yy} = \frac{E_y}{(1 - \nu_{xy}\nu_{yx})}(\nu_{xy}\varepsilon_x + \varepsilon_y) \tag{7.135}$$

$$\sigma_{xy} = 2G\varepsilon_{xy}$$

7.21 Consider the axisymmetrical thermal elasticity problem in two dimensions. Prove that the compatibility equation in polar coordinates is given by [10]

$$\nabla^2\left[\frac{1}{r}(r\varphi_{,r})_{,r}\right] + E\alpha\Delta T = 0, \tag{7.136}$$

where

$$\Delta T = T - T_0 \text{ as defined by } (4.41)$$ (7.137)

and

$$\phi = \text{Airy stress function.}$$ (7.138)

Hint: Generalize (7.8) based on (4.42).

7.22 For the case of plane stress and no body forces, show that

$$\phi = (Ae^y + Be^{-y} + Cye^y + Dye^{-y})\sin(x)$$ (7.139)

is a valid stress function.

7.23 For the large shear panel shown below, determine the stresses around the edges of the small hole.

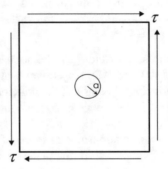

7.24 Consider the functionally graded beam discussed in Sect. 7.12.

(a) Noting that the normalized axial displacements and axial stresses correlate well between elasticity theory and beam theory for lower values of ξh, discuss why the results differ for higher values of this parameter.

(b) Discuss the variation of the transverse shear stresses between elasticity theory and beam theory, particularly the variations through the depth of the cross section

References

1. Love AEH (1944) A treatise on the mathematical theory of elasticity, 4th edn. Dover, New York
2. Timoshenko S, Goodier JN (1951) Theory of elasticity, 2nd edn. McGraw-Hill Book Company, Inc., New York
3. Popov EP (1978) Mechanics of materials, 2nd edn. Prentice-Hall, Englewood Cliffs
4. Little RW (1973) Elasticity. Prentice-Hall, Englewood Cliffs
5. Filonenko-Borodich M (1965) Theory of elasticity. Dover Publications, New York
6. Boresi AB, Sidebottom OM, Seeley FB, Smith J (1978) Advanced mechanics of materials, 3rd edn. John Wiley & Sons, New York

7. Trahair NS, Ansourian P (2016) In-plane behaviour of web-tapered beams. Eng Struct 108:47–52
8. Trahair NS, Ansourian P (2016) In-plane behaviour of mono-symmetric tapered beams. Eng Struct 108:53–58
9. Sankar BV (2001) An elasticity solution for functionally graded beams. Comp Sci Technol 61:689–696
10. Ugural AC, Fenster SK (1975) Advanced strength and applied elasticity. Elsevier-North Holland, New York
11. Barry JE, Ainso H (1983) Single-span deep beams. J Struct Eng ASCE 109(ST3):646–663
12. Boresi AP, Chong KP (1987) Elasticity in engineering mechanics. Elsevier, New York
13. Little GH (1989) Large deflections of orthotropic plates under pressure. J Eng Mech ASCE 115 (12):2601–2620

Chapter 8
Thin Plates and Shells

Abstract When a thin plate is loaded *transverse* to the plane, the plate bends and the deflection of the surface in the normal direction is predominant, similar to a gridwork of beams. Correspondingly a distributed transverse loading can be efficiently resisted by an initially curved thin surface structure such as a cylindrical or spherical tank. The initial curvature mobilizes in-plane or membrane resistance that can be far more efficient than the bending resistance provided by flat plates. Several classical problems for the bending of rectangular and circular plates and for the membrane behavior of thin shells of revolution are presented as an introduction to this extensive topic.

8.1 Introduction

In the preceding chapter, several cases of thin elastic plates loaded *in* the plane of the plate were analyzed using plane stress or plane strain equations. The resulting deformations are confined to the plane of the plate in accordance with the plane strain assumption. Now we consider a rectangular plate supported along all or part of the perimeter as shown in Fig. 8.1.

If such a plate is loaded *transverse* to the plane by a distributed load $q(x, y)$ applied to the upper surface, the plate bends and the deflection of the surface in the normal direction is predominant. The problem of plate bending is in part an extension of the elementary theory of beams. However, plate bending is more complex than beam bending or even the compatible bending of a network of orthogonal beams connected at their intersections. In effect, a solid plate is much stiffer than such a gridwork. This will be elaborated on at the end of Sect. 8.7.

Correspondingly, distributed transverse loading can be efficiently resisted by a thin initially curved surface structure in the manner of a cylindrical or spherical tank. The initial curvature mobilizes in-plane resistance that can be more efficient than the bending resistance provided by flat plates. In effect, a thin shell becomes a three-dimensional extension of an arch in a similar manner as the plate extends a beam.

Plates and shells are classified in terms of their thinness by a characteristic ratio h/l, where h is the thickness and l is a plan dimension such as the length of a side or a radius

© Springer International Publishing AG, part of Springer Nature 2018

P. L. Gould, Y. Feng, *Introduction to Linear Elasticity*,

https://doi.org/10.1007/978-3-319-73885-7_8

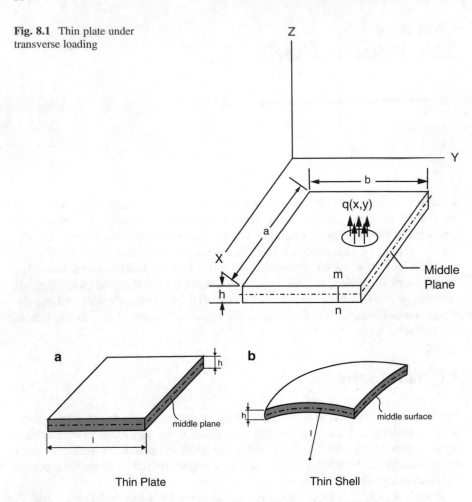

Fig. 8.1 Thin plate under transverse loading

Fig. 8.2 Thin plate and thin shell. (**a**) Thin plate. (**b**) Thin shell. (From Gould [3])

as shown in Fig. 8.2a. This classification is meaningful primarily for loading transverse to the surface. If the plate is very thin, $h/l < 0.001$, in-plane or membrane forces along with bending and twisting moments are needed to provide equilibrium mobilizing finite deformations. On the other hand, if the plate is thick, $h/l > 0.4$, three-dimensional effects are relevant. Between these bounds lie medium-thin plates which can be analyzed by a linear theory in reference to a middle plane, a two-dimensional analog to the familiar neutral axis of a beam. Similarly, a shell is characterized with respect to a defining radius of curvature as shown in Fig. 8.2b.

8.2 Assumptions

Several assumptions are introduced in order to relax some of the rigor of the theory of elasticity so as to facilitate solutions for thin plates and shells. Foremost these assumptions permit the analysis to be referred to a reference plane or surface as indicated on Fig. 8.2a, b. This is again analogous to the theory of beams that is generally formulated with respect to a reference or neutral axis. Here three assumptions are stated for plates, and the first two are relevant for shells as well. Further assumptions will be invoked for shells later:

1. Based on the relative thinness of the plate and the orientation of the loading, the plate is assumed to be approximately in both plane strain and plane stress, i.e.,

$$\varepsilon_{zz} = 0 \tag{8.1a}$$

and

$$\sigma_{zz} = 0 \tag{8.1b}$$

except directly under a surface load. The corresponding strain–stress Eqs. (7.12a, b, d) become

$$\varepsilon_{xx} = \frac{1}{E}\left(\sigma_{xx} - \nu\sigma_{yy}\right) \tag{8.2a}$$

$$\nu\varepsilon_{yy} = \frac{1}{E}\left(\sigma_{yy} - \nu\sigma_{xx}\right) \tag{8.2b}$$

$$\varepsilon_{xy} = \frac{1}{2G}\sigma_{xy}. \tag{8.2c}$$

Also, the plate thickness does not change so that the normal displacement is constant along a normal through the thickness, i.e.,

$$u_z = u_z(x, y). \tag{8.3}$$

2. The second assumption is stated with respect to the line *mn*, shown in Fig. 8.1 and again in the section normal to the *x*-axis, Fig. 8.3. Line *mn* represents all lines initially perpendicular to the undeformed middle plane and is presumed to remain *straight* and *normal* to the deformed middle plane. This is the *hypothesis of linear elements*, attributed to G. Kirchhoff by Filonenko-Borodich [1]. Referring to Fig. 8.3, the undeflected line *mn* is superimposed on the deflected line *m'n'* in the upper cross section. The rotation of the line is ω_{yz}. By the application of the hypothesis, this rotation is equal to the rotation of a tangent to the middle plane:

$$\omega_{yz} = u_{z,y}. \tag{8.4}$$

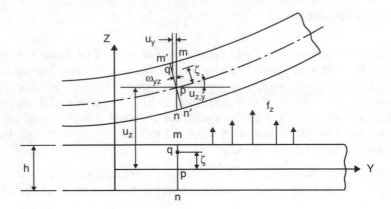

Fig. 8.3 Deformation of thin plate

3. The last essential assumption states that the middle plane deflects only in the z-direction. Thus, on the middle plane where the normal coordinate $\zeta = 0$,

$$u_x(x, y, 0) = 0 \tag{8.5a}$$

$$u_y(x, y, 0) = 0 \tag{8.5b}$$

$$u_z(x, y, 0) = f(x, y). \tag{8.5c}$$

Therefore, there can be no extensional deformations or shears on the middle plane and

$$\sigma_{xx}(x, y, 0) = 0$$
$$\sigma_{yy}(x, y, 0) = 0 \tag{8.6}$$
$$\sigma_{xy}(x, y, 0) = 0.$$

The preceding assumptions enable a classical displacement formulation for transversely loaded medium-thin plates. All stresses, strains, and in-plane displacements will be described in terms of the normal displacement $u_z(x, y)$.

8.3   Formulation of Plate Theory

8.3.1   Geometric Relationships

We focus on two points which lie on mn, the normal to the middle plane, in Fig. 8.3. Point p is also on the middle plane, while point q is initially located a distance ζ away from the plane in the positive z-direction.

On the displaced surface, point q remains ζ from point p by Assumption (1) and has the same normal displacement $u_z(x, y)$ from Assumption (3). So the in-plane displacement is given by

$$u_y = -\zeta\omega_{yz} = -\zeta u_{z,y} \tag{8.7}$$

as per Assumption (2), (8.4). The negative sign indicates that a *positive* $\zeta u_{z,y}$ produces a u_y in the *negative* y-direction. A similar section perpendicular to the y-axis would reveal the x–z plane and yield

$$u_x = -\zeta\omega_{xz} = -\zeta u_{z,x}. \tag{8.8}$$

Thus, the in-plane displacements u_x and u_y are functions of the normal displacement u_z.

8.3.2 Strains and Stresses

Using (8.7) and (8.8), we may write the in-plane strain–displacement relationships from (3.15a, b, d) with 1, 2, 3 = x, y, z:

$$\varepsilon_{xx} = u_{x,x} = -\zeta u_{z,xx} \tag{8.9a}$$

$$\varepsilon_{yy} = u_{y,y} = -\zeta u_{z,yy} \tag{8.9b}$$

$$\varepsilon_{xy} = \frac{1}{2}\left(u_{x,y} + u_{y,x}\right) = -\zeta u_{z,xy}. \tag{8.9c}$$

The corresponding stresses are formed by inverting (8.2) and substituting (8.9a, b, c).

$$\sigma_{xx} = \frac{E}{(1-\nu^2)}(\varepsilon_{xx} + \nu\varepsilon_{yy})$$
$$= -\frac{\zeta E}{(1-\nu^2)}\left(u_{z,xx} + \nu u_{z,yy}\right) \tag{8.10a}$$

$$\sigma_{yy} = \frac{E}{(1-\nu^2)}(\varepsilon_{yy} + \nu\varepsilon_{xx})$$
$$= -\frac{\zeta E}{(1-\nu^2)}\left(u_{z,yy} + \nu u_{z,xx}\right) \tag{8.10b}$$

$$\sigma_{xy} = 2G\varepsilon_{xy}$$

$$= \frac{E}{(1+\nu)}\varepsilon_{xy}$$

$$= -\frac{\zeta E}{(1+\nu)}u_{z,xy} \tag{8.10c}$$

$$= -\frac{\zeta E}{(1-\nu^2)}(1-\nu)u_{z,xy}.$$

It should be mentioned that efficient applications for the analysis of plates and shells are facilitated by the introduction of *stress resultants and couples* which are forces and moments per width of middle surface. However, to retain a strictly pointwise description consistent with elasticity theory, stress resultants are not used here for plate analysis but will be introduced in a limited fashion for shells later in the chapter.

We now calculate the remaining components of stress from the equilibrium Eq. (2.46), again with 1, 2, 3 = x, y, z, and from (8.10a, b, c). From (2.46a) and $f_x = 0$,

$$\sigma_{xz,z} = -\sigma_{xx,x} - \sigma_{xy,y}$$

$$= \frac{\zeta E}{(1-\nu^2)}[u_{z,xxx} + \nu u_{z,yyx} + (1-\nu)u_{z,xyy}]$$

$$= \frac{\zeta E}{(1-\nu^2)}(u_{z,xx} + u_{z,yy})_{,x} \tag{8.11}$$

$$= \frac{\zeta E}{(1-\nu^2)}(\nabla^2 u_z)_{,x}.$$

Similarly, from (2.46b) and $f_y = 0$,

$$\sigma_{yz,z} = \frac{\zeta E}{(1-\nu^2)}(\nabla^2 u_z)_{,y}. \tag{8.12}$$

The preceding equations may be integrated with respect to z. Only ζ is a function of z, so that

$$\int \zeta dz = \int \zeta d\zeta = \frac{\zeta^2}{2} + C. \tag{8.13}$$

Since σ_{xz} and σ_{yz} should vanish at the surfaces $\zeta = \pm\frac{h}{2}$,

$$C = -\frac{h^2}{8}. \tag{8.14}$$

The final expressions for the transverse shearing stresses are found from the integrals of (8.11) and (8.12), considering (8.13) and (8.14), as

$$\sigma_{xz} = \frac{E\left(\zeta^2 - \frac{h^2}{4}\right)}{2(1 - \nu^2)} (\nabla^2 u_z)_{,x} \tag{8.15a}$$

$$\sigma_{yz} = \frac{E\left(\zeta^2 - \frac{h^2}{4}\right)}{2(1 - \nu^2)} (\nabla^2 u_z)_{,y}. \tag{8.15b}$$

Now we consider the normal stress σ_{zz} given by (2.46c), with $f_z = 0$ since there are no body forces:

$$\sigma_{zz,z} = -\sigma_{xz,x} - \sigma_{yz,y}$$

$$= -\frac{E\left(\zeta^2 - \frac{h^2}{4}\right)}{2(1 - \nu^2)} (\nabla^4 u_z). \tag{8.16}$$

The stress σ_{zz} is of little practical consequence since, at most, it is equal to the magnitude of the surface loading. However, (8.16) is useful in establishing the governing equation for the normal displacement.

8.3.3 Plate Equation

We first consider the z-dependent terms of (8.16) $\left(\zeta^2 - \frac{h^2}{4}\right)$ integrated through the thickness.

$$\int_{-h/2}^{h/2} \left(\zeta^2 - \frac{h^2}{4}\right) dz = \int_{-h/2}^{h/2} \left(\zeta^2 - \frac{h^2}{4}\right) d\zeta$$

$$= -\frac{h^3}{6} \tag{8.17}$$

Substituting (8.17) into (8.16) gives

$$\sigma_{zz}\left(\frac{h}{2}\right) - \sigma_{zz}\left(-\frac{h}{2}\right) = \frac{Eh^3}{12(1 - \nu^2)} \nabla^4 u_z. \tag{8.18}$$

Since the load $q(x, y)$ is assumed to be applied to the top surface in the positive z-direction,

$$\sigma_{zz}\left(\frac{h}{2}\right) = q(x, y) \tag{8.19a}$$

and

$$\sigma_{zz}\left(-\frac{h}{2}\right) = 0 \qquad (8.19b)$$

while from (8.18)

$$\nabla^4 u_z = q(x, y)/D, \qquad (8.20a)$$

where

$$D = \frac{Eh^3}{12(1 - \nu^2)} \qquad (8.20b)$$

is termed the *flexural rigidity* and is roughly equivalent to E times the moment of inertia of a unit width of plate, in consort with the theory of flexure where EI is termed the bending or flexural rigidity. Equation (8.20a) is widely known as *the* plate equation and is of the same biharmonic form as the compatibility equations encountered in the two-dimensional elasticity formulation in Chap. 7.

Although the parallel between the plate equation and the bending of a beam of unit width is evident, (8.20a) represents more than one- or even two-directional flexure. Writing the equation in expanded form gives

$$u_{z,xxxx} + 2u_{z,xxyy} + u_{z,yyyy} = q(x, y)/D. \qquad (8.21)$$

While the first and third terms on the l.h.s indeed represent flexure in the respective directions, each presumably resisting a portion of the load $q(x, y)$, the second term involves the mixed partial derivative and describes relative twisting of parallel fibers in the plate. It is this torsional resistance that differentiates a plate from a system of intersecting beams and contributes significantly to the overall rigidity of a solid plate.

The goal of the formulation, to express all displacements, strains, and stresses in terms of a single displacement component u_z, has been achieved. Once (8.21) is solved, the remaining components of displacement are found from (8.7) and (8.8); the in-plane strains are given by (8.9a, b, c); and the stresses follow from (8.10a, b, c), (8.15a), and (8.15b).

8.3.4   Polar Coordinates

For a circular plate, it is expedient to use polar coordinates, Fig. 7.1. In as much as the "plate" equation, (8.20a), and the compatibility equation from two-dimensional elasticity, (7.18), are essentially the same biharmonic equation, we may directly use the information given in Chap. 7. We simply replace ϕ with u_z in (7.30) for the general case $u_z(r, \theta)$ to get

$$\nabla^4 u_z = q(r,\theta)/D \qquad (8.22)$$

and in (7.31e) for the axisymmetric case $u_z(r)$ to get

$$\nabla^4 u_z = q(r)/D, \qquad (8.23)$$

where the $\nabla^4(\ )$ operator has been transformed to polar coordinates. Of course, appropriate particular solutions will be needed for the plate problem.

8.4   Solution for Plates

8.4.1   Rectangular Plate

Since the theory of thin plates has an extensive specialized literature residing for the most part outside of elasticity, we shall not pursue the analysis in depth here. However, much insight into the process of obtaining analytical solutions may be gained by considering a rectangular plate such as that shown in Fig. 8.1, with plan dimensions a along the x-axis and b along the y-axis and subjected to an arbitrary distributed loading $q(x, y)$ perpendicular to the plane of the plate in the $+z$-direction. The origin for x and y is set at one corner of the plate.

If the plate is simply supported, which means that the normal displacement u_z and the bending moments M_x and M_y are equal to zero on the boundaries, harmonic functions in the form of a double-sine series

$$u_z = \sum_{j=1}^{\infty} \sum_{k=1}^{\infty} u_z^{jk} \sin j\pi\frac{x}{a} \sin k\pi\frac{y}{b} \qquad (8.24)$$

identically satisfy the displacement boundary conditions

$$u_z(0,y) = u_z(a,y) = u_z(x,0) = u_z(x,b) = 0. \qquad (8.25)$$

Now considering the bending moments on the boundary, they are computed from the integrals of the respective extensional stresses $\sigma_{(ii)}$ through the thickness, just as in the case of a beam; see (7.59b). However, for a plate, each stress is proportional to the curvatures in *both* coordinate directions, (8.10a) and (8.10b). This condition is therefore written as

$$M_x \sim (u_{z,xx} + \nu u_{z,yy}) = 0 \qquad (8.26a)$$

at $(0, y)$ and (a, y) and

$$M_y \sim (u_{z,yy} + \nu u_{z,xx}) = 0 \qquad (8.26b)$$

at $(x, 0)$ and (x, b). The double-sine function in (8.24) satisfies these conditions as well.

When the loading is expanded in a similar double series as used for u_z in (8.24),

$$q(x,y) = \sum_{j=1}^{\infty} \sum_{k=1}^{\infty} q^{jk} \sin j\pi \frac{x}{a} \sin k\pi \frac{y}{b}, \qquad (8.27)$$

standard Fourier analysis may be used to obtain the solution for a general harmonic jk [2],

$$u_z^{jk} = \frac{q^{jk}}{D} \left\{ \frac{1}{\pi^4 \left[\left(\frac{j}{a}\right)^2 + \left(\frac{k}{b}\right)^2 \right]^2} \right\}, \qquad (8.28a)$$

where

$$q^{jk} = \frac{4}{ab} \int_0^b \int_0^a q(x,y) \sin j\pi \frac{x}{a} \sin k\pi \frac{y}{b} \, dxdy. \qquad (8.28b)$$

as illustrated in the following example.

Then the complete solution for the displacement at any point (x, y) is obtained by superposition from (8.24) as

$$u_z = \frac{1}{\pi^4 D} \sum_{j=1}^{\infty} \sum_{k=1}^{\infty} q^{jk} \left\{ \frac{1}{\left[\left(\frac{j}{a}\right)^2 + \left(\frac{k}{b}\right)^2 \right]^2} \right\} \sin j\pi \frac{x}{a} \sin k\pi \frac{y}{b}. \qquad (8.29)$$

We may verify that (8.29) is indeed the solution by performing the operations indicated on the l.h.s. of (8.21). Since the solution is formed by superimposing the individual harmonic contributions, it is sufficient to consider a typical component $\bar{j}\bar{k}$. Common to each term is the function

$$F(x, y; \bar{j}, \bar{k}) = \frac{q^{\bar{j}\bar{k}}}{\pi^4 D} \left\{ \frac{1}{\left[\left(\frac{\bar{j}}{a}\right)^2 + \left(\frac{\bar{k}}{b}\right)^2 \right]^2} \right\} \sin \bar{j}\pi \frac{x}{a} \sin \bar{k}\pi \frac{y}{b} \qquad (8.30)$$

whereupon

$$u_{z,xxxx} = \left(\frac{\bar{j}\pi}{a}\right)^4 F$$

$$2u_{z,xxyy} = 2\left(\frac{\bar{j}\pi}{a}\right)^2 \left(\frac{\bar{k}\pi}{b}\right)^2 F$$

$$u_{z,yyyy} = \left(\frac{\bar{k}\pi}{b}\right)^4 F$$

and the sum is

$$\nabla^4 u_z = \left[\left(\frac{\bar{j}\pi}{a}\right)^4 + 2\left(\frac{\bar{j}\pi}{a}\right)^2\left(\frac{\bar{k}\pi}{b}\right)^2 + \left(\frac{\bar{k}\pi}{b}\right)^4\right]F$$

$$= \pi^4\left[\left(\frac{\bar{j}}{a}\right)^2 + \left(\frac{\bar{k}}{b}\right)^2\right]^2 F \tag{8.31}$$

$$= (q^{\bar{j}\bar{k}}/D)\sin \bar{j}\pi\frac{x}{a}\sin \bar{k}\pi\frac{y}{b}$$

$$= q(x,y)/D$$

from (8.20a) and (8.27).

It remains to examine the convergence of the series (8.29) so as to set appropriate truncation limits for j and k. To illustrate, we take the example of a uniformly loaded plate with a constant distributed loading

$$q(x,y) = q_0 \tag{8.32}$$

and set about to determine the Fourier coefficients q^{jk}. Starting with (8.27)

$$q_0 = \sum_{j=1}^{\infty}\sum_{k=1}^{\infty} q^{jk}\sin j\pi\frac{x}{a}\sin k\pi\frac{y}{b}. \tag{8.33}$$

Following (8.28b), the procedure is to multiply both sides of (8.33) by $\sin m\pi\frac{x}{a}\sin n\pi\frac{y}{b}$, where m and n represent arbitrary harmonics and integrate over the area:

$$\int_0^a\int_0^b q_0\sin m\pi\frac{x}{a}\sin n\pi\frac{y}{b}dxdy$$

$$= \sum_{j=1}^{\infty}\sum_{k=1}^{\infty} q^{jk}\int_0^a \sin m\pi\frac{x}{a}\sin j\pi\frac{x}{a}dx \int_a^b \sin n\pi\frac{y}{b}\sin k\pi\frac{y}{b}dy. \tag{8.34}$$

For the l.h.s. of (8.34),

$$\int_0^a\int_0^b q_0\sin m\pi\frac{x}{a}\sin n\pi\frac{y}{b}dxdy = q_0\frac{a}{m\pi}\left[-\cos\frac{m\pi x}{a}\right]_0^a\frac{b}{n\pi}\left[-\cos\frac{n\pi y}{b}\right]_0^b$$

$$= q_0\frac{a}{m\pi}[\cos m\pi - 1]\frac{b}{n\pi}[\cos n\pi - 1]. \tag{8.35}$$

If m and/or n is an even integer, the expression $= 0$. But, for m and n both odd,

$$q_0\frac{a}{m\pi}[\cos m\pi - 1]\frac{b}{n\pi}[\cos n\pi - 1] = \frac{4ab}{\pi^2 mn}q_0. \tag{8.36a}$$

Now considering the r.h.s. of (8.34),

$$
\int_0^a \sin\ m\pi\frac{x}{a}\sin j\pi\frac{y}{a}dx = \frac{a}{2} \quad m = j
$$
$$
= 0 \quad m \neq j
$$
$$
\int_0^b \sin\ n\pi\frac{y}{b}\sin k\pi\frac{y}{b}dy = \frac{b}{2} \quad n = k \tag{8.36b}
$$
$$
= 0 \quad n \neq k.
$$

Therefore the entire summation reduces to

$$
q^{jk}\frac{ab}{4}. \tag{8.37}
$$

Replacing m and n by j and k to denote a general but specific harmonic, we have by equating (8.36a) and (8.37)

$$
q^{jk} = \frac{16}{\pi^2 jk}q_0 \tag{8.38}
$$

and the complete displacement function follows from (8.29) as

$$
u_z = \frac{16q_0}{\pi^6 D}\sum_{j=1}^{\infty}\sum_{k=1}^{\infty}\frac{1}{jk}\left\{\frac{1}{\left[\left(\frac{j}{a}\right)^2+\left(\frac{k}{b}\right)^2\right]^2}\right\}\sin j\pi\frac{x}{a}\sin k\pi\frac{y}{b} \quad \begin{array}{l} j = 1,\ 3,\ 5,\ldots \\ k = 1,\ 3,\ 5,\ldots \end{array}
$$
$$
\tag{8.39}
$$

The series will obviously converge very rapidly since j and k are of the fifth power in the denominator. Also, it should be dominated by the first terms $j = k = 1; j = 1$, $k = 3; j = 3, k = 1$, since $j = k = 3$ and higher terms will carry at least $(3)^5$ in the denominator.

Once the displacement function is found, the various stresses are computed routinely. For the in-plane or membrane stresses, σ_{xx}, σ_{yy}, and σ_{xy}, we use (8.10a, b, c) which requires differentiating the series. For the transverse shearing stresses, σ_{xz} and σ_{yz}, (8.11) and (8.12), respectively, are integrated through the thickness. In considering the truncation of a series such as (8.39), we must be cautious. We remember that we seek not only the displacements for the plate but the stresses as well. We see from (8.10a, b, c) that the stresses σ_{ij} are given by various combinations of $u_{z,\ ij}$. For example,

$$
u_{z,xx} = -\frac{16\,q_0}{\pi^4\,D}\sum_{j=1}^{\infty}\sum_{k=1}^{\infty}\left(\frac{j}{a}\right)^2\frac{1}{jk}\left\{\frac{1}{\left[\left(\frac{j}{a}\right)^2+\left(\frac{k}{b}\right)^2\right]^2}\right\}\sin j\pi\frac{x}{a}\sin k\pi\frac{y}{b}, \tag{8.40}
$$

which will converge slower than (8.39). The expressions for $u_{z,yy}$ and $u_{z,xy}$ are similar.

The double series is known as Navier's solution and is recognized as a great accomplishment in solid mechanics. However if boundary conditions other than all sides simply supported are encountered or if the planform is other than rectangular or circular, the difficulty of choosing harmonic functions which identically satisfy the boundary conditions increases enormously. Therefore numerical methods such as those introduced in Sect. 11.7 are widely applied to the analysis of plates with boundaries that are other than simply supported.

8.4.2 Circular Plate

General Solution for Axisymmetric Loading

We consider a circular plate with planform as shown in Fig. 7.1 and radial coordinate r subject to a uniform load q_0 and initially without specifying boundary conditions. A plate so loaded will deform uniformly around the circumference so that the axisymmetric form of the governing equation may be considered. The solution to (8.23) may be written by analogy with (7.32) as

$$u_z(r) = C_1 r^2 \ln r + C_2 r^2 + C_3 \ln r + C_4 + u_{zp}, \tag{8.41}$$

where u_{zp} is a particular solution to account for the applied loading.

For the case of a uniform load, u_{zp} must be proportional to r^4, say Cr^4. Performing the operations indicated by (7.31e),

$$\nabla^4 (Cr^4) = \frac{1}{r} \left[r \left(\frac{1}{r} \left[r (Cr^4)_{,r} \right]_{,r} \right)_{,r} \right]_{,r} = 64C \tag{8.42}$$

With $64C = q_0/D$,

$$C = \frac{q_0}{64D} \tag{8.43}$$

and

$$u_{zp} = \frac{q_0}{64D} r^4. \tag{8.44}$$

A variety of support conditions and geometries may be solved with (8.41). First, it is convenient to form the first and second derivatives:

$$u_{z,r} = C_1 r (2 \ln r + 1) + 2C_2 r + C_3 \frac{1}{r} + \frac{q_0}{16D} r^3 \tag{8.45a}$$

$$u_{z,rr} = C_1 (2 \ln r + 3) + 2C_2 - C_3 \frac{1}{r^2} + \frac{3q_0}{16D} r^2. \tag{8.45b}$$

Next, second derivatives of u_z with respect to x and y, which are contained in the simply supported boundary conditions (8.26a) and (8.26b), are transformed into polar coordinates. We refer to Fig. 7.1 and the r-dependent portions of (7.22):

$$\frac{\partial(\ )}{\partial x} = \frac{\partial(\ )}{\partial r} r_{,x} = \frac{\partial(\ )}{\partial r} \cos\theta = \frac{\partial(\ )}{\partial r} \frac{x}{r} \tag{8.46a}$$

$$\frac{\partial(\ )}{\partial y} = \frac{\partial(\ )}{\partial r} r_{,y} = \frac{\partial(\ )}{\partial r} \sin\theta = \frac{\partial(\ )}{\partial r} \frac{y}{r} \tag{8.46b}$$

from which, by product differentiation. of (8.46a) and (8.46b), respectively,

$$u_{z,xx} = \frac{\partial^2 u_z}{\partial x^2} = \frac{x}{r}\frac{\partial^2 u_z}{\partial r^2} + \frac{\partial u_{z,r}}{\partial r}\left(\frac{r - xr_{,x}}{r^2}\right) \tag{8.47a}$$

$$u_{z,yy} = \frac{\partial^2 u_z}{\partial y^2} = \frac{y}{r}\frac{\partial^2 u_z}{\partial r^2} + \frac{\partial u_{z,r}}{\partial r}\left(\frac{r - yr_{,y}}{r^2}\right). \tag{8.47b}$$

Taking r along the x-axis in Fig. 7.1, $x = r$, $y = 0$, $r_{,x} = \cos\theta = 1$, and $r_{,y} = \sin\theta = 0$, leaving

$$u_{z,xx} = u_{z,rr} \tag{8.48a}$$

$$u_{z,yy} = \frac{1}{r}u_{z,r} \tag{8.48b}$$

as the transformed derivatives.

Solid Plate

We consider a circular plate with outside radius $r = R_0$. Since the terms containing $\ln r$ would be singular at $r = 0$, we set C_1 and $C_3 = 0$ in (8.41) which leaves

$$u_z(r) = C_2 r^2 + C_4 + \frac{q_0}{64D} r^4. \tag{8.49}$$

If the plate is clamped at the boundary, we have $u_z(R_0) = u_{z,r}(R_0) = 0$ or

$$u_z(R_0) = C_2 R_0^2 + C_4 + \frac{q_0}{64D} R_0^4 = 0 \tag{8.50a}$$

$$u_{z,r}(R_0) = 2C_2 R_0 + \frac{q_0}{16D} R_0^3 = 0 \tag{8.50b}$$

from which

$$C_2 = -\frac{q_0}{32D} R_0^2 \tag{8.51a}$$

$$C_4 = \frac{q_0}{64D} R_0^4, \tag{8.51b}$$

and

$$u_z = \frac{q_0}{64D}(-2R_0^2 r^2 + R_0^4 + r^4)$$
$$= \frac{q_0}{64D}(R_0^2 - r^2)^2. \tag{8.52}$$

If the plate is simply supported at the boundary, we have (8.50a) along with $M_r = 0$. With $M_r = M_x$, $M_r = 0$ is written in polar coordinates from (8.26a) and (8.48a) as

$$M_r = \left(u_{z,rr} + \frac{\nu}{r} u_{z,r}\right)_{r=R_0} = 0. \tag{8.53}$$

Substituting (8.45b) and (8.45a) with C_1 and $C_3 = 0$, respectively, in (8.53) gives

$$2(1 + \nu)C_2 + \frac{(3 + \nu)}{16D} q_0 R_0^2 = 0 \tag{8.54}$$

so that

$$C_2 = -2\left(\frac{3 + \nu}{1 + \nu}\right)\frac{q_0 R_0^2}{64D} \tag{8.55a}$$

From (8.50a)

$$C_4 = \frac{q_0 R_0^4}{64D}\left[\frac{2(3 + \nu)}{1 + \nu} - 1\right]$$
$$- \frac{q_0}{64D}\left[\frac{5 + \nu}{1 + \nu}\right]R_0^4 \tag{8.55b}$$

and then from (8.49)

$$u_z(r) = \frac{q_0}{64D}(R_0^2 - r^2)\left\{\left[\frac{5 + \nu}{1 + \nu}\right]R_0^2 - r^2\right\}. \tag{8.56}$$

Annular Plate

We may also consider an annular plate with planform as shown in Fig. 7.2a but subjected to a *transverse* uniform loading q_0. We again consider the clamped boundary condition corresponding to

$$u_z(R_0) = u_{z,r}(R_0) = 0 \tag{8.57}$$

at the outer ring and

$$u_z(R_1) = u_{z,r}(R_1) = 0 \tag{8.58}$$

on the inner ring. Since the singularity as $r \to 0$ is not an issue, we retain C_1 and C_3 and evaluate from (8.41) and (8.45a):

$$C_1 R_0^2 \ln R_0 + C_2 R_0^2 + C_3 \ln R_0 + C_4 + \frac{q_0}{64D} R_0^4 = 0 \tag{8.59a}$$

$$C_1 R_0 (2 \ln R_0 + 1) + 2 C_2 R_0 + \frac{C_3}{R_0} + \frac{q_0}{16D} R_0^3 = 0 \tag{8.59b}$$

$$C_1 R_1^2 \ln R_1 + C_2 R_1^2 + C_3 \ln R_1 + C_4 + \frac{q_0}{64D} R_1^4 = 0 \tag{8.59c}$$

$$C_1 R_1 (2 \ln R_1 + 1) + 2 C_2 R_1 + \frac{C_3}{R_1} + \frac{q_0}{16D} R_1^3 = 0. \tag{8.59d}$$

The equations are best solved using numerical values for R_0 and R_1, whereupon the determined constants are used in (8.41) and subsequent calculations.

It is noted that this clamped boundary condition may be difficult to attain physically, as the boundary is also required to be unrestrained in the plane of the plate so as not to develop in-plane restraining forces.

Stresses

After the appropriate boundary conditions are applied and the constants determined, the complete displacement function and derivatives are available from (8.41), (8.44), and (8.45a) and (8.45b). Returning to the stresses for the axisymmetric case, σ_{xx} and σ_{yy} in (8.10a) and (8.10b) become σ_{rr} and $\sigma_{\theta\theta}$. Substituting (8.48a) and (8.48b) for the curvatures gives

$$\sigma_{rr} = -\frac{\zeta E}{1 - \nu^2} \left(u_{z,rr} + \frac{\nu}{r} u_{z,r} \right) \tag{8.60a}$$

$$\sigma_{\theta\theta} = -\frac{\zeta E}{1 - \nu^2} \left(\frac{1}{r} u_{z,r} + \nu u_{z,rr} \right). \tag{8.60b}$$

The evaluation of the stresses, in particular the maximum value through the thickness at $\zeta = \pm \frac{h}{2}$, may proceed routinely. It is important to recognize that the axisymmetric loading condition produces circumferential stresses $\sigma_{\theta\theta}$ as well as the expected radial stresses σ_{rr}. The circumferential stresses and resultant bending moments are not directly subject to the boundary conditions and can be critical for some loading cases.

8.4.3 Commentary on Plate Theories

It should also be mentioned that more refined linear theories of plates, between the elementary theory presented here and a three-dimensional elasticity solution, are available. Most prominent is a theory which relaxes part of Assumption (2). In effect, the line mn remains straight but not necessarily normal to the deformed middle surface. This negates (8.4) and gives independent rotations ω_{yz} and ω_{xz} in addition to $u_z(x, y)$. The difference between the two deformation fields is due to the transverse shearing strains. Although this difference is seldom significant in magnitude, the inclusion of these shearing terms often facilitates numerically based solutions [2]. Also, the extension of plate theory to include nonlinear geometric behavior, as introduced in Sect. 3.8, is popular due to the inherent flexibility of a thin plate and the difficulty in attaining the ideal boundary conditions, as noted at the end of the previous section.

8.5 Formulation of Shell Theory

8.5.1 Curvilinear Coordinates

The middle surface of the shell outlined in Fig. 8.2b is expanded in Fig. 8.4 where the surface is defined in the X–Y–Z Cartesian coordinates by

$$z = f(X, Y) \tag{8.61}$$

Then a set of coordinates α and β on the middle surface are related to the Cartesian system by

$$X = f_1(\alpha, \beta) \tag{8.62a}$$

$$Y = f_2(\alpha, \beta) \tag{8.62b}$$

$$Z = f_3(\alpha, \beta), \tag{8.62c}$$

where f_i are continuous, single-valued functions. Each ordered pair (α, β) corresponds to one point and uniquely describes the surface. They are called *curvilinear coordinates*. As shown on the figure, incrementing one of the coordinates, say $\beta = \beta_1, \beta = \beta_2, ..., \beta = \beta_n$, defines a series of parametric curves on the surface along which only α varies. These are called s_α coordinate lines. Similarly, varying α defines s_β coordinate lines. If the surface coordinate lines s_α and s_β are mutually perpendicular, the curvilinear coordinates are called orthogonal. We deal exclusively with orthogonal curvilinear coordinates in this introductory treatment. Several advanced theories are based on non-orthogonal coordinates and formulated in general tensor notation [2].

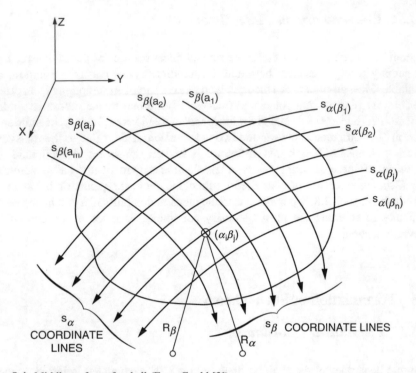

Fig. 8.4 Middle surface of a shell (From Gould [3])

8.5.2   *Middle-Surface Geometry*

Equation (8.61) for the middle surface may be described in terms of a position vector **r**

$$\mathbf{r} = X\mathbf{e}_x + Y\mathbf{e}_y + Z\mathbf{e}_z \tag{8.63}$$

as shown in Fig. 8.5, where $\mathbf{e}_i$ (i=x,y,z) are unit vectors along the X_i axes.

Substituting (8.62a, b, c) into (8.63) defines the position vector **r** in terms of the curvilinear coordinates.

$$\mathbf{r}(\alpha, \beta) = f_i(\alpha, \beta)\mathbf{e}_i \tag{8.64}$$

Derivatives of **r** with respect to the curvilinear coordinates $\mathbf{r},_\alpha$ and $\mathbf{r},_\beta$ are tangent to the respective coordinate lines as shown in the figure and are orthogonal as well so that $\mathbf{r},_\alpha \cdot \mathbf{r},_\beta = 0$.

The vector **ds** joining two points on the middle surface (α,β) and $(\alpha + d\alpha, \beta + d\beta)$ shown in Fig. 8.5 is

$$\mathbf{ds} = \mathbf{r},_\alpha d\alpha + \mathbf{r},_\beta d\beta \tag{8.65}$$

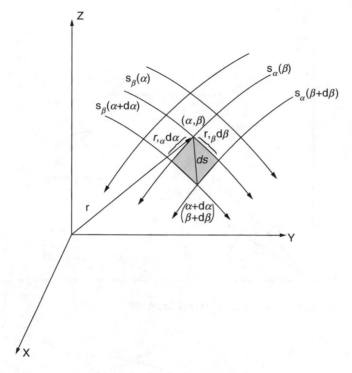

Fig. 8.5 Position vector to middle surface (From Gould [3])

and

$$\mathbf{ds} \cdot \mathbf{ds} = ds^2 = (\mathbf{r}_{,\alpha} \cdot \mathbf{r}_{,\alpha})d\alpha^2 + (\mathbf{r}_{,\beta} \cdot \mathbf{r}_{,\beta})d\beta^2 \qquad (8.66)$$

Defining $A^2 = \mathbf{r}_{,\alpha} \cdot \mathbf{r}_{,\alpha}$ and $B^2 = \mathbf{r}_{,\beta} \cdot \mathbf{r}_{,\beta}$,

$$ds^2 = A^2 d\alpha^2 + B^2 d\beta^2 \qquad (8.67)$$

which is known as the first quadratic form of the theory of shells.

The quantities A and B are called *Lamé parameters or measure numbers* and are fundamental to the understanding of curvilinear coordinates. To interpret their physical meaning, consider two cases where each of the coordinates is varied individually and independently. Then $ds_\alpha = A\, d\alpha$ and $ds_\beta = B d\beta$, respectively. Thus ds_α is the change in arc length along coordinate line s_α when α is incremented by $d\alpha$, and ds_β is the same along s_β. In general a Lamé parameter relates the change in arc length on the surface to the change in the corresponding curvilinear coordinate hence, the alternate name, measure number.

As a simple example, consider a circle with radius R_β in the Y–Z plane as shown in Fig. 8.6. If the polar angle β is selected as the curvilinear coordinate, $ds_\beta = R_\beta d\beta$ and the Lamé parameter is R_β. However there could be other perhaps less obvious

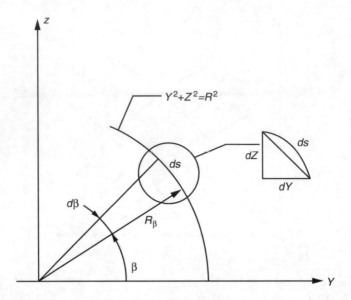

Fig. 8.6 Lamé parameter for circular arc (From Gould [3])

Fig. 8.7 Unit tangent
vectors (From Gould [3])

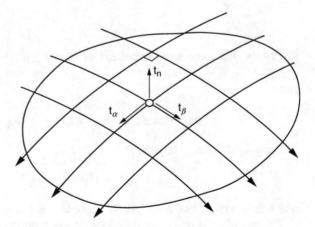

choices for the curvilinear coordinate, such as the axial coordinate Z or even the arc
length itself. In the latter case, the Lamé parameter is just 1.

It is convenient to define a reference set of unit vectors on the surface as shown in
Fig. 8.7. Based on the tangent vectors defined as $r_{,\alpha}$ and $r_{,\beta}$, we have the *unit* vectors

$$\mathbf{t}_\alpha = \frac{\mathbf{r},_\alpha}{\pm |\mathbf{r},_\alpha|} = \frac{\mathbf{r},_\alpha}{A} \tag{8.68a}$$

$$\mathbf{t}_\beta = \frac{\mathbf{r},_\beta}{\pm |\mathbf{r},_\beta|} = \frac{\mathbf{r},_\beta}{B} \tag{8.68b}$$

and the normal vector

$$\mathbf{t_n} = \mathbf{t_\alpha} \times \mathbf{t_\beta} = \frac{1}{AB}\left(\mathbf{r}_{,\alpha} \times \mathbf{r}_{,\beta}\right) \tag{8.68c}$$

The unit vectors $\mathbf{t}_i$ $(i = \alpha,\beta,n)$ differ from the Cartesian unit vectors $\mathbf{e}_i$ in that they change orientation at each point on the surface.

8.6 Shells of Revolution

8.6.1 Definitions

A surface of revolution is formed by rotating a plane curve generator about an axis of revolution to form a closed surface as shown in Fig. 8.8. The orthogonal coordinate lines are called the *meridians*, formed by planes containing the axis of rotation, and the *parallel circles*, traced by planes perpendicular to the axis of rotation. In the theory of surfaces, the meridians and the parallel circles correspond to lines of principal curvature. These lines are defined by principal radii of curvature, as we will show in the next section. The principal radii of curvature at any point are the maximum and minimum of all possible values at the point. We will restrict our development to this special class of shells since it is the most common in engineering applications.

Fig. 8.8 Shell of revolution
(From Gould [3])

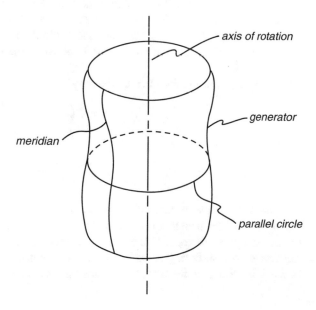

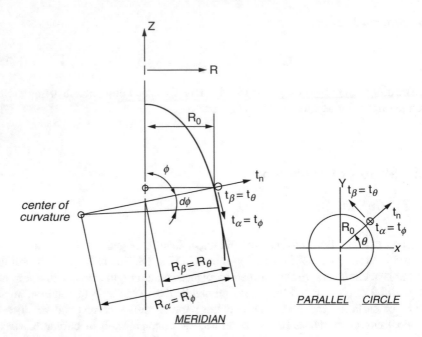

Fig. 8.9 Geometry for shell of revolution (From Gould [3])

8.6.2   *Curvilinear Coordinates*

The meridian of a shell of revolution is shown in Fig. 8.9 where the definitive parameters are indicated for the curvilinear coordinates $\alpha = \phi$ and $\beta = \theta$. ϕ is the meridional angle formed by the extended normal to the surface and the axis of rotation, and θ is the circumferential angle between the radius of the parallel circle and the x-axis.

The equation of the meridian is given by

$$Z = Z(R) \tag{8.69a}$$

where

$$R = \sqrt{(X^2 + Y^2)} \tag{8.69b}$$

Note that the meridional radius of curvature R_ϕ defining the s_ϕ coordinate line emanates from the center of curvature and extends to the surface; the circumferential radius of curvature R_θ defining the s_θ coordinate line extends from the axis of rotation to the surface. The radius of the parallel circle, which is equal to R in (8.69b), is termed as the horizontal radius and denoted as R_o. R_o is *not* a principal radius of curvature because it is not normal to the surface; rather it is the projection of R_θ on the horizontal plane, i.e.,

$$R_\mathrm{o} = R_\theta \sin\phi. \tag{8.70a}$$

We also add, initially without proof, the differentiation formula [3]

$$(R_\mathrm{o})_{,\phi} = (R_\theta \sin\phi)_{,\phi} = R_\phi \cos\phi \tag{8.70b}$$

which is known as the Gauss–Codazzi relation and is renown in the theory of surfaces. The relationship is confirmed in Sect. 8.8 and is applied to avoid the term $R_{\theta,\phi}$ in the equilibrium equations. For shells defined by (8.69a), R_ϕ and R_θ may be determined using well-known formulas from the theory of surfaces. For many common curves, the radii of curvature are available from elementary mathematics texts. We will treat only elementary geometric forms in this introductory presentation.

A closed shell of revolution is called a dome and the peak is the pole. The pole introduces some mathematical complications because at this point, $R_\mathrm{o} \to 0$. For completeness, we note that Fig. 8.9 shows only the case where R_ϕ and R_θ lie on the same side of the meridian. If they lie on opposite sides, the meridian is of hyperbolic form, and the shell cannot be closed. The former case is called positive Gaussian curvature and the latter negative Gaussian curvature.

With the curvilinear coordinates selected as ϕ and θ,

$$ds_\phi = R_\phi d\phi \tag{8.71}$$

and

$$ds_\theta = R_\mathrm{o} d\theta \tag{8.72}$$

so that $A = R_\phi$ and $B = R_\mathrm{o}$.

8.7 Stress Resultants and Couples

An element of a shell is shown in Fig. 8.10a bound by the s_α and s_β coordinate line defined on Fig. 8.5. The geometry of the middle surface of such an element was considered in Sect. 8.5.2. Here we show the entire thickness h with the coordinate ζ defined in the direction of $\mathbf{t_n}$ and highlight a differential volume element with thickness $d\zeta$ parallel to and displaced from the middle surface by a distance ζ. The element is bound by coordinate lines s_α and s_β and their respective differential increments.

On Fig. 8.10b the stresses acting on the volume element are indicated following the conventions of the theory of elasticity introduced in Chap. 2. As mentioned in Sec. 8.3.2, plate and shell theories deal with a simplification of the stress-at-a-point

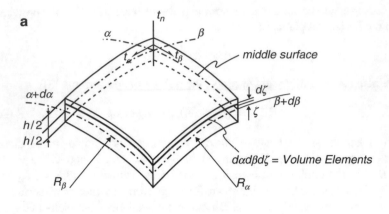

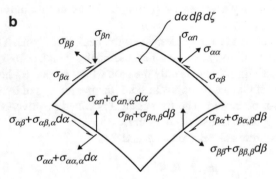

Fig. 8.10 Shell volume element and stresses (From Gould [3])

problem to solve for the total forces and moment intensities per length of middle surface, known as *stress resultants* and *stress couples*, respectively. The stresses are then evaluated, sometimes approximately, by back-substitution as we will discuss later.

The stress resultants and stress couples are shown on an element of the middle surface in Fig. 8.11. The extensional forces N_α and N_β, transverse shear forces Q_α and Q_β, and bending moments M_α and M_β are represented by singly subscripted variables indicating the surface on which the force or moment acts.

The in-plane shear forces $N_{\alpha\beta}$ and $N_{\beta\alpha}$ and twisting moments $M_{\alpha\beta}$ and $M_{\beta\alpha}$ carry double subscripts corresponding to the surface acted upon and the direction, respectively. While this subscripting is not completely consistent with elasticity notation, it has been widely adopted.

Figure 8.12 is a view of Fig. 8.10a normal to t_β. An arc of the middle surface along the s_α coordinate line is measured by

$$ds_\alpha = A d\alpha \tag{8.73}$$

Fig. 8.11 Stress resultants and stress couples (From Gould [3])

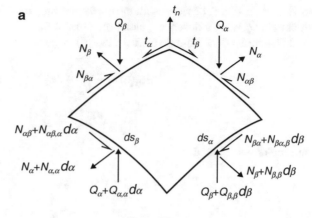

Stress Resultants

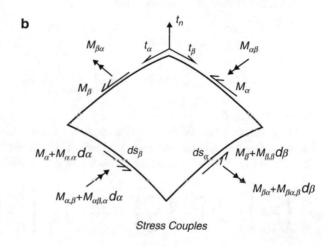

Stress Couples

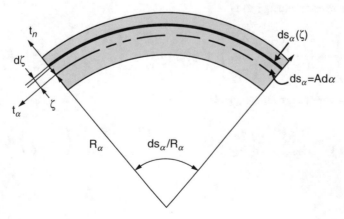

Fig. 8.12 Section normal to $\mathbf{t_\beta}$ (From Gould [3])

as shown on Fig. 8.12 along with the corresponding arc of the volume element with thickness $d\zeta$ lying a distance ζ along the normal $\mathbf{t_n}$.

The length of the volume element $ds_\alpha\,(\zeta)$ is given by

$$ds_\alpha(\zeta) = \frac{R_\alpha + \zeta}{R_\alpha} ds_\alpha$$
$$= ds_\alpha\left(1 + \frac{\zeta}{R_\alpha}\right) \qquad (8.74)$$

and the projected area is

$$da_\beta(\zeta) = ds_\alpha(\zeta)\, d\zeta$$
$$= ds_\alpha\left(1 + \frac{\zeta}{R_\alpha}\right) d\zeta \qquad (8.75)$$

A similar section normal to $\mathbf{t_\alpha}$ gives

$$ds_\beta(\zeta) = ds_\beta\left(1 + \frac{\zeta}{R_\beta}\right)$$
$$da_\alpha(\zeta) = ds_\beta\left(1 + \frac{\zeta}{R_\beta}\right) d\zeta, \qquad (8.76)$$

where $ds_\beta = B d\beta$.

Now we can calculate the magnitudes of the stress resultants and stress couples, respectively, forces, and moments per length of middle surface, in terms of the corresponding stresses. The stress resultants on the section normal to $\mathbf{t_\alpha}$ are

$$\begin{Bmatrix} N_\alpha \\ N_{\alpha\beta} \\ Q_\alpha \end{Bmatrix} = \frac{1}{ds_\beta} \int_{-h/2}^{h/2} \begin{Bmatrix} \sigma_{\alpha\alpha} \\ \sigma_{\alpha\beta} \\ \sigma_{\alpha n} \end{Bmatrix} da_\alpha(\zeta) = \int_{-h/2}^{h/2} \begin{Bmatrix} \sigma_{\alpha\alpha} \\ \sigma_{\alpha\beta} \\ \sigma_{\alpha n} \end{Bmatrix}\left(1 + \frac{\zeta}{R_\beta}\right) d\zeta \qquad (8.77)$$

and on the section normal to $\mathbf{t_\beta}$

$$\begin{Bmatrix} N_\beta \\ N_{\beta\alpha} \\ Q_\beta \end{Bmatrix} = \frac{1}{ds_\alpha} \int_{-h/2}^{h/2} \begin{Bmatrix} \sigma_{\beta\beta} \\ \sigma_{\beta\alpha} \\ \sigma_{\beta n} \end{Bmatrix} da_\beta(\zeta) = \int_{-h/2}^{h/2} \begin{Bmatrix} \sigma_{\beta\beta} \\ \sigma_{\beta\alpha} \\ \beta n \end{Bmatrix}\left(1 + \frac{\zeta}{R_\alpha}\right) d\zeta. \qquad (8.78)$$

The corresponding stress couples are

$$\begin{Bmatrix} M_\alpha \\ M_{\alpha\beta} \end{Bmatrix} = \frac{1}{ds_\beta} \int_{-h/2}^{h/2} \begin{Bmatrix} \sigma_{\alpha\alpha} \\ \sigma_{\alpha\beta} \end{Bmatrix} \zeta da_\alpha(\zeta) = \int_{-h/2}^{h/2} \begin{Bmatrix} \sigma_{\alpha\alpha} \\ \sigma_{\alpha\beta} \end{Bmatrix}\zeta\left(1 + \frac{\zeta}{R_\beta}\right) d\zeta \qquad (8.79)$$

and

$$\begin{Bmatrix} M_\beta \\ M_{\beta\alpha} \end{Bmatrix} = \frac{1}{ds_\alpha} \int_{-h/2}^{h/2} \begin{Bmatrix} \sigma_{\beta\beta} \\ \sigma_{\beta\alpha} \end{Bmatrix} \zeta \, da_\beta(\zeta) = \int_{-h/2}^{h/2} \begin{Bmatrix} \sigma_{\beta\beta} \\ \sigma_{\beta\alpha} \end{Bmatrix} \zeta \left(1 + \frac{\zeta}{R_\alpha} \right) d\zeta. \quad (8.80)$$

At this point we recognize that once the plate or shell problem has been solved for the stress resultants and stress couples, the recovery of the pointwise stresses would necessitate the inversion of the preceding equations. This is straightforward for plates but difficult for curved cross sections. In practice the stresses at a point are usually recovered by assuming that the cross section is rectangular, which is tantamount to neglecting the terms of magnitude $\zeta/R{:}1$ in the above equations and assuming that all stresses except the transverse shears $\sigma_{\alpha\beta}$ and $\sigma_{\beta\alpha}$ vary linearly over the cross section. Of course $\sigma_{\alpha\beta}$ and $\sigma_{\beta\alpha}$ are equal from the theory of elasticity, but some small differences may arise in reducing the corresponding stress resultants and couples. The preceding arguments imply

$$\sigma_{(ii)}(\zeta) = \sigma_{(ii)}(0) + \frac{[\sigma_{(ii)}(h/2) - \sigma_{(ii)}(0)]\zeta}{(h/2)} \quad (i = \alpha, \beta)$$

$$\sigma_{ij}(\zeta) = \sigma_{ij}(0) + \frac{[\sigma_{ij}(h/2) - \sigma_{ij}(0)]\zeta}{(h/2)} \quad (i = \alpha, \beta; j = \beta, \alpha) \quad (8.81)$$

It is consistent with elementary beam theory as described in Sect. 6.3.1 to assume that the transverse shear stresses are distributed parabolically through the thickness with a maximum at the middle surface, (6.20c). Thus

$$\sigma_{in}(\zeta) = \sigma_{in}(0) \left[1 - \left(\frac{\zeta}{h/2} \right)^2 \right] \quad (i = \alpha, \beta). \quad (8.82)$$

When these relationships are introduced into (8.77), (8.78), (8.79), and (8.80), the maximum stresses on the cross section are

$$\sigma_{(ii)}\left(\pm \frac{h}{2} \right) = \frac{N_i}{h} \pm \frac{6M_i}{h^2} \quad (i = \alpha, \beta) \quad (8.83a)$$

$$\sigma_{(ij)}\left(\pm \frac{h}{2} \right) = \frac{N_{ij}}{h} \pm \frac{6M_{ij}}{h^2} \quad \begin{pmatrix} i = \alpha, & \beta \\ j = \beta, & \alpha \end{pmatrix} \quad (8.83b)$$

$$\sigma_{in}(0) = \frac{3}{2} \frac{Q_i}{h} \quad (i = \alpha, \beta) \quad (8.83c)$$

Then the stresses at any other level within the cross section are found from (8.81) to (8.82).

At this point we have introduced the highlights of a vector-based linear theory of thin shells that has been developed in Ref. [3]. Before pursuing some modest applications, it is appropriate to comment on the stress resultants and couples derived in (8.77), (8.78), (8.79), and (8.80). These are cross-section quantities as opposed to

the stresses-at-a-point associated with elasticity and thus can illustrate the global
load resistance mechanisms of the plate and shell forms.

First we reconsider transversely loaded plates treated in the first part of this
chapter. Two-dimensional beam resistance is obviously provided by a plate and
represented by the bending stress couples M_α, M_β, and the accompanying transverse
shears, Q_α and Q_β. But there is another formidable resistance mechanism provided
by the twisting stress couples $M_{\alpha\beta}$ and $M_{\beta\alpha}$ that add considerable rigidity to solid
plates beyond a two-way grid of beams. The plate analyses in the earlier sections of
the chapter could have been formulated in terms of these stress resultants and
couples, but it was unnecessary for the basic examples where the stress-at-a-point
approach was sufficient.

Next we consider transversely loaded shells that would obviously provide
two-dimensional arch action represented by the in-plane direct stress resultants N_α
and N_β along with shear resistance similar to that provided by plates against in-plane
loading and represented by in-plane shear stress resultants $N_{\alpha\beta}$ and $N_{\beta\alpha}$. At first
glance it would appear that a shell also has the same bending and twisting resistances
as the plate. However the utility of shells is largely based on providing *in-plane*
resistance to *transverse* loading, which is *far* more efficient than the bending
resistance of thin plane or curved surface structures.

If the shell is loaded by forces *distributed* as opposed to concentrated, has
geometry that is relatively *smooth* and *continuous*, and has appropriate *boundary
conditions* to develop the in-plane stress resultants, then the bending and twisting
effects may be neglected except perhaps locally. The resulting formulation is termed
the *membrane theory* and is widely used for the analysis of shells [4, 5]. We will use
the membrane theory and the accompanying stress resultants for examples to follow.
Because the membrane theory is statically determinate, displacements are not
required to solve for the stresses, so they are not discussed in this abbreviated
presentation.

8.8   Membrane Theory Equilibrium Equations for Shells of Revolution

The shell of revolution described in Sect. 8.6 is a widely used geometry and provides
a sufficiently general yet easily visualized model for the derivation and interpretation
of membrane theory equilibrium equations. A middle surface element of a shell of
revolution is shown in Fig. 8.13.

The curvilinear coordinates are ϕ and θ as indicated on Fig. 8.8, and the Lamé
parameters are R_ϕ and R_o, respectively. On Fig. 8.13, note that the meridians are not
parallel. For an increment of $d\theta$ on the circumference of the parallel circle, adjacent
tangents to the respective meridians $\mathbf{t}_\phi$ (ϕ,θ) and $\mathbf{t}_\phi$ $(\phi, \theta + d\theta)$ can be extended to
intersect at an angle $d\chi$ as shown on the figure.

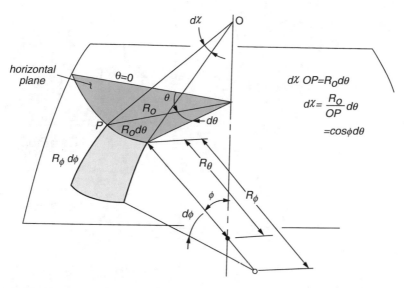

Fig. 8.13 Middle surface element and geometrical relationships for a shell of revolution. (From Gould [3])

On Fig. 8.14, the middle surface element is shown with distributed loadings in the ϕ, θ, and n directions; also the in-plane stress resultants that act on the middle surface element of a shell of revolution are shown in several views. All loading and stress resultants are multiplied by the respective length of middle surface on which they act, $R_\phi\,d\phi$, or $R_o d\theta$, to convert them to forces.

The equilibrium equations for the shell can be derived elegantly using vector mechanics [3]. However this requires differentiation of the unit tangent vectors $\mathbf{t_i}$ (ϕ,θ) defined in (8.68a, b, c) in the three directions ϕ, θ, n, which is somewhat involved and beyond the scope of this introductory presentation. Therefore, equilibrium will be determined using the free-body diagram approach summing forces in the parallel to the $\mathbf{t_i}$ (ϕ,θ,n) unit vectors, i.e., in the ϕ, θ, and n directions.

$\mathbf{t_\phi}$ direction: Referring to the Force Element in Fig. 8.14, we have the meridional force $N_\phi R_o d\theta$ changing over the distance $R_\phi d\phi$ by $(N_\phi R_o d\theta)_{,\phi}d\phi$, and the shear force $N_{\theta\phi}R_\phi d\phi$ changing over the distance $R_o d\theta$ by $(N_{\theta\phi}R_\phi d\phi)_{,\theta}d\theta$. Less obvious is the contribution of the circumferential force $N_\theta R_\phi d\phi$. Passing to view HV1 and dropping the higher-order differential term, we see that this force develops a radial component $N_\theta R_\phi d\phi d\theta$. Then moving to view MV1, the radial component is resolved in the meridional ($\mathbf{t_\phi}$) and normal ($\mathbf{t_n}$) directions. We retain the normal component for equilibrium in the $\mathbf{t_n}$-direction while adding $-N_\theta R_\phi d\phi d\theta \cos\phi$ in the $\mathbf{t_\phi}$ direction, with the minus sign indicating that it is acting opposite to the change in ϕ. Finally we include the q_ϕ loading from the Load Element, note that R_ϕ is not a function of θ and divide through by $d\phi d\theta$ to produce the equilibrium equation:

$$(N_\phi R_o)_{,\phi} + N_{\theta\phi,\theta}R_\phi - N_\theta R_\phi \cos\phi \ + \ q_\phi R_o R_\phi = 0. \qquad (8.84)$$

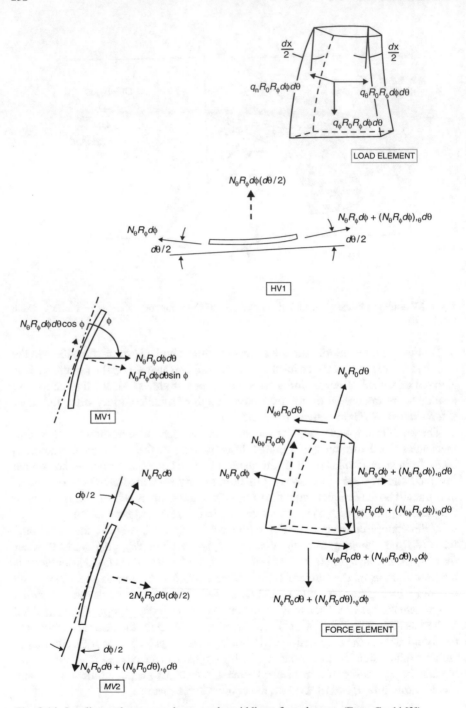

Fig. 8.14 Loading and stress resultants on the middle surface element. (From Gould [3])

t_θ-direction: From the Force Element in Fig. 8.14, we have the circumferential force $N_\theta R_\phi d\phi$ changing over the distance $R_o d\theta$ by $(N_\theta R_\phi d\phi)_{,\theta}$ and the shear force $N_{\phi\theta} R_o d\theta$ changing over the distance $R_\phi d\phi$ by $(N_{\phi\theta} R_o d\phi)_{,\phi} d\phi$. Additionally we consider the shear force $N_{\theta\phi} R_\phi d\phi$. Referring back to Fig. 8.13, we see that this force generates a component in the t_θ-direction because of the converging meridians. Neglecting the higher-order differential term, we have $2 N_{\theta\phi} R_\phi d\phi (d\chi/2) = N_{\theta\phi} R_\phi d\phi$ $\cos\phi$ from Fig. 8.13. Collecting the contributions, adding the q_θ loading, and dividing by $d\phi d\theta$ produce the equilibrium equation:

$$N_{\theta,\theta} R_\phi + (N_{\phi\theta} R_o)_{,\phi} + N_{\theta\phi} R_\phi \cos\phi + q_\theta R_o R_\phi = 0. \tag{8.85}$$

It is of interest to compare the first two terms of the preceding two equations. The two terms with R_0 are differentiated as products since R_0 is a function of ϕ. However the terms with R_ϕ have only the stress resultants differentiated since R_ϕ is not a function of θ. This is reflected in the final form (8.88).

t_n-direction: First we return to view MV1 in Fig. 8.14, previously discussed for the t_ϕ-direction, and recall the term $N_\theta R_\phi d\phi d\theta$ sin ϕ acting in the negative t_n-direction. We use a similar resolution for the meridional force $N_\phi R_o d\theta$ in view MV2 to get the normal component $N_\phi R_o d\phi d\theta$, again in the negative t_n-direction. Adding the loading term q_n and dividing by $d\phi d\theta$ give the equilibrium equation:

$$-N_\theta R_\phi \sin\phi - N_\phi R_o + q_n R_o R_\phi = 0. \tag{8.86}$$

Equation (8.86) is of paramount importance in that it demonstrates the primary load mechanism of shells whereby *transverse* loading is resisted by *in-plane* forces. As noted earlier, this mechanism is, in general, far more efficient than the bending and twisting moments and corresponding transverse shears required for thin plates. The equation may be called "*the*" shell equation in the manner that (8.20a) was named "*the*" plate equation.

Equations (8.84), (8.85), and (8.86) will be refined and simplified somewhat to complete the membrane theory equations for shells of revolution. In refining the equations, liberal use is made of the Gauss–Codazzi relationship, (8.70b). A geometric interpretation is shown in Fig. 8.15. Referring to points B and D, $R_o(\phi) = AB$ and $R_o (\phi + \Delta\phi) = CD$, respectively, so that $\Delta R_o = CD - AB = BD \sin\left(\frac{\pi}{2} - \phi\right) = BD \cos\phi$, where BD is the straight line approximating the arc BD over the differential arc $\Delta\phi$. Since $BD = R_\phi \Delta\phi$,

$$\Delta R_o = R_\phi \cos\phi \Delta\phi. \tag{8.87}$$

Dividing both sides by $\Delta\phi$ and taking lim $\Delta\phi \to 0$ give (8.70b).

We may now collect the three equations derived from equilibrium considerations using the Gauss–Codazzi relationships and the further assumption that $N_{\theta\phi} \cong N_{\phi\theta} = S$ (8.84, 8.85, 8.86). This is an extension of the symmetry of the stress tensor and is accurate within $O(h/R{:}1)$ as shown by the definitions in (8.77) and (8.78).

Fig. 8.15 Geometric
interpretation of Gauss–
Codazzi equation (From
Gould [3])

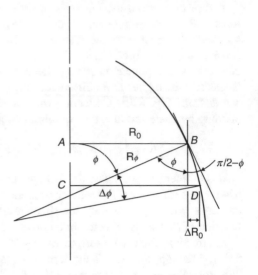

Fig. 8.16 Resolution of
distributed loading in
z-direction along curvilinear
coordinates. (From Gould
[3])

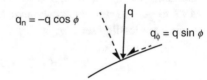

$$\frac{1}{R_\phi}N_{\phi,\phi} + \frac{\cot\phi}{R_\theta}\left(N_\phi - N_\theta\right) + \frac{1}{R_0}S_{,\theta} + q_\phi = 0 \tag{8.88a}$$

$$\frac{1}{R_\phi}S_{,\phi} + \frac{2\cot\phi}{R_\theta}S + \frac{1}{R_0}N_{\theta,\theta} + q_\theta = 0 \tag{8.88b}$$

$$\frac{N_\phi}{R_\phi} + \frac{N_\theta}{R_\theta} = q_n \tag{8.88c}$$

At this point it is appropriate to comment on the distributed loading that is shown resolved into components q_ϕ, q_θ, and q_n on the Load Element in Fig. 8.14. If a shell of revolution has loadings q_ϕ and/or q_n, the linear response will only give stress resultants $N_\theta(\phi)$ and $N_\phi(\phi)$ but not S, and ϕ is the single independent variable. Consequently (8.88b) is not needed, and the loading and response are termed axisymmetric. If there is also q_θ loading, all three of the equations participate; the loading and response are termed non-axisymmetric and are a function of two variables, ϕ and θ. The distinction is important because shells of revolution are commonly called axisymmetric shells based on their geometry when, in fact, the loading may or may not be axisymmetric.

From here on, we restrict the discussion to *axisymmetric* loading. The most common loads are gravity or dead load and live load.

On Fig. 8.16 we show a load q in the z-direction distributed over the shell middle surface. In accordance with the choice of the meridional coordinate as ϕ, q is expressed as $q(\phi)$. For dead load q_{DL}, $q = q_{DL}(\phi)$ with components q_ϕ and q_n as shown on the figure. Live load, for example, snow loading, is generally based on the projected surface defined by ΔR_o in Fig. 8.15 and is designated as $q = q_{LL}(R)$. Since the components on the Load Element are based on loading distributed on the middle surface, $q = q_{LL} \Delta R_o/BD = q_{LL} \cos \phi$. The load is then resolved as shown on the figure. The negative sign on q_n follows from the Load Element on Fig. 8.14 where the positive direction is shown as outward from the center of curvature. Another common loading is external or internal pressure p_e or p_i as introduced in Fig. 7.2. In the present notation, these loads correspond to q_n on Fig. 8.14 with the appropriate sign.

8.9   Solution of Membrane Theory Equations

We now consider (8.88a) and (8.88c) dropping the S term since there are no in-plane shearing stresses with the symmetric loading. An obvious reduction is to solve (c) for N_θ and substitute into (a). This would produce a first-order linear differential equation in N_ϕ with a variable coefficient.

A more efficient approach after Novozhilov is to introduce an integrating function that facilitates the solution [6]

$$\psi = N_\phi R_\theta \sin^2 \phi \qquad (8.89)$$

while replacing N_θ in (8.88a) by

$$N_\theta = R_\theta \left(q_n - \frac{N_\phi}{R_\phi} \right) \qquad (8.90)$$

Then (8.88a) becomes

$$\frac{R_\theta^2 \sin \phi}{R_\phi} \psi_{,\phi} = (q_n \cos \phi - q_\phi \sin \phi) R_\theta^3 \sin^2 \phi \qquad (8.91)$$

which reduces to

$$\psi_{,\phi} = (q_n \cos \phi - q_\phi \sin \phi) R_\phi R_\theta \sin \phi \qquad (8.92)$$

with the solution

$$\psi(\phi) = \int (q_n \cos \phi - q_\phi \sin \phi) R_\phi R_\theta \sin \phi \, d\phi \qquad (8.93)$$

In many cases, an alternate form of (8.93) is convenient by recognizing that evaluating the integral for a particular geometry will produce a single integration constant that is determined by the boundary condition on ψ or N_ϕ. If we designate this boundary as ϕ' and the corresponding boundary condition as $\psi(\phi')$, we may write the indefinite integral in the form

$$\psi(\phi) = \psi(\phi') + \int_{\phi'}^{\phi} (q_n \cos\phi - q_\phi \sin\phi)R_\phi R_\theta \sin\phi\, d\phi \qquad (8.94)$$

Of course at $\phi = \phi'$, $\psi = \psi(\phi') = N_\phi(\phi')R_\theta(\phi')\sin^2\phi'$ as specified.
The solution for N_ϕ follows from (8.89) and (8.94) as

$$N_\phi(\phi) = \frac{N_\phi(\phi')R_\theta(\phi')\sin^2\phi'}{R_\theta\sin^2\phi}$$
$$\qquad\qquad (8.95)$$
$$+ \frac{1}{R_\theta\sin^2\phi}\int_{\phi'}^{\phi}(q_n\cos\phi - q_\phi\sin\phi)R_\phi R_\theta\sin\phi\, d\phi$$

and then N_θ is computed from (8.90).

Before we turn to the solution of (8.95) for specific geometries, we consider a closed shell as shown in Fig. 8.17.

Referring to Fig. 8.9, the point where the coordinate $R_o \to 0$, corresponding to $\phi = 0$ in Fig. 8.17a and $\phi = \phi_t$ in Fig. 8.17b, is called the pole. In either case, (8.89) becomes $\psi = N_\phi R_o(\phi)\sin\phi = 0$ since $R_o = 0$ regardless of $\sin\phi$. Then the first terms of both (8.94) and (8.95) vanish.

We may make several observations for closed shells based on the remaining (8.95) and (8.88c):

1. The pole condition does not necessarily imply that $N_\phi = 0$ at that point.
2. For the case of a dome, Fig. 8.17a, where $\phi = 0$, the integrated expressions may produce an indeterminate form $\left(\frac{0}{0}\right)$ that must be evaluated by a limiting operation.
3. Since the available boundary condition on N_ϕ are implied by the pole, there is no opportunity to specify any further boundary conditions on the stress resultants within the membrane theory. To maintain equilibrium, the remaining boundary of the shell *must* develop *whatever* value of N_ϕ is computed from (8.95). Ideally the physical boundary will not also impose constraints that develop transverse shears and bending moments.

Fig. 8.17 Closed shells.
(From Gould [3])

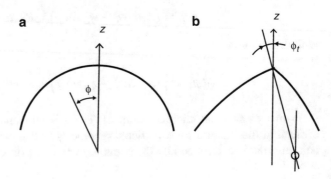

4. For a dome, all meridians meet at the pole, and any direction is parallel to one meridian and perpendicular to another, i.e., the curvilinear coordinates are interchangeable and indistinguishable at the pole. That is, $R_\phi(0) = R_\theta(0) = R_p$ and $N_\phi(0) = N_\theta(0)$. Then from (8.88c),

$$N_\phi(0) = N_\theta(0) = q_n(0)R_p/2 \qquad (8.96)$$

8.10   Spherical Shell Under Dead Load

A spherical shell of revolution with radius a, constant thickness h, mass density ρ, and bound by angles and ϕ_t and ϕ_b is shown in Fig. 8.18.

The load q is

$$q = q_{DL} \qquad (8.97)$$

The weight of the shell, known as the self-weight, is taken as the total dead load q_{DL} and resolved into components q_n and q_ϕ as shown on the figure. Then the principal radii of curvature are easily evaluated. The meridional radius R_ϕ is a, and since the horizontal radius $R_o = a \sin\phi$ through (8.70a), $R_\theta = R_o/\sin\phi = a$.

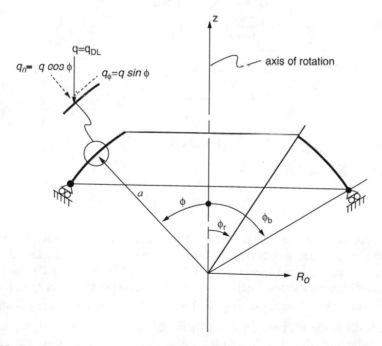

Fig. 8.18 Spherical shell under dead load. (From Gould [3])

The boundaries are assumed to satisfy ideal membrane theory assumptions. The upper boundary is taken as stress-free, and the lower boundary is constrained to be unyielding in the ϕ or meridional direction so as to develop N_ϕ and free to displace in the normal direction $\mathbf{t_n}$ and rotate about an axis along $\mathbf{t_\theta}$. The latter two releases correspond to the transverse shear Q_ϕ and the meridional moment M_ϕ defined in (8.77) and (8.79), respectively, with $\alpha = \phi$.

We substitute these values into (8.95) taking $\phi' = \phi_t$ so that $N_\phi(\phi_t) = 0$. Then we have

$$
\begin{aligned}
N_\phi(\phi) &= \frac{1}{a\sin^2\phi} \int_{\phi'}^{\phi} (-q\cos^2\phi - q\sin^2\phi)a^2\sin\phi \, d\phi \\
&= \frac{qa}{\sin^2\phi}(\cos\phi - \cos\phi_t).
\end{aligned}
\tag{8.98}
$$

Then from (8.90),

$$
N_\theta = a\left(-q\cos\phi - \frac{N_\phi}{a}\right)
\tag{8.99}
$$

We may now investigate a spherical dome by setting $\phi_t = 0$ in (8.98). Using an elementary trigonometric identity, Eq. (8.98) becomes

$$
\begin{aligned}
N_\phi &= \frac{qa}{(1 + \cos\phi)(1 - \cos\phi)}(\cos\phi - 1) \\
&= \frac{-qa}{1 + \cos\phi}
\end{aligned}
\tag{8.100}
$$

and (8.99) follows asd

$$
\begin{aligned}
N_\theta &= a\left(-q\cos\phi - \frac{q}{1 + \cos\phi}\right) \\
&= qa\left(\frac{1}{1 + \cos\phi} - \cos\phi\right)
\end{aligned}
\tag{8.101}
$$

At the pole with $\phi = 0$,

$$
N_\phi = N_\theta = \frac{-qa}{2}
\tag{8.102}
$$

It is instructive to consider the behavior of stress resultants N_ϕ and N_θ along the meridian of a dome as the angle defining the base ϕ_b increases, i.e., the shell becomes deeper. It is apparent from (8.100) or from Fig. 8.19 that the meridional stress continues to increase proportional to the increased mass but spread over a larger circumference until $\phi_b = \frac{\pi}{2}$. In fact it can be shown that N_ϕ is directly proportional to the total weight of the shell. However N_θ given by (8.99) behaves quite differently. Basically N_θ is not subject to any boundary constraints but is

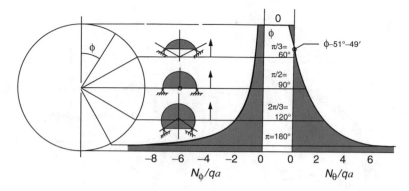

Fig. 8.19 Self-weight stress resultants for spherical dome. (From Gould [3])

determined by the *difference* of two terms, one containing the surface loading q and the other the meridional stress. In Fig. 8.19, (8.101) and (8.102) are plotted for increasing values of ϕ, so the graphs represent the solution for *any* selected constant thickness dome with the lower boundary defined by the corresponding value of ϕ. As shown, N_θ may vary from tension to compression over a shell surface depending on the depth. Note that the ordinate is ϕ so that the curve distorts the vertical distribution of N_θ. Nevertheless, shells that are fairly deep ($\phi' > 51°$) will have circumferential tension in the lower portion under dead load, a concern for unreinforced concrete or masonry shells.

8.11   Lower Boundary for a Shell of Revolution

We now examine the lower boundary of a shell of revolution with the form shown in Fig. 8.18.

Recall that the lower support must develop the calculated value of N_ϕ at $\phi = \phi_b$. In Fig. 8.20a an idealized support is shown where a circumferential ring beam is employed to resist the thrust that is assumed to act at the centroid of the beam. While the vertical component **V** is transmitted directly to the foundation, the horizontal reaction $H = -N_\phi(\phi_b)\cos\phi_b$ is applied around the ring as shown in Fig. 8.20b and resisted by the hoop tension T. Summing forces along $\mathbf{t}_x$,

$$2\mathbf{T} = \int_{-\pi/2}^{\pi/2} -N_\phi(\phi_b)\cos\phi_b\cos\theta(bd\theta)\mathbf{t}_x \tag{8.103}$$

with magnitude

$$T = -N_\phi(\phi_b)b\cos\phi_b \tag{8.104}$$

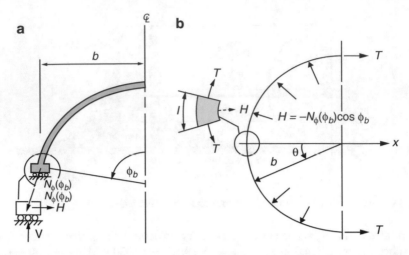

Fig. 8.20 Ring beam for a shell of revolution. (From Gould [3])

For a spherical shell,

$$b = a \sin \phi_b \tag{8.105}$$

so that from (8.104)

$$T = -N_\phi(\phi_b)a \cos \phi_b \sin \phi_b \tag{8.106}$$

Substituting the solution for the self-weight on a spherical dome (8.100), the tension is

$$T_{DL} = qa^2 \frac{\cos \phi_b \sin \phi_b}{1 + \cos \phi_b} \tag{8.107}$$

Dividing T_{DL} by the area of the ring beam gives the circumferential stress in the ring beam, equivalent to $\sigma_{\theta\theta}$ in the shell. Obviously there would be an incompatibility of the corresponding circumferential strains at the junction unless these quantities coincide by chance. To enforce a compatible condition at the junction requires consideration of bending forces; from a practical standpoint, it is generally assumed that the effects of the mismatch are confined to a narrow region near the boundary. This is in accordance with St. Venant's principle used extensively in Chap. 7. For concrete shells where tension is undesirable, prestressing is commonly used to reduce the incompatibility.

8.12   Spherical Shell Under Internal Pressure

Now we consider the case of uniform internal pressure $qn = +p$ in a spherical dome. From the second term in (8.95) and the spherical geometry,

$$
\begin{aligned}
N_\phi(\phi) &= \frac{1}{a\sin^2\phi} \int_0^\phi p\cos\phi\, a^2\sin\phi\, d\phi \\
&= \frac{pa^2}{a\sin^2\phi} \left[\frac{1}{2}\sin^2\phi\right]_0^\phi \\
&= \frac{pa}{2}
\end{aligned}
\tag{8.108}
$$

and from (8.90)

$$
\begin{aligned}
N_\theta(\phi) &= a\left(p - \frac{N_\phi}{a}\right) \\
&= \frac{pa}{2}
\end{aligned}
\tag{8.109}
$$

so that there is a uniform state of stress $N_\phi = N_\theta = pa/2$ over the entire shell.

8.13   Cylindrical Shell Under Normal Pressure

Cylindrical shells have been mentioned briefly in Sect. 7.5 related to Fig. 7.2b, where values of $\sigma_{\theta\theta}$ are plotted to demonstrate the differences between thick and thin shell solutions. The thin shell solution is easily found from (8.88c) with $R_\phi = \infty$, $R_\theta = r$ denoting the middle-surface radius of the cylinder, and $q_n = p$ as

$$
N_\theta = pr
\tag{8.110}
$$

with $i=\theta$, giving from (8.83a)

$$
\sigma_{\theta\theta} = pr/h
\tag{8.111}
$$

for a shell of thickness h. This is equivalent to the stress $(\sigma_{\theta\theta})_{\mathrm{av}}$ plotted in Fig. 7.2 and discussed in Sect. 7.5.

8.14   Use of MuPAD to Solve Plate Problems (Sect. 8.4)

If a thin plate has an elliptical planform, following the strategy used for the torsion problem, the solution is assumed proportional to the equation of the boundary: $u_z = C\left(\frac{x^2}{D^2} + \frac{y^2}{B^2} - 1\right)^2$. Show for a uniform load q_0 the computation of the constant C.

To solve this problem, we use the plate Eq. (8.20). Using MuPAD, we can avoid the long manual derivation of the term $\nabla^4 u_z$ and solve the problem easily with the following commands.

- We first calculate the term $\nabla^4 u_z$.

```
uz := C*(x^2/D^2 + y^2/B^2 - 1)^2;
```
$$C \left(\frac{x^2}{D^2} + \frac{y^2}{B^2} - 1 \right)^2$$

```
del4u := diff(uz,x,x,x,x) + diff(uz,y,y,y,y) + 2*diff(uz,x,x,y,y);
```
$$\frac{24\,C}{B^4} + \frac{24\,C}{D^4} + \frac{16\,C}{B^2\,D^2}$$

- Then, we use the plate equation to solve C.

```
Simplify(q0/(Elas*h^3) * 12*(1-v^2)/(Elas*h^3)/((24)/B^4 + (24)/D^4 + (16)/(B^2*D^2)))
```
$$-\frac{12\,q0\,(v^2-1)}{Elas^2\,h^6 \left(\frac{24}{D^4} + \frac{16}{B^2\,D^2} + \frac{24}{B^4} \right)}$$

- We can also visualize the 2-D function u_z.

```
plot(plot::Function3d(subs(uz,C=1,B=1,D=2), x = -2..2, y=-1..1))
```

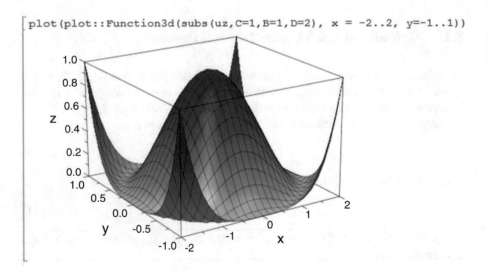

8.15   Exercises

8.1 Consider a thin plate having the elliptical planform shown in Fig. 6.7a. Following the strategy used for the torsion problem, the solution is assumed proportional to the equation of the boundary [1]

$$u_z = C\left(\frac{x^2}{D^2} + \frac{y^2}{B^2} - 1\right)^2.$$

(a) Show that this solution satisfies the clamped boundary conditions equivalent to (8.57).
(b) For a uniform load, q_0, compute the constant C.
(c) Evaluate the maximum displacement.

8.2 Consider the solution for a simply supported plate, (8.39), and particularly the displacement at the center of the plate.

(a) Compare the values after one, three, and five terms.
(b) Repeat the computation for the stress σ_{xx} at the center.

8.3 Consider a clamped solid circular plate with radius R_0 under a "sand heap" loading

$$q(r) = q_0(r - R_0)$$

and determine the deflection function.

8.4 Consider the solutions for the solid circular plate with both clamped and simply supported boundary conditions.

(a) Evaluate $\sigma_{rr}(0)$ for the two boundary conditions. How does this compared to the difference between a clamped and simply supported beam if $v = 0.3$?
(b) Evaluate $\sigma_{\theta\theta}$ for both cases.
(c) Why are the values of σ_{rr} and $\sigma_{\theta\theta}$ equal at $r = 0$ in each case?

8.5 Considering the equilibrium equations for a shell of revolution (8.84), verify the reduced form given by (8.88). Note that terms requiring derivatives with respect to θ, mainly $R_{\theta,\theta}$, should be differentiated using the Gauss–Codazzi relationship (8.70b), so that it may be necessary to multiply top and bottom of a term containing R_θ by $\sin \phi$ in some cases.

8.6 For the spherical shell considered in Sect. 8.10, derive a solution for N_ϕ and N_θ for the case of a live load $q = q_{LL}\cos\phi$ as discussed in Sect. 8.8. Then on Fig. 8.18 the components are $q_n = -q\,\cos^2\phi$ and $q_\phi = q\,\cos\phi\,\sin\phi$. Also evaluate the ring beam tension T_{LL} for a dome with a base angle of $45°$.

8.7 Consider the shell shown in Fig. 8.18

(a) Show the meridional stress resultant at the base $N_\phi(\phi_b)$.
(b) Assume that the total weight of the shell above the base is Q acting in the negative z-direction.
(c) Develop a relationship between Q and $N_\phi(\phi_b)$ in terms of the radius a and the base angle ϕ_b.

8.8 Consider a cylindrical tank with middle-surface radius r, thickness h, and height H. The Z coordinate is measured from the base of the shell.

(a) Compute the membrane theory stress resultants for a hydrostatic loading with fluid of density γ for a full tank.
(b) Discuss in general the requirements for the lower boundary of the shell to maintain membrane action.

8.9 Consider an annular plate with planform as shown in Fig. 7.2a.

The outside boundary $r = R_1$ is clamped, while the inside boundary $R = R_o$ is free to displace but restrained against rotation. State the boundary conditions in terms of the displacement u_z and its derivatives if possible.

References

1. Filonenko-Borodich M (1965) Theory of elasticity. Dover, New York
2. Green AE, Zerna W (1968) Theoretical elasticity. Oxford University Press, Oxford
3. Gould PL (1988) Analysis of shells and plates. Springer, New York
4. Timoshenko S, Woinowski-Kreiger S (1959) Theory of plates and shells, 2nd edn. McGraw-Hill, New York
5. Billington DD (1982) Thin shell concrete structures, 2nd edn. Mc Graw-Hill, New York
6. Novozhilov VV (1964) Thin shell theory [Translated from 2nd Russian ed. by P.G. Lowe] Noordhoff, Gronigen

Chapter 9
Dynamic Effects

Abstract The strains in an elastic body may be computed from a specified displacement field using the equations of compatibility, regardless of whether the displacements arise from static or dynamic excitation. The corresponding stresses and, indeed, the displacements themselves may be dependent on the rate characteristics of the loading function. Therefore, the time derivatives of the displacements, i.e., velocities and accelerations, enter into these equations. In this chapter, the field equations of linear elasticity are generalized to include rate-dependent terms. Further, in the case of dynamic excitation, two important system characteristics are the *natural frequencies*, or their reciprocal the *periods* of vibration, and the corresponding *mode* shapes. A related quantity, which is dependent only on the material properties, is the *velocity* at which an elastic wave *propagates* through a medium or a body. Damping, which represents internal energy dissipation, is also an important consideration in dynamic response computations and may be dependent on the displacement, the velocity, or other rate quantities.

9.1 Introduction

The strains in an elastic body may be computed from a specified displacement field using the equations of compatibility, regardless of whether the displacements arise from static or dynamic excitation. The corresponding stresses and indeed the displacements themselves may be dependent on the rate characteristics of the loading function. Therefore the time derivatives of the displacements, i.e., velocities and accelerations, enter into these equations.

It is often expedient to separate the response of a linear system into a product of the effects of the *system* and the effects of the *load*. This concept is useful in static cases and is usually based on identifying the Green's function for the system, the response (displacement, strain, stress) due to the application of a *unit* load, and integrating that function over a specified load range to evaluate the corresponding total response. For example, (7.93a–c) with $P = 1$ would be Green's functions for a vertical load on a semi-infinite plate.

In the case of dynamic excitation, two important system characteristics are the *natural frequencies* or their inverse, the *periods* of vibration, and the corresponding *mode* shapes. Once these are found by considering the homogeneous part of the governing differential equation, the response to a forcing function may be readily obtained as a particular solution. Also even if the loading is not specified, these properties are broadly utilized for system characterization. For example, structures are said to be stiff or flexible based on the relative natural frequencies. A related quantity which is dependent only on the material properties is the *velocity* at which an elastic wave *propagates* through a medium or a body. Wave velocities are widely used for comparative classifications of media and materials.

Damping, which represents internal energy dissipation, is an important consideration in dynamic response computations and may be dependent on the displacement, the velocity, or other rate quantities. In effect, damping is a quantity that is commonly considered at the macro-level of linear elasticity but which really represents microlevel molecular processes. Obviously, there are a variety of models that attempt to depict physical reality.

In this chapter, the field equations of linear elasticity will be generalized to include rate-dependent terms. We will first focus on the propagation of waves through the medium and then describe the motion of the particles comprising the medium. The transmission of energy by elastic waves through a semi-infinite medium and also along the surface is the basis for understanding the effects of a rupture occurring kilometers beneath the earth's surface on objects possibly thousands of kilometers distant during an earthquake.

9.2   Vibrations in an Infinite Elastic Medium

9.2.1   Equilibrium Equations

We begin with Eq. (5.7) and consider the body forces to be the inertial forces corresponding to the acceleration, by D'Alembert's principle. Thus, we have

$$\mu u_{j,ii} + (\lambda + \mu)u_{i,ij} - \rho \ddot{u}_j = 0, \tag{9.1}$$

where ρ is the mass density, $(\cdot)$ indicates $\frac{\partial(\ )}{\partial t}$, and t is the time variable.

Initially, we are not concerned with boundary conditions on u_i since the medium is assumed to be infinite. The general derivations are carried out in indicial notation with augmentation using explicit notation [1]. We consider separately longitudinal vibrations, which correspond to the decomposition of u_i into an irrotational component, and transverse vibrations, which correspond to a divergence-free or non-volume change component, as discussed in Chap. 5.

Fig. 9.1 Plane in Cartesian
space

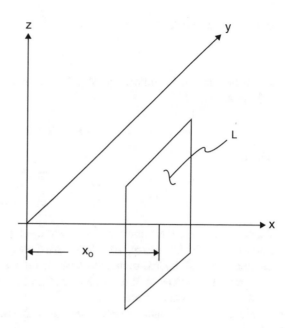

9.2.2   *Longitudinal Vibrations*

We refer to a Cartesian coordinate system and consider time-dependent displacements along one axis, say the *x*-axis in Fig. 9.1. For the moment, we do not dwell on
the source of the motion but only the response. The displacement field is

$$u_x = u_x(x, t) \tag{9.2a}$$

$$u_y = 0 \tag{9.2b}$$

$$u_z = 0. \tag{9.2c}$$

Considering the plane L parallel to the y–z plane at a distance x_0 from the origin, it
will assume a position $x = x_0 + u_x(x_0, t)$ at any time $t > 0$ during the motion. If
excitations along the x-axis are present, the plane will travel back and forth along the
axis. Other planes initially parallel to L will move in a like manner. While the planes
will remain parallel, the distance between any two will alternately increase and
decrease creating a state of uniform longitudinal vibration. Two planes initially at
$x = x_0$ and $x = x_1$ will be separated by a distance

$$
\begin{aligned}
d(x_0, x_1, t) &= [x_1 + u_x(x_1, t)] - [x_0 + u_x(x_0, t)] \\
&= x_1 - x_0 + u_x(x_1, t) - u_x(x_0, t)
\end{aligned}
\tag{9.3}
$$

at any time t. For this case, (9.1) with $i = x$, y, z becomes

$$\mu u_{x,xx} + (\lambda + \mu)u_{x,xx} - \rho \ddot{u}_x = 0, \tag{9.4}$$

since the motion is uniform and only a function of x. Equation (9.4) may be written as

$$\ddot{u}_x = c_P^2 u_{x,xx} \tag{9.5a}$$

where

$$c_P = \sqrt{\frac{\lambda + 2\mu}{\rho}} = \sqrt{\frac{E(1 - \nu)}{\rho(1 + \nu)(1 - 2\nu)}}, \tag{9.5b}$$

which is known as the one-dimensional wave equation. Thus, longitudinal vibrations described by (9.2a), (9.2b), and (9.2c) are possible if $u_x(x, t)$ satisfies (9.5a). The parameter c_P will be interpreted later. Since the waves generated oscillate along the axis producing extension and contraction, they are dilation waves that are commonly called P (for pressure) waves.

The preceding equation may be derived from another approach. The condition of no rotation

$$\omega_{ij} = \frac{1}{2}\left(u_{i,j} - u_{j,i}\right) = 0, \tag{9.6}$$

from (3.31b) is satisfied by defining a potential function Φ such that

$$u_i = \Phi_{,i} \tag{9.7a}$$

and

$$u_j = \Phi_{,j} \tag{9.7b}$$

from which

$$u_{i,i} = \Phi_{,ii}. \tag{9.8}$$

Then (9.1) becomes

$$\mu \Phi_{,jii} + (\lambda + \mu)\Phi_{,iij} - \rho \ddot{u}_j = 0, \tag{9.9}$$

Using (9.7b) to replace $\Phi_{,j}$ we get

$$(\lambda + 2\mu)u_{j,ii} - \rho \ddot{u}_j = 0 \tag{9.10a}$$

or

$$(\lambda + 2\mu)\nabla^2 u_j - \rho \ddot{u}_j = 0, \tag{9.10b}$$

which reduces to (9.5a) for the one directional motion described by (9.2a), (9.2b), and (9.2c). Thus, the specialization which produces P waves is known as irrotational wave propagation. Also, the form of (9.10a) and (9.10b) is more general in that the motion need not be completely uniform throughout the medium so long as it is irrotational.

9.2.3 Transverse Vibrations

Now we choose a displacement field normal to the x-axis, say in the z-direction,

$$u_x = 0 \tag{9.11a}$$

$$u_y = 0 \tag{9.11b}$$

$$u_z = u_z(x, t). \tag{9.11c}$$

In this case, the plane L and all parallel planes will move perpendicular to the x-axis. All points on L move the same distance u_z, but the planes move relative to one another in the z-direction setting up a state of uniform transverse vibrations. Equation (9.1) reduces to

$$\mu u_{z,xx} - \rho \ddot{u}_z = 0 \tag{9.12}$$

since the motion is still only a function of x or

$$\ddot{u}_z = c_s^2 u_{z,xx} \tag{9.13a}$$

where

$$c_s = \sqrt{\frac{\mu}{\rho}} = \sqrt{\frac{E}{2\rho(1+\nu)}}. \tag{9.13b}$$

Thus, transverse vibrations described by (9.11a), (9.11b), and (9.11c) may occur if $u_z\,(x,\,t)$ satisfies (9.13a). Again the parameter c_S will be interpreted later. The waves generated are known as shear or S waves since they produce transverse motion but no extensional deformation.

We may also derive the preceding equations by noting that the transverse vibrations cause no volume change. From (3.41), the dilation

$$\Delta = \varepsilon_{ii} = u_{i,i} = 0, \tag{9.14}$$

Setting $u_{i,i} = 0$ in (9.1) gives

$$\mu u_{j,ii} - \rho \ddot{u}_j = 0 \tag{9.15a}$$

or

$$\mu \nabla^2 u_j - \rho \ddot{u}_j = 0, \tag{9.15b}$$

which reduces to (9.12) for the motion described by (9.11a), (9.11b), and (9.11c). Therefore, the specialization which produces S waves is known as incompressible wave propagation.

It should also be noted that an identical solution could have been constructed for the displacement in the y-direction, i.e., $u_y = u_y(x, t)$, so that two orthogonal sets of shear waves, **SH** and **SV** waves, may be generated. Again, the form of (9.15a) and (9.15b) is more general than the reduced (9.12).

9.2.4   Transverse Isotropic Material

The Lamé equation (9.1) may be generalized in principle for a general anisotropic medium as characterized by (4.6) [2]. This is beyond the scope of the present treatment, but some insight may be gained by considering the case of transverse isotropy as developed in Sect. 4.4.

The transverse isotropic model is suitable for biological applications because it adequately describes the elastic properties of bundled fibers aligned in one direction as shown in Fig. 4.5 [3]. The five material constants in $[C_{TI}]$ (4.31) are grouped as the two shear moduli $\mu_{\parallel}$ and $\mu_{\perp}$ that describe the shear wave propagation perpendicular and parallel to the direction of the fibers and the Lamé constants $\lambda_{\perp} \lambda_{\parallel}$ and λ_M that describe the propagation of the longitudinal waves in the respective directions. Wave propagation is discussed in the next section.

After some minor simplifications and the introduction of the constitutive law (4.31), the generalized Lamé equation is written in the form similar to (9.1) for $\boldsymbol{u} = \{u_x u_y u_z\}$ as [3]

$$\mu_{\perp} \nabla^2 \boldsymbol{u} + (\lambda + \mu_{\perp}) \nabla (\nabla \cdot \boldsymbol{u}) + \tau \begin{bmatrix} u_{x,zz} + u_{z,xz} \\ u_{y,zz} + u_{z,yz} \\ \nabla^2 u_z + (\nabla \cdot \boldsymbol{u})_{,z} \end{bmatrix} = \rho \ddot{\boldsymbol{u}} - \zeta \nabla^2 \dot{\boldsymbol{u}}, \tag{9.16}$$

where $\tau = \mu_{\parallel} - \mu_{\perp}$ and where it is assumed that all elastic moduli related to the longitudinal wave are equal, i.e., $\lambda = \lambda_{\perp} = \lambda_{\parallel} = \lambda_M$. The latter assumption is strictly justified for incompressible materials where $\lambda \gg \mu$ [3], but other applications may alter this assumption. Also introduced in this equation is the shear viscosity ζ that produces energy dissipation proportional to the velocity $\dot{\boldsymbol{u}}$. Note that the τ term, representing the difference of the shear moduli between the longitudinal and transverse directions, results in a coupling of the displacements in the equilibrium equations. Also the operator $\nabla (\nabla \cdot \mathbf{u}) = \nabla^2 \mathbf{u} + \nabla \times (\nabla \times \mathbf{u})$ [4], which is useful for decoupling the irrotational and incompressible components derived in Sects. 9.2.2 and 9.2.3, respectively, is utilized.

Of course when $\tau = 0$ and ζ is neglected, we retrieve the uncoupled isotropic (9.1) and proceed to further separate the longitudinal and transverse components. In the general case where $\tau \neq 0$, it is noted that because soft tissue is nearly incompressible with $\nu \rightarrow 0.5$, the orders of magnitude of the shear moduli $\mu_\parallel$ and $\mu_\perp$ are far smaller than the λ values (See 4.24b). Then the transverse and longitudinal components of the vibrating wave may be uncoupled under controlled conditions [3].

From this slight extension of the idealized isotropic material assumption, it is apparent that anisotropy in a medium complicates the description of the propagation of waves considerably.

9.2.5   Harmonic Vibrations

Longitudinal Motion

We now consider a sinusoidal form for the function u_x in (9.2a)

$$u_x(x, t) = A \sin 2\pi \left(\frac{x}{l_P} - \frac{t}{T_P} \right). \tag{9.17}$$

The normalizing terms l_P and T_P will be interpreted later. Equation (9.17) is substituted into (9.5a) to get

$$\frac{1}{T_P^2} = \frac{c_P^2}{l_P^2}. \tag{9.18}$$

Therefore, the conditions on l_P and T_P to produce the motion described by (9.17) are

$$\frac{l_P}{T_P} = c_P = \sqrt{\frac{\lambda + 2\mu}{\rho}}. \tag{9.19}$$

The amplitude of the vibration, A, is arbitrary.

This motion is represented by the trace of the planes such as L in the x–z plane in Fig. 9.2. The term T_P in (9.17) is called the *period* of the traveling waves. For a point at $x = \bar{x}$, which represents all points on a single plane, the displacement $u_x(\bar{x}, t)$ evaluated at a specific time $\bar{t}$ is

$$u_x(\bar{x}, \bar{t}) = A \left[\sin \left(2\pi \frac{\bar{x}}{l_P} \right) \cos \left(2\pi \frac{\bar{t}}{T_P} \right) - \cos \left(2\pi \frac{\bar{x}}{l_P} \right) \sin \left(2\pi \frac{\bar{t}}{T_P} \right) \right] \tag{9.20a}$$

Fig. 9.2 Traces of longitudinally vibrating planes

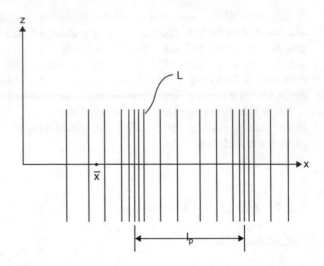

while at a time $\bar{t} + T_P$

$$
u_x(\bar{x}, \ \bar{t} + T_P) = A\left[\sin\left(2\pi\frac{\bar{x}}{l_P}\right)\cos\left(2\pi\frac{\bar{t} + T_P}{T_P}\right) - \cos\left(2\pi\frac{\bar{x}}{l_P}\right)\sin\left(2\pi\frac{\bar{t} + T_P}{T_P}\right)\right]
$$

$$
= A\left[\sin\left(2\pi\frac{\bar{x}}{l_P}\right)\cos\left(2\pi + 2\pi\frac{\bar{t}}{T_P}\right) - \cos\left(2\pi\frac{\bar{x}}{l_P}\right)\sin\left(2\pi + 2\pi\frac{\bar{t}}{T_P}\right)\right]
$$

$$
= u_x(\bar{x}, \ \bar{t})
$$

(9.20b)

since $\cos(2\pi + t) = \cos t$ and $\sin(2\pi + t) = \sin t$. The same would be true at $t = \bar{t} + 2T_P$ or $t = \bar{t} + nT_P$, where n is an integer. Thus, the motion is *repeated* after a duration of T_P.

To interpret parameter l_P, we focus on the axial strain (3.15a)

$$
\varepsilon_{xx}(x, \ t) = u_{x,x} = A\frac{2\pi}{l_P}\cos 2\pi\left(\frac{x}{l_P} - \frac{t}{T_P}\right).
$$

(9.21)

At an instant in time $t = \bar{t}$, a representative sampling of planes that were equally spaced when the system was at rest is selected. It may be shown that at $t = \bar{t}$, a plane located at $x + l_P$ has the same strain as the plane at x, i.e.,

$$
\varepsilon_{xx}(x + l_P, \ \bar{t}) = \varepsilon_{xx}(x, \ \bar{t}),
$$

(9.22a)

or, in general,

$$
\varepsilon_{xx}(x + nl_P, \ \bar{t}) = \varepsilon_{xx}(x, \ \bar{t}).
$$

(9.22b)

The relationship between equally strained planes is shown in Fig. 9.2. The separation l_P is known as the *wave length*. The maximum contraction strain is in the region where the planes are most dense, while the maximum extension is in the least dense region. Referring to (9.21), these maxima occur when

$$\cos 2\pi \left(\frac{x}{l_p} - \frac{t}{T_p} \right) = \pm 1 \qquad (9.23a)$$

or

$$\frac{x}{l_P} - \frac{t}{T_P} = \frac{n}{2} \qquad (9.23b)$$

where n is an integer. Also, (9.23b) indicates that the coordinates of these points of extreme strain $x = x_e$ move linearly in time with a velocity

$$\dot{x}_e = \frac{l_P}{T_P}. \qquad (9.24)$$

But, from (9.18),

$$\frac{l_P}{T_P} = c_P = \sqrt{\frac{\lambda + 2\mu}{\rho}} \qquad (9.25)$$

so that the parameter c_P encountered originally in (9.5a) and (9.5b) has finally been identified as the velocity of propagation of the pressure waves produced by longitudinal vibrations.

An important application is for sonic vibrations and the associated strains, which occur in the range $\frac{1}{16}s > T_P > 2 \times 10^{-5}$ s [1]. Such waves propagate through an infinite medium with a velocity dependent on the material properties λ, μ, and ρ which enter into (9.25). Some closely related applications in the strength of materials are the longitudinal vibration of a bar and longitudinal wave motion in a fluid column.

Transverse Motion

If we assume an analogous sinusoidal form for u_z in (9.11c)

$$u_z = B \sin 2\pi \left(\frac{x}{l_S} - \frac{t}{T_S} \right) \qquad (9.26)$$

Fig. 9.3 Traces of
transversely vibrating planes

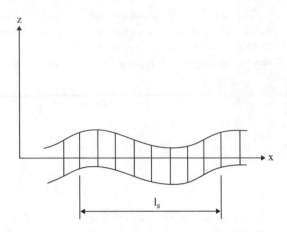

the traces for the transverse vibration of the waves are shown in Fig. 9.3. The
wavelength is l_S and the velocity of propagation is

$$c_S = \sqrt{\frac{\mu}{\rho}}, \tag{9.27}$$

which is smaller than c_P. Here, we have a shear strain computed from (3.15e) as

$$\varepsilon_{xz} = \frac{1}{2}u_{z,x}$$
$$= B\frac{\pi}{l_S}\cos 2\pi \left(\frac{x}{l_S} - \frac{t}{T_S}\right). \tag{9.28}$$

The shear strain propagates with the same velocity c_S, just as in the strength of
materials where the transverse vibration of a shear beam is described by the response
to waves propagating with velocity c_S.

To compare the longitudinal and transverse motions, we form the ratio

$$\frac{c_S}{c_P} = \sqrt{\frac{\mu}{\lambda + 2\mu}}. \tag{9.29}$$

Both μ and λ may be expressed in terms of Poisson's ratio v and Young's modulus
E through (4.24a, b). Then,

$$\frac{c_S}{c_P} = \sqrt{\frac{1 - 2v}{2 - 2v}}. \tag{9.30}$$

For $v = \frac{1}{3}, \frac{c_S}{c_P} = \frac{1}{2}$.

While the velocities c_P and c_S are quite high, on the order of 5 km/s through metal
[1], the difference in velocities can often be distinguished and the P and S waves thus

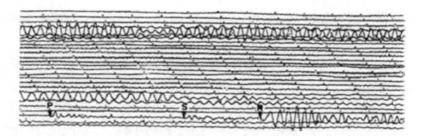

Fig. 9.4 Seismogram from a long-period seismograph showing the vertical component of elastic wave motion recorded at Oroville, California, during part of 1 day. The third trace from the bottom is from a magnitude 5 earthquake in Alaska. The time between breaks in the trace is 1 min (Bolt [10], © 1970, p. 5. Reprinted by permission of Prentice Hall, Englewood Cliffs, NJ)

identified. In Fig. 9.4, a seismogram which records periodic wave motion is shown [5]. The marked record is an earthquake in Alaska recorded in California. The peaks corresponding to the faster-moving P waves followed by the slower S waves are shown along with peaks due to another even slower moving type of wave, the L or Love waves that are caused by surface effects.

Again we note the possibility of a second orthogonal set of S waves acting in the x–y plane and the multidirectional forms of the wave propagation equations, (9.10a), (9.10b), (9.15a), and (9.15b). However, in any case, the velocity of propagation remains a constant dependent only on the properties of the medium. The difference in velocity between the P and S waves is used in many geophysical techniques.

9.3   Free Vibration

9.3.1   Equations of Motion

Vibration of elastic bodies is very important in practical engineering and is most often applied to specialized structural forms such as beams and plates. However, the governing equations can be developed to a certain extent within the present context of elasticity.

While the previous sections have focused on the propagation of waves through a medium, we now focus on the motion of particles within the medium. Again, the source of the motion is not germane at this point. We simply assume that the body is vibrating in a normal mode, which means that all particles are moving synchronously and pass through the zero or rest position simultaneously [6]. The deformation may be visualized as a series of independently vibrating orthogonal modes, such as sine or cosine curves, which may be superposed to approximate a more complicated motion. This is, of course, the physical interpretation of a Fourier series and is the basis of one of the most powerful analytical tools in mathematical physics, harmonic analysis.

Referring to (9.1), the displacement is taken as the product of a function A_j of the spatial coordinates x_j and a function f of the time coordinate t, i.e.,

$$u_j(x_j, \ t) = A_j(x_j)f(t). \tag{9.31}$$

Substituting into (9.1) gives

$$\mu\left[A_{j,ii} + \left(\frac{\lambda + \mu}{\mu}\right)A_{i,ij}\right]f = \rho A_j \ddot{f} \tag{9.32}$$

or

$$\frac{\mu}{A_j}\left[A_{j,ii} + \left(\frac{\lambda + \mu}{\mu}\right)A_{i,ij}\right] = \rho\frac{\ddot{f}}{f}. \tag{9.33}$$

With the l.h.s. being a function of x_j and the r.h.s. being a function of $t, \dfrac{\ddot{f}}{f}$ is at most a constant, i.e.,

$$\frac{\ddot{f}}{f} = -\omega^2. \tag{9.34}$$

Thus, we take

$$f = C\cos\omega t + D\sin\omega t \tag{9.35}$$

but since the time origin is arbitrary, we retain

$$f = D\sin\omega t. \tag{9.36}$$

Then the equilibrium (9.1) becomes, in view of (9.34),

$$\mu\left[u_{j,ii} + \left(\frac{\lambda + \mu}{\mu}\right)u_{i,ij} + \frac{\rho}{\mu}\omega^2 u_j\right] = 0 \tag{9.37}$$

or by comparing (5.1) and (5.7) and reintroducing the stress tensor, we can write (9.37) as

$$\sigma_{ij,i} + \rho\omega^2 u_j = 0. \tag{9.38}$$

Equation (9.37) may also be stated in terms of the engineering constants G and v using (4.24) as

$$u_{j,ii} + \frac{1}{1 - 2v}u_{i,ij} + \frac{\rho}{G}\omega^2 u_j = 0. \tag{9.39}$$

Equations (9.38) or (9.39) are eigenvalue equations, similar to (2.55) for the principal stresses in Chap. 2. For a set of boundary conditions prescribed on the surface of the body, (9.38) or (9.39) may be satisfied only for certain values of ω, which are the *natural frequencies* of vibration in this case. The corresponding u_j are the *mode shapes*. Within the context of our general treatment, we may proceed with a general proof of the orthogonality condition that is an important component to this analysis.

9.3.2 Orthogonality Conditions

We seek to derive certain conditions on the solutions of (9.39). We select two such solutions, say $u_j^{(1)}$ and $u_j^{(2)}$, and first specialize the equation for $u_j^{(1)}$:

$$u_{j,ii}^{(1)} + \frac{1}{1 - 2\nu} u_{i,ij}^{(1)} + \frac{\rho}{G} \omega_1^2 u_j^{(1)} = 0 \qquad (9.40)$$

or in terms of the stresses, as given by (9.38),

$$\sigma_{ij,i}^{(1)} + \rho \omega_1^2 u_j^{(1)} = 0. \qquad (9.41)$$

Multiplying (9.41) by $u_j^{(2)}$ and integrating over the volume gives

$$\int_V u_j^{(2)} \sigma_{ij,i}^{(1)} dV = -\rho \omega_1^2 \int_V u_j^{(2)} u_j^{(1)} dV. \qquad (9.42)$$

Next, expanding the l.h.s. using integration by parts on a volume integral [6] produces

$$\int_V u_j^{(2)} \sigma_{ij,i}^{(1)} dV = \int_A u_j^{(2)} \sigma_{ij}^{(1)} n_i dA - \int_V u_{j,i}^{(2)} \sigma_{ij}^{(1)} dV. \qquad (9.43)$$

Note that the first term on the r.h.s contains n_i similar to the divergence theorem (1.14). This term is written in terms of the traction T_j as

$$\int_A u_j^{(2)} \sigma_{ij}^{(1)} n_i dA = \int_A u_j^{(2)} T_j^{(1)} dA \qquad (9.44)$$

using (2.11). The second term on the r.h.s. of (9.43) may be manipulated into a form that is symmetric in (1) and (2) by first rearranging the integrand as

$$u_{j,i}^{(2)} \sigma_{ij}^{(1)} = \frac{1}{2} \sigma_{ij}^{(1)} \left(u_{i,j}^{(2)} + u_{j,i}^{(2)} \right) \qquad (9.45)$$

recognizing the summation convention. Continuing, from (5.2),

$$\frac{1}{2} \sigma_{ij}^{(1)} \left(u_{i,j}^{(2)} + u_{j,i}^{(2)} \right) = \varepsilon_{ij}^{(2)} \sigma_{ij}^{(1)}. \qquad (9.46a)$$

We use the constitutive law (4.30) to replace $\varepsilon_{ij}^{(2)}$ and produce

$$\varepsilon_{ij}^{(2)}\sigma_{ij}^{(1)} = \frac{1}{E}\left[(1+\nu)\sigma_{ij}^{(2)}\sigma_{ij}^{(1)} - \nu\delta_{ij}\sigma_{kk}^{(2)}\sigma_{ij}^{(1)}\right]$$
$$= \frac{1}{E}\left[(1+\nu)\sigma_{ij}^{(2)}\sigma_{ij}^{(1)} - \nu\sigma_{kk}^{(2)}\sigma_{jj}^{(1)}\right] \tag{9.46b}$$

as the second term on the r.h.s. of (9.43).

Obviously, (9.46b) is symmetric. Finally using (9.44) and (9.46b), (9.42) becomes

$$\int_A u_j^{(2)}T_j^{(1)}dA - \int_V \frac{1}{E}\left[(1+\nu)\sigma_{ij}^{(1)}\sigma_{ij}^{(2)} - \nu\sigma_{kk}^{(2)}\sigma_{jj}^{(1)}\right]dV$$
$$= -\rho\omega_1^2\int_V u_j^{(2)}u_j^{(1)}dV \tag{9.47}$$

If the calculations on (9.39) are repeated for $u_j^{(2)}$, the resulting equation will be identical to (9.47) with the indices 1 and 2 interchanged. Then subtracting the two equations, we find

$$\int_A \left(u_j^{(2)}T_j^{(1)} - u_j^{(1)}T_j^{(2)}\right)dA = -\rho(\omega_1^2 - \omega_2^2)\int_V u_j^{(2)}u_j^{(1)}dV \tag{9.48}$$

recognizing the symmetry. We refer to Betti's law, in integral form [7], which is to be discussed in Sect. 11.5.4, to set the l.h.s. = 0, whereupon

$$-\rho(\omega_1^2 - \omega_2^2)\int_V u_j^{(2)}u_j^{(1)}dV = 0 \tag{9.49}$$

Since ω_1 and ω_2 are presumed to be different,

$$\int_V u_j^{(2)}u_j^{(1)}dV = 0 \tag{9.50}$$

is the statement of orthogonality.

9.3.3 Rayleigh's Quotient

It is often useful to rewrite (9.39) in a form in which ω^2 is expressed explicitly. Taking the first two terms in operator form

$$u_{j,ii} + \frac{1}{1-2\nu}u_{i,ij} = L_j(u_j), \tag{9.51}$$

we have

$$L_j\left(u_j\right) = -\frac{\rho}{G}\omega^2 u_j. \tag{9.52}$$

We multiply by u_j and integrate to get

$$\omega^2 = -\frac{G}{\rho}\frac{\int_V u_j L_j\left(u_j\right)dV}{\int_V u_j u_j dV}, \tag{9.53}$$

which is called Rayleigh's quotient [6].

This form allows the mode shapes to be approximated in the manner of the Rayleigh–Ritz method, Sect. 11.7.2. So long as the assumed functions satisfy the boundary conditions, good approximations for the eigenvalues, which are the natural frequencies of the system, may be obtained.

9.3.4 Axial Vibration of a Bar

As an example, we consider a slender bar of length H shown in Fig. 9.5, which is oriented along the x-axis in Fig. 9.1 and vibrates in an axial mode as discussed in Sect. 9.2.2. A similar problem was analyzed in Sect. 6.2 using 3-D elasticity equations but with a uniformly distributed self-weight loading instead of the constant axial load P producing the initial elongation shown on Fig. 9.5 [see (6.19c)].

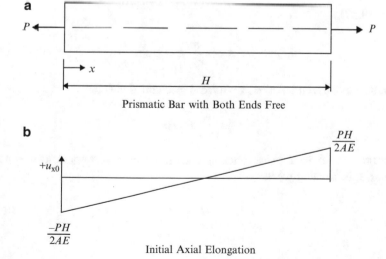

a

Prismatic Bar with Both Ends Free

b

Initial Axial Elongation

Fig. 9.5 Longitudinal vibration of elastic bar

In accordance with the elementary theory, we neglect the transverse displacements u_y and u_z. Then (9.37) becomes

$$(\lambda + 2\mu)u_{x,xx} + \rho\omega^2 u_x = 0 \tag{9.54}$$

in the manner of (9.4). Equation (9.54) is written compactly as

$$u_{x,xx} + \frac{\omega^2}{c_P^2} u_x = 0 \tag{9.55}$$

following (9.5a) and (9.5b). Also, the assumption of one-dimensional motion implies that Poisson's ratio $v = 0$ so that

$$c_P = c_{P1} = \sqrt{\frac{E}{\rho}}. \tag{9.56}$$

Then the spatial solution is taken as

$$u_x(x) = A \sin\frac{\omega}{c_{P1}}x + B \cos\frac{\omega}{c_{P1}}x. \tag{9.57}$$

Now the boundary conditions on the bar must be considered to evaluate ω. For a bar that is free to displace at both ends, the displacement oscillates between $+\bar{u}_x$ and $-\bar{u}_x$ at each end. Since the stress-free condition implies that the axial strain is zero, $u_{x,\,x} = 0$. Thus, we have

$$u_{x,x}(0,\ t) = u_{x,x}(H,\ t) = 0. \tag{9.58}$$

From (9.57),

$$u_{x,x} = \frac{\omega}{c_{P1}}\left[A \cos\frac{\omega}{c_{P1}}x - B \sin\frac{\omega}{c_{P1}}x\right] \tag{9.59}$$

so that, for a nontrivial solution, we have $A = 0$ and $B \neq 0$ if

$$\sin\frac{\omega}{c_{P1}}H = 0. \tag{9.60}$$

Equation (9.60) is a transcendental equation which is typical of eigenvalue problems. It is satisfied when

$$\frac{\omega_n}{c_{P1}}H = n\pi \tag{9.61a}$$

or

$$\omega_n = \frac{n\pi}{H}c_{P1}$$

$$= \frac{n\pi}{H}\sqrt{\frac{E}{\rho}}. \tag{9.61b}$$

This result is identical to that obtained from the specialized formulation for a one-dimensional bar [7].

The complete solution for the displacement $u_x(x, t)$ is found by substituting ω_n into (9.57) and (9.35) and then recombining these equations into (9.31). Thus,

$$u_x(x) = B\cos\frac{n\pi}{H}x \tag{9.62}$$

$$f(t) = C\cos\omega_n t + D\sin\omega_n t \tag{9.63}$$

and

$$u_{xn}(x, \ t) = \cos\frac{n\pi}{H}x\left(C_n\cos\omega_n t + D_n\sin\omega_n t\right). \tag{9.64}$$

In the last equation, the constants have been combined and the solution written for a general harmonic n.

Free vibration and the propagation of elastic waves as discussed in the preceding sections may be initiated by a specified displacement or velocity distribution at $t = 0$. This distribution may be in the form of a single harmonic, i.e., proportional to $\cos\frac{n\pi}{H}x$ or $\sin\frac{n\pi}{H}x$, or may be a more general pattern requiring several terms of the form of (9.64) to describe. Thus, the complete solution is

$$u_x(x, \ t) = \sum_{n=1,2,}^{\infty}\cos\frac{n\pi}{H}x\left(C_n\cos\omega_n t + D_n\sin\omega_n t\right). \tag{9.65}$$

Correspondingly, the initial conditions are specified in terms of a similar Fourier series, and the constants C_n and D_n are evaluated harmonic by harmonic using the orthogonality property discussed in Sect. 9.3.2. At time $t = 0$,

$$u_{x0}(x) = \sum_{n=1,2,}^{\infty}C_n\cos\frac{n\pi}{H}x \tag{9.66a}$$

$$\dot{u}_{x0}(x) = \omega_1\sum_{n=1,2,}^{\infty}nD_n\cos\frac{n\pi}{H}x \tag{9.66b}$$

because $\omega_n = n\omega_1$. Both sides of each of (9.66a) and (9.66b) are multiplied by $\cos\frac{n\pi x}{H}$ and integrated between the limits $x = 0, H$ using the orthogonality condition similar to (8.36b)

$$\int_0^H \cos\frac{m\pi x}{H}\cos\frac{n\pi x}{H}dx = 0, \quad m \neq n,$$

$$= \frac{H}{2}, \quad m = n \tag{9.67}$$

to get the following expressions

$$C_n = \frac{2}{H}\int_0^H u_{x0}(x)\cos\frac{n\pi x}{H}dx \tag{9.68a}$$

$$D_n = \frac{2}{\omega_1 n H}\int_0^H \dot{u}_{x0}(x)\cos\frac{n\pi x}{H}dx. \tag{9.68b}$$

Continuing the example, we consider the bar initially stretched by static axial forces P as shown in Fig. 9.5a and released at time $t = 0$. These initial conditions

$$u_{x0}(x) = -\frac{PH}{2AE}\left(1 - 2\frac{x}{H}\right) \tag{9.69a}$$

and

$$\dot{u}_{x0}(x) = 0 \tag{9.69b}$$

are substituted into (9.68a) and (9.68b) to get

$$C_n = 0 \qquad \text{for } n = 2, 4, 6$$
$$= \frac{-4PH}{AEn^2\pi^2} \qquad \text{for } n = 1, 3, 5 \tag{9.70}$$

and

$$D_n = 0.$$

The displacement of any cross section of the bar is found by substituting C_n into (9.65) as

$$u_x(x, \ t) = \frac{-4PH}{\pi^2 AE}\sum_{n=1,3,5}^{\infty}\frac{1}{n^2}\cos\frac{n\pi}{H}x\cos\omega_n t \tag{9.71}$$

since only the odd-numbered modes enter into the vibration, reflecting the antisymmetry about the center line $x = \frac{H}{2}$.

The stresses are determined from (9.71) using the elementary stress–strain law consistent with this one-dimensional formulation $\sigma_{xx} = E\varepsilon_{xx} = Eu_{x,\,x}(x,t)$ as

$$\sigma_{xx}(x,t) = \frac{4P}{\pi A} \sum_{n=1,3,5}^{\infty} \frac{1}{n} \sin \frac{n\pi}{H} x \cos \omega_n t. \tag{9.72}$$

Note that in the stress series, the spatial terms are multiplied by coefficients $\frac{1}{n}$, where as in the displacement series the coefficients are $\frac{1}{n^2}$ indicating that the convergence is faster as n increases or that the higher modes are relatively more important for the stress solution than for the displacement.

Additional details on the vibration of prismatic bars with various boundary conditions are presented in Jacobsen and Ayre [7].

9.4   Uniform Rotation of a Beam

9.4.1   Equilibrium Equations

We consider a beam with an arbitrary cross section, as shown Fig. 9.6 which rotates about the x-axis with a constant angular velocity Ω. The cross-sectional x and y are chosen to be principal axes in order to preclude torsion due to asymmetric body forces. The beam is subjected to extension along the z-axis and bending in the y–z plane and is essentially a combination of the problems discussed in Sects. 6.2 and 6.3.

Fig. 9.6   Rotating beam

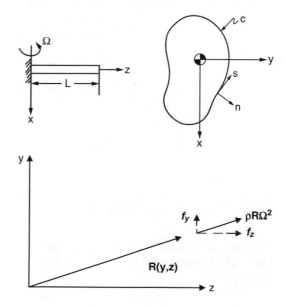

This model represents an idealized turbine blade which is of great practical interest. The self-weight of the member is neglected, as well as the time-dependent effects in the transient phase while the blade is brought up to speed. Essentially each point in the member is subjected to a centripetal acceleration $R\Omega^2 \mathbf{e}_R$, where $R(y, z)$ is the projection of the position vector emanating from the origin onto the y–z plane as shown on the figure. The components of the inertial force at a reference point along the axis of the beam designated by z and within the cross section (x, y) are

$$f_y = \rho R\Omega^2 (y/R) = \rho \Omega^2 y \tag{9.73a}$$

and

$$f_z = \rho R\Omega^2 (z/R) = \rho \Omega^2 z, \tag{9.73b}$$

respectively, and Eq. (2.46) for (x, y, z) become

$$\sigma_{xx,x} + \sigma_{yx,y} + \sigma_{zx,z} = 0 \tag{9.74a}$$

$$\sigma_{xy,x} + \sigma_{yy,y} + \sigma_{zx,z} + \rho \Omega^2 y = 0 \tag{9.74b}$$

$$\sigma_{xz,x} + \sigma_{yz,y} + \sigma_{zz,z} + \rho \Omega^2 z = 0. \tag{9.74c}$$

9.4.2   Boundary Conditions

All lateral boundaries and the end $z = L$ are presumed to be traction-free. For the lateral boundaries, we write the components of the traction from (2.44) as

$$T_x = \sigma_{xx}n_x + \sigma_{yx}n_y + \sigma_{zx}n_z = 0 \tag{9.75a}$$

$$T_y = \sigma_{xy}n_x + \sigma_{yy}n_y + \sigma_{zy}n_z = 0 \tag{9.75b}$$

$$T_z = \sigma_{xz}n_x + \sigma_{yz}n_y + \sigma_{zz}n_z = 0. \tag{9.75c}$$

For the end $z = L$,

$$\sigma_{zx} = \sigma_{zy} = \sigma_{zz} = 0. \tag{9.76}$$

9.4.3   Semi-inverse Solution

Employing the semi-inverse method of St. Venant, which was described in Sect. 5.7 and applied extensively in Chap. 6, Stephen and Wang [8] stated the following assumptions:

(a) Shearing stresses and strains in the z-direction are zero:

$$\sigma_{xz} = \sigma_{yz} = \varepsilon_{xz} = \varepsilon_{yz} = 0. \tag{9.77}$$

(b) Planar stresses and strains are independent of z:

$$\sigma_{xx}, \sigma_{xy}, \sigma_{yy} \text{ and } \varepsilon_{xy} \neq f(z). \tag{9.78}$$

(c) Longitudinal strain is taken in the form

$$\varepsilon_{zz} = \varepsilon_0 + \frac{1}{2}\frac{\rho}{E}\Omega^2\left(L^2 - z^2\right) - \kappa_0 x - \kappa_0' y + \varepsilon_1\left(x^2 + y^2\right), \tag{9.79}$$

where ε_0, ε_1, κ_0, and κ_0' are constants to be determined. In (9.79), the first two terms are similar to those found in the solution for the extension problem in Sect. 6.2, (6.5c), but one degree higher since the loading here is linear in z rather than constant. Similarly, the last term represents the distortion of the cross section into a paraboloid of revolution as described by (6.19c). The third and fourth terms are associated with bending since they are linear in the cross-sectional coordinates.

With these assumptions, the equilibrium (9.74a) reduce to

$$\sigma_{xx,x} + \sigma_{yx,y} = 0 \tag{9.80a}$$

$$\sigma_{xy,x} + \sigma_{yy,y} + \rho\Omega^2 y = 0 \tag{9.80b}$$

$$\sigma_{zz,z} + \rho\Omega^2 z = 0 \tag{9.80c}$$

and the St. Venant compatibility equations, (3.59), for (x, y, z) become

$$\varepsilon_{xx,yy} + \varepsilon_{yy,xx} = 2\varepsilon_{xy,xy} \tag{9.81a}$$

$$\varepsilon_{yy,zz} + \varepsilon_{zz,yy} = 0 \tag{9.81b}$$

$$\varepsilon_{zz,xx} + \varepsilon_{xx,zz} = 0 \tag{9.81c}$$

$$\varepsilon_{xx,yz} = 0 \tag{9.81d}$$

$$\varepsilon_{yy,xz} = 0 \tag{9.81e}$$

$$\varepsilon_{zz,xy} = 0. \tag{9.81f}$$

Note that (9.79) satisfies (9.81f), and a stress function $\phi(x, y)$ such that

$$\sigma_{xx} = \phi_{,yy} \tag{9.82a}$$

$$\sigma_{xy} = -\phi_{,xy} \tag{9.82b}$$

$$\sigma_{yy} = \phi_{,xx} - \frac{1}{2}\rho\Omega^2 y^2 \tag{9.82c}$$

satisfies the equilibrium Eqs. (9.80a) and (9.80b).

We now consider Hooke's law specialized from the first three of (4.29b):

$$\varepsilon_{xx} = (1/E)\sigma_{xx} - (\nu/E)(\sigma_{yy} + \sigma_{zz}) \tag{9.83a}$$

$$\varepsilon_{yy} = (1/E)\sigma_{yy} - (\nu/E)(\sigma_{xx} + \sigma_{zz}) \tag{9.83b}$$

$$\varepsilon_{zz} = (1/E)\sigma_{zz} - (\nu/E)(\sigma_{xx} + \sigma_{yy}). \tag{9.83c}$$

The first two equations are differentiated with respect to z, recalling (9.78), to produce

$$\varepsilon_{xx,zz} = \varepsilon_{yy,zz} = -(\nu/E)\sigma_{zz,zz} = (\nu/E)\rho\Omega^2. \tag{9.84}$$

The last equality is stated after considering (9.80c). Returning to compatibility, Eqs. (9.81b) and (9.81c) gives

$$\varepsilon_{zz,yy} = \varepsilon_{zz,xx} = -(\nu/E)\rho\Omega^2 \tag{9.85a}$$

from which we may assign the value

$$\varepsilon_1 = -\frac{1}{2}(\nu/E)\rho\Omega^2 \tag{9.85b}$$

to the constant in (9.79) since it is the only term which multiplies quadratic functions of x and y.

The remaining extensional stress–strain Eq. (9.83c) is now solved for σ_{zz} in terms of the remaining unknown constants as shown in (9.86):

$$\sigma_{zz} = E\varepsilon_0 + \frac{1}{2}\rho\Omega^2(L^2 - z^2) - E\kappa_0 x - E\kappa_0' y - \frac{1}{2}\nu\rho\Omega^2(x^2 + y^2) + \nu(\sigma_{xx} + \sigma_{yy}). \tag{9.86}$$

At this point, the evaluation of these constants requires consideration of the resultant force and bending moment on the cross section.

We first form the resultant axial force on the cross section as

$$T(z) = \int_A \sigma_{zz}dx\,dy$$

$$= EA\varepsilon_0 + \frac{1}{2}\rho\Omega^2 A(L^2 - z^2) - \frac{1}{2}\nu\rho\Omega^2(I_x + I_y) \tag{9.87}$$

$$+ \nu\int_V (\sigma_{xx} + \sigma_{yy})dx\,dy,$$

where

$$I_x = \int_A y^2 dx\,dy \tag{9.88a}$$

and

$$I_y = \int_A x^2 dx\,dy \tag{9.88b}$$

are the principal moments of inertia. Note that the first moment terms in (9.79) do not contribute to the integral (9.87).

Next, we focus on the last term of $T(z)$ and recall the useful relationship between contour and area integrals known as Green's theorem. From (1.13),

$$\oint_C (P\,dx + Q\,dy) = \int_A (Q_{,x} - P_{,y})\,dx\,dy \tag{9.89}$$

where $P(x, y)$ and $Q(x, y)$ are C^1 continuous functions. Now we express

$$\int_A (\sigma_{xx} + \sigma_{yy})\,dx\,dy = I_1 + I_2 \tag{9.90}$$

where

$$I_1 = \int_A \left\{ (x\sigma_{xx} + y\sigma_{xy})_{,x} + (x\sigma_{yx} + y\sigma_{yy})_{,y} \right\} dx\,dy \tag{9.91a}$$

and

$$I_2 = -\int_A \left\{ x(\sigma_{xx,x} + \sigma_{xy,y}) + y(\sigma_{yx,x} + \sigma_{yy,y}) \right\} dx\,dy. \tag{9.91b}$$

The separation of the integral into two parts may be easily verified by expanding the derivatives. Invoking Green's theorem (9.89) with

$$Q = (x\sigma_{xx} + y\sigma_{xy}) \tag{9.92a}$$

and

$$P = -(x\sigma_{yx} + y\sigma_{yy}) \tag{9.92b}$$

$$\begin{aligned} I_1 &= \oint_C (-x\sigma_{yx}dx - y\sigma_{yy}dx + x\sigma_{xx}dy + y\sigma_{xy}dy) \\ &= \oint_C \{x(\sigma_{xx}dy - \sigma_{yx}dx) + y(\sigma_{xy}dy - \sigma_{yy}dx)\}. \end{aligned} \tag{9.93}$$

Further, we may write the contour integral in terms of the differential contour increment ds by noting the n–s coordinates on the contour in Fig. 9.6 and referring to (2.18) with $i = y$, which gives

$$\frac{dy}{ds} = s_y = \mathbf{e}_y \cdot \mathbf{s} = n_x \tag{9.94a}$$

and

$$\frac{dx}{ds} = s_x = \mathbf{e}_x \cdot \mathbf{s} = -n_y. \tag{9.94b}$$

Replacing dy and dx in (9.93) by the relationships in terms of ds produces

$$I_1 = \oint_C \left\{ x(\sigma_{xx} n_x + \sigma_{yx} n_y) + y(\sigma_{xy} n_x + \sigma_{yy} n_y) \right\} ds \tag{9.95}$$

which vanishes by virtue of the lateral bounding conditions, (9.75a) and (9.75b). Thus, we have remaining in (9.90)

$$I_2 = \int_A (\sigma_{xx} + \sigma_{yy}) dx \, dy$$
$$= \int_A \rho \Omega^2 y^2 dx \, dy \tag{9.96}$$

by comparison of (9.91b) to the equilibrium equations, (9.80a) and (9.80b).

Finally, returning to the last original term on the r.h.s. of (9.87) and noting (9.88a),

$$\nu \int_A (\sigma_{xx} + \sigma_{yy}) dx \, dy = \nu \rho \Omega^2 I_x \tag{9.97}$$

and the resultant axial force becomes

$$T(z) = EA\varepsilon_0 + \frac{1}{2}\rho\Omega^2 A(L^2 - z^2) + \frac{1}{2}\nu\rho\Omega^2(I_x - I_y). \tag{9.98}$$

The stress-free end condition

$$T(L) = 0 \tag{9.99}$$

requires that

$$\varepsilon_0 = -\frac{1}{2}\frac{\nu\rho\Omega^2}{EA}(I_x - I_y) \tag{9.100}$$

from which

$$T(z) = \frac{1}{2}\rho\Omega^2 A(L^2 - z^2). \tag{9.101}$$

The constants κ_0 and κ_0' in (9.86) remain to be evaluated by considering the bending moments on the cross section

$$M_x = -\int_A \sigma_{zz} y \, dx \, dy \tag{9.102a}$$

$$M_y = \int_A \sigma_{zz} x \, dx \, dy, \tag{9.102b}$$

First consider M_x

$$M_x(z) = -\int_A \sigma_{zz} y \, dx \, dy$$

$$= -\int_A \left\{ E\varepsilon_0 y + \frac{1}{2}\rho\Omega^2(L^2 - x^2)y - E\kappa_0 xy - E\kappa_0' y^2 \right.$$

$$\left. \times -\frac{1}{2}\nu\rho\Omega^2(x^2 + y^2)y + \nu(\sigma_{xx} + \sigma_{yy})y \right\} dx \, dy.$$

The first three terms do not contribute since x and y are principal axes, so that

$$M_x(z) = E\kappa_0' I_x + \frac{1}{2}\nu\rho\Omega^2 \int_A (x^2 + y^2)y \, dx \, dy - \nu \int_A (\sigma_{xx} + \sigma_{yy})y \, dx \, dy. \tag{9.103}$$

Again we have an area integral to be transformed,

$$\int_A (\sigma_{xx} + \sigma_{yy})y \, dx \, dy = I_3 + I_4, \tag{9.104}$$

where

$$I_3 = \int_A \left\{ \left[\frac{1}{2}(y^2 - x^2)\sigma_{yy} + xy\sigma_{yx} \right]_{,y} \right.$$

$$\left. + \left[xy\sigma_{xx} + \frac{1}{2}(y^2 - x^2)\sigma_{xy} \right]_{,x} \right\} dx \, dy \tag{9.105a}$$

$$I_4 = -\int_A \left\{ xy(\sigma_{xx,x} + \sigma_{xy,y}) + \frac{1}{2}(y^2 - x^2)(\sigma_{yx,x} + \sigma_{yy,y}) \right\} dx \, dy \tag{9.105b}$$

which is easily verified. Then following (9.89) with

$$P = -\left[\frac{1}{2}(y^2 - x^2)\sigma_{yy} + xy\sigma_{yx} \right] \tag{9.106a}$$

and

$$Q = xy\sigma_{xx} + \frac{1}{2}(y^2 - x^2)\sigma_{xy} \tag{9.106b}$$

and applying the transformation to (9.105a),

$$I_3 = \int_A \left\{ -\frac{1}{2}(y^2 - x^2)\sigma_{yy} dx - xy\sigma_{yx} dx + xy\sigma_{xx} dy + \frac{1}{2}(y^2 - x^2)\sigma_{xy} dy \right\}$$

$$= \int_A \left\{ xy(\sigma_{xx} dy - \sigma_{yx} dx) + \frac{1}{2}(y^2 - x^2)(\sigma_{xy} dy - \sigma_{yy} dx) \right\}. \tag{9.107}$$

The terms in parentheses in (9.107) are identical to those in (9.93), so that $I_3 = 0$ by (9.95) and

$$
\begin{aligned}
I_4 &= \int_A (\sigma_{xx} + \sigma_{yy}) y \, dx \, dy \\
&= \frac{1}{2} \rho \Omega^2 \int_A (y^2 - x^2) y \, dx \, dy
\end{aligned}
\tag{9.108}
$$

from (9.105b), (9.73a), and (9.73b). Continuing from (9.103) after including (9.108) multiplied by $-\nu$,

$$
M_x(z) = EI_x \kappa_0' + \nu \rho \Omega^2 \int_A x^2 y \, dx \, dy.
\tag{9.109}
$$

Similarly, we may compute

$$
M_y(z) = -EI_y \kappa_0 + \frac{1}{2} \nu \rho \Omega^2 \int_A (y^2 - x^2) x \, dx \, dy.
\tag{9.110}
$$

These expressions are independent of z and are required to vanish at $z = 0$; hence,

$$
\kappa_0 = \frac{1}{2} \frac{\nu \rho}{EI_y} \Omega^2 \int_A (y^2 - x^2) x \, dx \, dy
\tag{9.111}
$$

and

$$
\kappa_0' = -\frac{\nu \rho}{EI_x} \Omega^2 \int_A x^2 y \, dx \, dy.
\tag{9.112}
$$

This completes the general solutions for σ_{zz} (9.86) in terms of ε_0, κ_0, and κ_0', which are functions of the cross-sectional geometry and the stress function ϕ. It is also useful to have expressions for the direct strains as defined in (9.83a), (9.83b), and (9.83c), incorporating ε_0 from (9.100) and κ_0 and κ_0' from (9.111) and (9.112) into the equation for σ_{zz}. The results are given by Stephen and Wang [8] and are used along with the shear stain ε_{xy}, which may be obtained from (4.29b), to derive a two-dimensional approximation.

9.4.4 Two-Dimensional Problem

A solution independent of the axial coordinate z is sought by considering the first of the compatibility equations, (9.81a), in the form

$$
\varepsilon_{xx,yy} + \varepsilon_{yy,xx} - 2\varepsilon_{xy,xy} = 0.
\tag{9.113}
$$

The resulting equation is found from (9.83a), (9.83b), (9.83c), (9.82a), (9.82b), and (9.82c) as

$$\nabla^4 \phi = -\nu \left(\frac{1+3\nu}{1-\nu^2}\right)\rho\Omega^2. \tag{9.114}$$

The rest of the compatibility Eqs. (9.81a), (9.81b), (9.81c), (9.81d), (9.81e), and (9.81f) are satisfied identically.

The applicable boundary conditions, (9.75a) and (9.75b), may be written in terms of ϕ and the differential along the contour, ds, by using (9.82a), (9.82b), and (9.82c) for the stresses and (9.94a) and (9.94b) for the components of the normal to the contour. The relationships are

$$
\begin{aligned}
T_x &= \phi_{,yy}\frac{dy}{ds} + \phi_{,xy}\frac{dx}{ds} \\
&= \frac{d}{dy}(\phi_{,y})\frac{dy}{ds} + \frac{d}{dx}(\phi_{,y})\frac{dx}{ds} \\
&= \frac{d}{ds}(\phi_{,y}) \\
&= 0
\end{aligned} \tag{9.115a}
$$

and

$$
\begin{aligned}
T_y &= -\phi_{,xy}\frac{dy}{ds} - \left[\phi_{,xx} - \frac{1}{2}\rho\Omega^2 y^2\right]\frac{dx}{ds} \\
&= \frac{d}{ds}(\phi_{,x}) + \frac{1}{2}\rho\Omega^2 y^2\frac{dx}{ds} \\
&= 0.
\end{aligned} \tag{9.115b}
$$

9.4.5   Circular Cross Section

Stephen and Wang [8] give a solution for the stress function for a member with a circular cross section, Fig. 7.1, with the contour equation

$$x^2 + y^2 - r^2 = 0 \tag{9.116}$$

in the form

$$
\begin{aligned}
\phi(x,y) &= -\left(\frac{\nu+3\nu^2}{1-\nu^2}\right)\frac{\rho\Omega^2}{64}(x^2+y^2-r^2)^2 \\
&= \frac{\rho\Omega^2}{64}\left(y^4 - \frac{5}{3}x^4 + 2x^2y^2 + 14r^2x^2 - 2r^2y^2\right).
\end{aligned} \tag{9.117}
$$

As in previous examples, the equation of the contour is included in the stress function. This function is easily shown to satisfy (9.114) by differentiation and substitution. Also, it must satisfy (9.115a) and (9.115b). In order to compute the derivatives along the boundary when $\phi = \phi(x, y)$, the chain rule

$$\frac{d(\ )}{ds} = (\ )_{,x}\frac{dx}{ds} + (\ )_{,y}\frac{dy}{ds} \tag{9.118}$$

is required. For a circular contour,

$$\frac{dx}{ds} = \frac{dx}{d\theta}\frac{d\theta}{ds}; \quad \frac{dy}{ds} = \frac{dy}{d\theta}\frac{d\theta}{ds}. \tag{9.119}$$

With $x = r\cos\theta$ and $y = r\sin\theta$, (7.20a, b), and noting from Fig. 7.1 that $ds = r\,d\theta$,

$$\frac{d\theta}{ds} = \frac{1}{r}; \quad \frac{dx}{ds} = -\sin\theta = \frac{-y}{r}; \quad \frac{dy}{ds} = \cos\theta = \frac{x}{r} \tag{9.120}$$

so that the differentiation formula, (9.118), becomes

$$\frac{d(\ )}{ds} = -(\ )_{,x}\frac{y}{r} + (\ )_{,y}\frac{x}{r}. \tag{9.121}$$

This may be directly applied to (9.117) to check (9.115a) and (9.115b). To illustrate, consider (9.115a) and form

$$\phi_{,y} = -\left(\frac{\nu + 3\nu^2}{1 - \nu^2}\right)\frac{\rho\Omega^2}{64}4y\left(x^2 + y^2 - r^2\right)$$
$$+ \frac{\rho\Omega^2}{64}\left(4y^3 + 4x^2 y - 4r^2 y\right) \tag{9.122a}$$

$$\frac{d}{ds}\left(\phi_{,y}\right) = -\left(\phi_{,y}\right)_{,x}\frac{y}{r} + \left(\phi_{,y}\right)_{,y}\frac{x}{r} \tag{9.122b}$$

We expand the two terms on the r.h.s. of (9.122b) as

$$-\left(\phi_{,y}\right)_{,x}\frac{y}{r} = \left(\frac{\nu + 3\nu^2}{1 - \nu^2}\right)\frac{\rho\Omega^2}{16}(2xy)\frac{y}{r} - \frac{\rho\Omega^2}{64}(8xy)\frac{y}{r} \tag{9.123a}$$

$$+\left(\phi_{,y}\right)_{,y}\frac{x}{r} = -\left(\frac{\nu + 3\nu^2}{1 - \nu^2}\right)\frac{\rho\Omega^2}{16}(x^2 + 3y^2 - r^2)\frac{x}{r}$$
$$+ \frac{\rho\Omega^2}{64}4(3y^2 + x^2 - r^2)\frac{x}{r}. \tag{9.123b}$$

Now combining the variable terms with common coefficients in (9.123a) and (9.123b), the first set of terms becomes

$$x\left(2y^2 - x^2 - 3y^2 + r^2\right) \to 0 \quad \text{on } C. \tag{9.124}$$

while the second set is

$$x\left(-8y^2 + 12y^2 + 4x^2 - 4r^2\right) \to 0 \quad \text{on } C \qquad (9.125)$$

so that (9.115a) is satisfied. The stresses σ_{xx}, σ_{yy}, and σ_{xy} are evaluated from (9.82a), (9.82b), and (9.82c) in view of (9.116) as

$$\sigma_{xx} = \frac{\rho\Omega^2}{64}\left[12y^2 + 4x^2 - 4r^2 - 4\left(\frac{\nu + 3\nu^2}{1 - \nu^2}\right)\left(x^2 + 3y^2 - r^2\right)\right] \qquad (9.126a)$$

$$\sigma_{yy} = \frac{\rho\Omega^2}{64}\left[12x^2 + 4y^2 - 4r^2 - 4\left(\frac{\nu + 3\nu^2}{1 - \nu^2}\right)\left(3x^2 + y^2 - r^2\right)\right] \qquad (9.126b)$$

$$\sigma_{xy} = -\frac{1 - \nu - 4\nu^2}{8(1 - \nu^2)}\rho\Omega^2 xy. \qquad (9.126c)$$

To compute σ_{zz}, we must first consider the geometric constants ε_0, κ_0, and κ_0'. For the circular cross section, $I_x = I_y$ so that $\varepsilon_0 = 0$, (9.100). Likewise, the functions κ_0 and κ_0', when evaluated with x and y from (9.120), vanish for this cross section. Therefore, from (9.86)

$$\sigma_{zz} = \frac{1}{2}\rho\Omega^2[L^2 - z^2 - \nu(x^2 + y^2)] + \nu(\sigma_{xx} + \sigma_{yy}). \qquad (9.127)$$

Bezhad and Bastami [9] studied the vibration characteristics of a rotating circular shaft by determining the change in the lateral natural frequency due to an axial force. In this case, the rotation is about the z-axis shown in Fig. 9.6. The axial force is not a direct result of the centrifugal force produced by the gyroscopic rotation but comes from the Poisson's effect. The Bernoulli–Euler beam theory, which is the basis of the solution presented in Sect. 6.3.1, is extended to include an axial load, which is calculated from an elasticity solution similar to that producing (9.127) and takes the form

$$EI\frac{d^4y}{dx^4} - P\frac{d^2y}{dx^2} - \rho A\omega^2 y = 0 \qquad (9.128)$$

where E is the Young's modulus, I is the moment of inertia, P is the axial force, A is the cross-sectional area, and ω is the vibration frequency. The result is

$$\frac{\omega^2 - \omega_0^2}{\omega_0^2} = \frac{2\nu\rho}{\pi^2 E}L^2\Omega^2 \qquad (9.129)$$

where ω_0 is the first natural frequency of the non-rotating shaft, ν is the Poisson's ratio, and L is the length of the shaft. The solution is interesting in that:

1. The influence of the axial stresses on the frequency resulting from the gyroscopic effect and Poisson's ratio increases rapidly with length of the shaft.
2. The change in the lateral natural frequency is independent of the shaft diameter.

This is another example of the combining features from the theory of elasticity with a less rigorous strength-of-materials approach.

9.5   Use MuPAD to Derive Axial Vibration Solutions (Sect. 9.3.4)

In this section, we use MuPAD to derive the axial vibration solutions.

- First, for spatial solutions:

$$
\begin{bmatrix}
\texttt{ux := A*sin(w/cpl*x) + B*cos(w/cpl*x)} \\[4pt]
B \cos\left(\dfrac{w\,x}{\mathrm{cpl}}\right) + A \sin\left(\dfrac{w\,x}{\mathrm{cpl}}\right)
\end{bmatrix}
$$

- The corresponding $u_{x,\,x}$ can be derived:

$$
\begin{bmatrix}
\texttt{ux\_x := diff(ux,x)} \\[4pt]
\dfrac{A\,w \cos\left(\frac{w\,x}{\mathrm{cpl}}\right)}{\mathrm{cpl}} - \dfrac{B\,w \sin\left(\frac{w\,x}{\mathrm{cpl}}\right)}{\mathrm{cpl}}
\end{bmatrix}
$$

- Based on boundary conditions, we solve the transcendental equations

$$
\begin{bmatrix}
\texttt{solve(sin(w/cpl*H)=0, w, IgnoreSpecialCases)} \\[4pt]
\left\{ \dfrac{\pi\,\mathrm{cpl}\,k}{H} \;\middle|\; k \in \mathbb{Z} \right\}
\end{bmatrix}
$$

- Therefore, at $t = 0$, we can have the summation terms derived as follows

$$
\begin{bmatrix}
\texttt{uxn := cos(n*PI/H*x)*(Cn*cos(wn*t) + Dn*sin(wn*t))} \\[4pt]
\cos\left(\dfrac{\pi\,n\,x}{H}\right)(Cn \cos(t\,wn) + Dn \sin(t\,wn))
\end{bmatrix}
$$

$$
\begin{bmatrix}
\texttt{ux0 := simplify(subs(uxn, t=0))} \\[4pt]
Cn \cos\left(\dfrac{\pi\,n\,x}{H}\right)
\end{bmatrix}
$$

$$
\begin{bmatrix}
\texttt{ux0p := simplify(subs(diff(uxn,t), t=0))} \\[4pt]
Dn\,wn \cos\left(\dfrac{\pi\,n\,x}{H}\right)
\end{bmatrix}
$$

- Note that we use EE rather than E to represent the elastic modulus. This is because E is a reserved variable in MuPAD. To verify the results, we can substitute $n = 2$ and 3:

$$\begin{bmatrix} \texttt{simplify(subs(Cn , n=2))} \\ 0 \end{bmatrix}$$

$$\begin{bmatrix} \texttt{simplify(subs(Cn , n=3))} \\ -\dfrac{4\,H\,P}{9\,A\,\text{EE}\,\pi^2} \end{bmatrix}$$

9.6 Exercises

9.1 Perform the calculations in Sect. 9.3.2 using $u_j^{(2)}$ in place of $u_j^{(1)}$.

9.2 For the free-free bar solved in Sect. 9.3.4, verify the solution for the constants of integration and the displacement u_x.

9.3 For the bar solved in Problem 9.2, verify the formula for the evaluation of the stress $\sigma_{xx} = P/A$, and plot the stress at $x = H/2$ and $x = H/4$ as a function of time. Hint: it may be convenient to use a time axis incremented in terms of $H/c_{\text{P}1}$.

9.4 Consider the vibration of an elastic bar as solved in Sect. 9.3.4 but with one end fixed and the other free. Determine ω_n.

9.5 For the boundary conditions of Problem 9.4, solve for the displacement u_x.

9.6 For the boundary conditions of Problem 9.4, solve for σ_{xx} and plot the stress at the fixed end as a function of time. Hint: it may be convenient to use a time axis incremented in terms of $H/c_{\text{P}1}$.

9.7 In Ref. [8], a solution for the uniform rotation of a shaft with an elliptical cross section with the equation

$$\frac{x^2}{D^2} + \frac{y^2}{B^2} - 1 = 0$$

is given. The cross section is shown in Fig. 6.7a. The stress function is

$$\phi(x,y) = \frac{\Lambda}{8}D^2B^2\left(\frac{x^2}{D^2} + \frac{y^2}{B^2} - 1\right)^2 - \frac{\rho\Omega}{24}\left(\frac{B}{D}\right)^2 x^4 + \frac{\rho\Omega}{4}B^2x^2,$$

where

$$\Lambda = \rho\Omega^2\left[\left\langle\left(\frac{B}{D}\right)^2 - \left(\frac{\nu + 3\nu^2}{1 - \nu^2}\right)\right\rangle\left\langle\frac{D^2B^2}{3D^4 + 2D^2B^2 + 3B^4}\right\rangle\right].$$

Verify that:

(a) ϕ satisfies (9.114), (9.115a), and (9.115b).
(b) κ_0 and κ_0' are zero.
(c) Compute the stresses.

9.8 Verify (9.39).

9.9 Consider a cylindrical shaft with an inner radius a, outer radius b, and length L, which rotates with a constant rotational speed Ω as shown in the figure above. The shaft is supported on two bearings at shaft ends which suppress axial movement in the shaft so that the shaft can be considered stationary. Due to D'Alembert principle, the inertia force per unit volume $\rho\Omega^2 r$, originating from centrifugal acceleration, is applied in the radial direction. The inner and outer surfaces of the shaft are free surfaces [9].

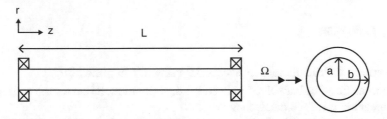

1. Verify the equilibrium equation in the radial direction

$$r\sigma_{rr,r} + \sigma_{rr} - \sigma_{\theta\theta} + \rho\Omega^2 r^2 = 0$$

where ρ is the shaft density.
Hint: use axisymmetric conditions to simplify the equilibrium equation
And solve for
2. Axial stress σ_{zz} induced by the centrifugal force produced radial stress.
3. Axial force P generated from the axial stress in (1).
Hint: The axisymmetric shaft has a length much larger than its diameter and the bearings suppress axial movement of the shaft; therefore, the shaft is under plain strain.

References

1. Filonenko-Borodich M (1965) Theory of elasticity. Dover Publications, Inc., New York
2. Royer D, Dieulesaint E (1996) Elastic waves in solids, vol 1. Springer, New York
3. Sinkus R et al (2005) Imaging anisotropic and viscous properties of breast tissue by magnetic resonance-Elastography. Magn Reson Med 53:372–387. Wiley Interscience
4. Chou PC, Pagano NJ (1992) Elasticity, tensor, dyadic and engineering approaches. Dover Publications, Inc, New York
5. Bolt BA (1970) In: Weigel RL (ed) Earthquake engineering. Prentice-Hall, Englewood Cliffs
6. Pearson CE (1959) Theoretical elasticity. Harvard University Press, Cambridge, MA
7. Jacobsen LS, Ayre RS (1958) Engineering vibrations. McGraw-Hill Book Co, New York
8. Stephen NG, Wang PJ (1986) Stretching and bending of rotating beam. J Appl Mech ASME 53:869–872
9. Behzad M, Bastami AR (2004) Effect of centrifugal force on natural frequency of lateral vibration of rotating shafts. J Sound Vib 274:985–995
10. Bolt BA (1970) Elastic waves in the vicinity of the earthquake source. In: Weigel R (ed) Earthquake engineering. Prentice Hall, Englewood Cliffs, p 5

Chapter 10
Viscoelasticity

Abstract The stress–strain relationships derived in Chap. 4 pertain to elastic materials that respond *immediately* to the load applied. Viscous materials also resist loading but undergo strain with a time lag. For such viscoelastic systems, the stress at a particular time depends on the *strain rate* in addition to the strain itself. Viscoelastic materials exhibit several important characteristics, including creep, stress relaxation, hysteresis, and stiffness that depends on the strain rate. These characteristics are explained in the context of several standard models with applications to biological tissues.

10.1 Introduction

While the stress–strain relationships derived in Chap. 4 define elastic materials that respond immediately to the load applied, viscous materials also resist loading and undergo strain, but the stress depends on the strain rate in addition to the strain itself. Many real-world materials, both man-made and biological, demonstrate both viscous and elastic properties and are thus deemed viscoelastic materials. Since the stress depends on both strain and strain rate, the stress history affects the current state of the material as well as the instantaneous stress. Viscoelastic materials exhibit several important characteristics, including creep, stress relaxation, hysteresis, and stiffness that depends on the strain rate. The stereotypical example of a viscoelastic material is Silly Putty™ (Fig. 10.1), which deforms differently depending on the rate of strain. If Silly Putty™ is quickly stretched, it will break instantly. But it will rebound elastically if dropped, will deform under its own weight, and can be stretched slowly for a long time without breaking as shown on the pictures. The introductory treatment in this chapter is based on a widely circulated public domain course note set by Roylance, augmented by some recent biomechanical extensions.

Viscoelastic materials are incorporated into many mechanical devices for the purpose of energy dissipation. The most familiar is perhaps the piston shock absorber, enlarged as shown in Fig. 10.2 for use in earthquake protection. An innovative modern device is the viscous wall damper (Fig. 10.3), that is, used to reduce forces

© Springer International Publishing AG, part of Springer Nature 2018
P. L. Gould, Y. Feng, *Introduction to Linear Elasticity*,
https://doi.org/10.1007/978-3-319-73885-7_10

Fig. 10.1 Silly Putty™, the stereotypical viscoelastic substance (Copyright Cambridge Polymer Group. Reproduced with permission)

Fig. 10.2 Piston dampers (Copyright Taylor Devices, Inc. Reproduced with permission)

and displacements in buildings under earthquake forces. The damper is composed of a viscoelastic fluid, and a baffle system enclosed in a wall panel cladding.

10.2   Properties of Viscoelastic Materials

10.2.1   Rate Dependence

As stated earlier, stresses in viscoelastic materials depend on strain rate as well as strain so that that the strain history has an effect on the current state of the material.

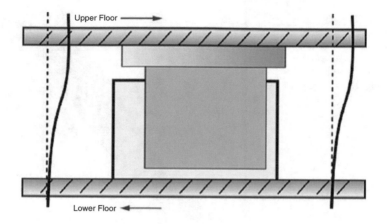

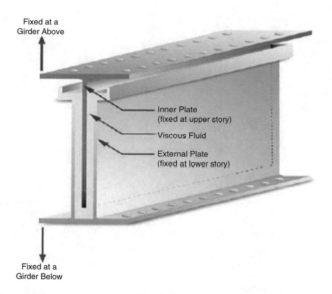

Fig. 10.3 Viscous wall damper (Courtesy Dynamic Isolation Systems, Inc.)

Many viscoelastic materials exhibit "fading memory" meaning that recent strain history has a greater effect than older loading history.

10.2.2   Hysteresis

Viscoelastic materials show different loading and unloading curves in terms of stress and strain. This occurs because of the energy that viscoelastic materials dissipate in loading and unloading. The area of the hysteresis loop formed by the loading and

Fig. 10.4 Hysteresis loop in a viscoelastic material. The area of the loop is proportional to the energy lost in loading and unloading [1]

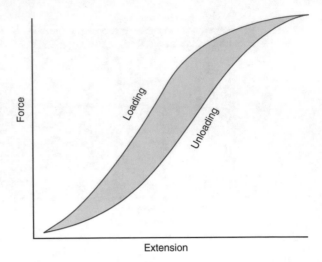

Fig. 10.5 These stress curves correspond to three different levels of strain. Notice how the stress relaxes as time t increases, after an initial peak [1]

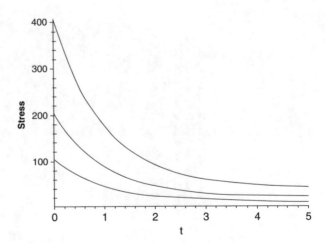

unloading curves shown in Fig. 10.4 equals the amount of energy lost per unit volume, as first mentioned in Sect. 4.2.

10.2.3 Stress Relaxation

In viscoelastic materials exposed to a step increase in strain, the stress peaks and then relaxes. This relaxation occurs as the viscous response of the material takes effect as shown in Fig. 10.5.

Fig. 10.6 This shows the strains measured at three different stress levels. Notice how the strain cannot instantly increase to its maximum value due to the viscous properties of the material [1]

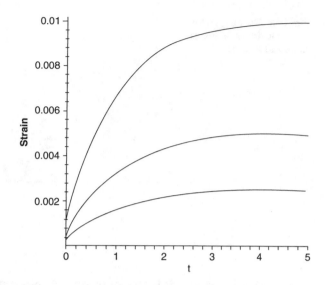

10.2.4   Creep

When a viscoelastic material is exposed to a constant stress, the strain will increase asymptotically with time but not instantaneously as shown in Fig. 10.6. Continuously increasing strain over time, or creep, is often observable in many engineered components.

10.2.5   Dynamic Loading

While stress relaxation and creep tests are convenient for studying the response of materials at longer times, dynamic tests are well suited for the 1 s or less time period. Upon application of a sinusoidally varying stress, a viscoelastic material will reach a steady state in which the resulting strain will also be sinusoidal with the same circular frequency ω but retarded by a phase angle δ [1].

It is convenient to select the origin on the time axis as the time when the strain passes through the maximum, thus transferring the δ to the stress function. That is

$$\varepsilon(t) = \varepsilon_0 \cos \omega t \qquad (10.1)$$

and

$$\sigma(t) = \sigma_0 \cos(\omega t + \delta), \qquad (10.2)$$

where ε_0 and σ_0 are the peak values of the stress and strain.

Fig. 10.7 Rotating vector representation of harmonic stress and strain [1]

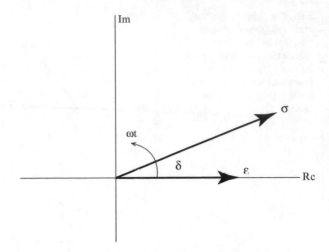

In harmonic analysis, it is helpful to separate phase-separated quantities using complex variables. Here a stress function σ^* is defined as a complex quantity with the real part in phase with the strain and the imaginary part 90° out-of-phase:

$$\sigma^* = \sigma'_o \cos \omega t + i\sigma''_o \sin \omega t. \tag{10.3}$$

Referring to Fig. 10.7, the stress and strain may be visualized as the projection of stress and strain vectors that are rotating in the complex plane at a frequency ω onto the real axis, Re. The strain vector $\boldsymbol{\varepsilon}$ is shown just passing the real axis, and the stress vector $\boldsymbol{\sigma}$ is ahead by the phase angle δ.

Now the terms in (10.3) may be interpreted using Fig. 10.7 with $\boldsymbol{\sigma} = \sigma^*$:

$$\tan \delta = \sigma''_o / \sigma'_o \tag{10.4a}$$

$$|\sigma^*| = \sigma_o = \left[\left(\sigma'_o \right)^2 + \left(\sigma''_o \right)^2 \right]^{1/2} \tag{10.4b}$$

$$\sigma'_o = \sigma_o \cos \delta \tag{10.4c}$$

$$\sigma''_o = \sigma_o \sin \delta \tag{10.4d}$$

Using the complex stress function in (10.3), two dynamic elastic moduli that have different physical interpretations are defined. First, the real or "storage" modulus is the ratio of the in-plane stress to the strain

$$E' = \sigma'_o / \varepsilon_0. \tag{10.5}$$

Then the imaginary or "loss" modulus is the ratio of the out-of-plane stress to the strain

$$E'' = \sigma''_o / \varepsilon_0. \tag{10.6}$$

The harmonic stress and strain functions are written in exponential form as

$$\sigma = \sigma_0^* e^{i\omega t} \tag{10.7a}$$

and

$$\varepsilon = \varepsilon_0^* e^{i\omega t} \tag{10.7b}$$

based on the Euler relation $e^{i\theta} = \cos\theta + i\sin\theta$ and expressing both the stress and strain as complex numbers. This removes the restriction of placing the origin at the time of maximum strain by incorporating the phase angle δ. Finally, the widely used complex elastic modulus is written as

$$E^*(i\omega) = \sigma_0^*/\varepsilon_0^* \tag{10.8}$$

which may be written in the form

$$E^*(i\omega) = E'(\omega) + E''(i\omega) \tag{10.9}$$

as illustrated in Sect. 10.4.4.

Similarly, the complex shear modulus is

$$G^*(i\omega) = G'(\omega) + G''(i\omega) \tag{10.10}$$

where $G'(\omega)$ is called the storage (shear) modulus and $G''(i\omega)$ is termed the loss (shear) modulus.

10.3   Constitutive Models for Linear Viscoelasticity

It is illustrative to model linearly viscoelastic materials as combinations of springs and dashpots as shown in Fig. 10.8. The elastic component of these materials is represented by "Hookean" springs which are defined by the stress–strain relationship:

$$\sigma = E\varepsilon \tag{10.11}$$

In this case, the Young's modulus E is analogous to the spring constant k. The viscous component is represented by a "Newtonian" dashpot with the stress–strain relationship

$$\sigma = \eta\dot{\varepsilon}. \tag{10.12}$$

Fig. 10.8 Maxwell model
[1]

This equation demonstrates how the stress on the material depends on strain rate. Various spring and dashpot combinations result in models of viscoelastic materials with different properties [1].

10.3.1 Maxwell Model

The Maxwell model of a viscoelastic material consists of a spring and a dashpot arranged in series. In this arrangement both components of the model, the spring and the dashpot, experience the same stress, but the total strain is the sum of the strain on each component. The model is arranged in series as shown in Fig. 10.8 and represents a typical viscoelastic fluid. Taking $E = k$ and $\tau = \eta/k$, the single constitutive equation is

$$\dot{\varepsilon}_{\text{Total}} = \dot{\varepsilon}_{\text{D}} + \dot{\varepsilon}_{\text{S}} = \frac{\sigma}{\eta} + \frac{\dot{\sigma}(t)}{k} \tag{10.13a}$$

or

$$k\dot{\varepsilon}_{\text{Total}} = \frac{\sigma}{\tau} + \dot{\sigma}(t) \tag{10.13b}$$

Even though this is a model of a linearly viscoelastic material, there is no single constant of proportionality between stress and strain due to the time derivatives in the constitutive equation. The modulus, or stress–strain ratio, is time dependent, and it must be determined through testing procedures such as the stress relaxation test. When put under a constant strain, the stresses in the Maxwell model gradually relax. Under a constant stress, the elastic component stretches immediately, while the viscous component stretches gradually. The dashpot in this model resembles the action of the piston damper shown in Fig. 10.2.

10.3.2 Kelvin–Voigt Model

The Kelvin–Voigt model for a viscoelastic material consists of a spring and a dashpot in parallel with each other. This model represents a typical viscoelastic solid. The diagram for the model is shown in Fig. 10.9.

Fig. 10.9 Kelvin–Voigt model [1]

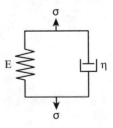

In this arrangement, the spring and dashpot experience the same strain, but the total stress is the sum of the stress on each component. The constitutive equation for this model is

$$\sigma(t) = k\varepsilon(t) + \eta\dot{\varepsilon}(t), \qquad (10.14)$$

where for the linear spring, $k = E$.

Under a constant stress, this model deforms at a decreasing rate as the portion carried by the linear spring increases. Once the stress is released, the material gradually relaxes. This model might be suitable to characterize the viscous wall damper shown in Fig. 10.3. Some recent studies have generalized this relationship by replacing ε by a so-called fractional derivative operator designed to capture the frequency dependence of the material properties [2, 3].

10.3.3   Standard Linear Solid Model

This viscoelastic model generalizes the Maxwell model with spring stiffness k_1 by placing it in parallel with another Hookean spring k_e as shown in Fig. 10.10.

The extra spring provides a rubber stiffness that most polymers exhibit. This stiffness prevents the unrestricted flow under constant stress allowed by the Maxwell model. When exposed to a constant stress, this model deforms instantly to a strain and then gradually continues to deform at a decreasing rate [1].

The constitutive equation for this model is derived by adding the series spring element k_1 to that for the model in Fig. 10.9 and evaluating the total stress as the sum of the contributions from the two components. The solution is complicated by the presence of the time derivative of the stress term and the stress term itself. The development of the constitutive equation is facilitated by using the Laplace transform solution technique as developed in Sect. 10.4.1.

Fig. 10.10 Linear
viscoelastic model [1]

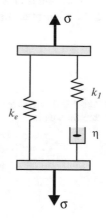

10.4   Response of Viscoelastic Materials

10.4.1   Laplace Transforms

As seen in (10.13a) and (10.13b), the time derivatives of the strain ε and the stress σ are present in the governing equations of viscoelastic models. These equations are best handled by Laplace transform that reduce differential equations to algebraic equations. We present a brief introductory summary of this technique before proceeding.

The basic definition of the Laplace transform $\mathcal{L}\{\ \}$ is

$$\mathcal{L}\{f(t)\} = f^*(s) = \int_0^\infty e^{-st} f(t) dt \qquad (10.15)$$

so that

$$f^*(s) = \mathcal{L}\{f(t)\} \qquad (10.16a)$$

and

$$f(t) = \mathcal{L}^{-1}\{f^*(s)\} \qquad (10.16b)$$

where $\mathcal{L}^{-1}\{\ \}$ is the inverse Laplace transform. As we will see, following the transformation of the original differential equation $f(t)$ into the Laplace variable $f^*(s)$, the resulting algebraic equation is solved for $f^*(s)$. Then, the solution in terms of the original variable t is recovered through the inverse transform. The definitive mathematics is not pursued here; however, we list a few inverse transforms needed to treat some elementary cases.

In Table 10.1, $u(t)$ is the Heaviside step function defined as

$$u(t) = 0, t < 0 \quad \text{and} \quad u(t) = 1, t \geq 0 \qquad (10.17)$$

This function is convenient for representing a suddenly applied constant amplitude effect measured from $t = 0$ or from another initial time.

Table 10.1 Laplace transform pairs

$f(t)$	$f^*(s)$
1	$1/s$
e^t	$1/(s-1)$ $(s > 1)$
t^n	$n!/s^{n+1}$
$u(t)$	$1/s$
$e^{-\alpha t}$	$1/(s + \alpha)$
$\dot{f}(t)$	$-f(0) + s\mathcal{L}\{f(t)\}$

Fig. 10.11 Stress and strain
histories in a stress
relaxation test [1]

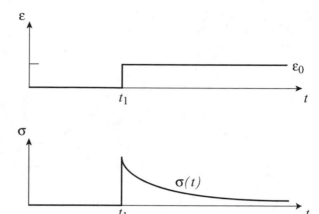

10.4.2   Stress Relaxation

We consider the phenomenon of stress relaxation discussed briefly in Sect. 10.2.3. In
Fig. 10.11, a constant strain ε_0 is applied to the material at $t = t_1$. This could be
represented using the slightly generalized Heaviside step function, defined in
(10.17), as $\varepsilon(t) = \varepsilon_0\, u(t - t_1)$. The resulting time-dependent stress $\sigma\,(t)$ is found
from (10.13a) and (10.13b).

Taking (10.13b) with $\dot{\varepsilon}_{\text{Total}} = 0$,

$$\frac{\sigma}{\tau} + \dot{\sigma}\,(t) = 0. \tag{10.18}$$

Separating variables and integrating from $\sigma_0 = k\varepsilon_0$

$$\int_{\sigma_0}^{\sigma} \frac{d\sigma}{\sigma} = -\frac{1}{\tau}\int_0^t dt$$

$$\ln\sigma - \ln\sigma_0 = -\frac{t}{\tau}$$

or

$$\sigma(t) = \sigma_0 e^{-t/\tau} \tag{10.19}$$

$\tau = \eta/k$ is a characteristic relaxation parameter representing the time needed for the
stress to fall to $1/e$ of its original value. Then, the relaxation modulus E_{rel} is obtained
directly from (10.11) as

$$\begin{aligned}
E_{\text{rel}}(t) &= \frac{\sigma(t)}{\varepsilon_0} \\
&= \frac{\sigma_0}{\varepsilon_0}e^{-t/\tau} \\
&= ke^{-t/\tau}.
\end{aligned} \tag{10.20}$$

Fig. 10.12 Relaxation modulus for the Maxwell model

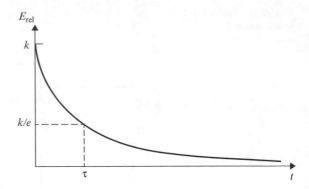

Equation (10.20) is plotted in Fig. 10.12 where the two parameters k and τ can be adjusted to force the model to match an experimental plot at two points. Ordinarily, k would be set to the initial modulus, and τ would be selected for a convenient point on the curve. The relaxation parameter τ may be strongly dependent on temperature [1].

10.4.3   Standard Linear Solid Model

While the Maxwell model may be adequate for some materials that accommodate unrestricted flow, the standard linear viscoelastic model (S.L.S.) shown in Fig. 10.10 provides a residual stiffness k_e that remains after the stresses in the Maxwell arm have relaxed as the dashpot extends. Both the Maxwell arm and the parallel spring arm k_e experience the same strain ε so that the stress is the sum

$$\sigma = \sigma_e + \sigma_m. \tag{10.21}$$

The stress in the Maxwell arm σ_m given in (10.13b) is awkward to solve because of the time derivative and is addressed by the Laplace transform technique. The transforms of the time derivatives are the Laplace variables times the transforms of the functions so that

$$\mathcal{L}\{\dot{\varepsilon}\} = s\bar{\varepsilon} \tag{10.22a}$$

$$\mathcal{L}\{\dot{\sigma}\} = s\bar{\sigma}, \tag{10.22b}$$

where the overline denotes the transformed functions. Proceeding with (10.13b) with $\sigma = \sigma_m$ and $k = k_1$, the transformed equation is

$$k_1 s\bar{\varepsilon} = (1/\tau)\bar{\sigma}_m + s\bar{\sigma}_m \tag{10.23}$$

yielding

$$\bar{\sigma}_m = [k_1 s/(s+1/\tau)]\bar{\varepsilon}. \tag{10.24}$$

Then adding the stress in the parallel spring, the total stress is

$$\bar{\sigma} = [k_e + k_1 s/(s+1/\tau)]\bar{\varepsilon} \tag{10.25a}$$

which may be written as

$$\bar{\sigma}(s) = \mathcal{E}\bar{\varepsilon}(s) \tag{10.25b}$$

where the associated viscoelastic modulus in the Laplace plane is

$$\mathcal{E} = k_e + k_1 s/(s+1/\tau). \tag{10.25c}$$

We may consider two cases based on (10.25b) and (10.25c), strain input and stress input.

For a given input strain function $\varepsilon(t)$, the resulting stress function is obtained by the following steps:

1. Transform the strain function to $\bar{\varepsilon}(s)$.
2. Form $\bar{\sigma}(s) = \mathcal{E}\bar{\varepsilon}(s)$.
3. Inverse transform $\bar{\sigma}(s)$ to yield $\sigma(t)$.

This process is applied to the stress relaxation test shown in Fig. 10.11 and solved previously for the Maxwell model. The strain function $\varepsilon(t)$ commencing at $t = 0$ is represented by the Heaviside function, (10.17) as

$$\varepsilon(t) = \varepsilon_0 u(t) \tag{10.26a}$$

with Laplace transform from Table 10.1

$$\bar{\varepsilon}(s) = \varepsilon_0/s. \tag{10.26b}$$

Substituting in (10.25a) and dividing by ε_0 gives

$$\bar{\sigma}/\varepsilon_0 = k_e/s + k_1/(s+1/\tau) \tag{10.27}$$

Noting from Table 10.1 that

$$\mathcal{L}^{-1}\left\{\frac{1}{s}\right\} = 1 \text{ and } \mathcal{L}^{-1}\left\{\frac{1}{s+\alpha}\right\} = e^{-\alpha t} \text{ and taking } \alpha = 1/\tau,$$

(10.27) is inverse transformed to give

$$\frac{\sigma(t)}{\varepsilon_0} = E_{rel}(t) = k_e + k_1 e^{-t/\tau} \tag{10.28}$$

which is the value for the Maxwell model augmented by k_e.

The curve plotted in Fig. 10.13 was generated by (10.28). Note that plotting E_{rel} against log t expands the graph in the early phases of the deformation. The upper

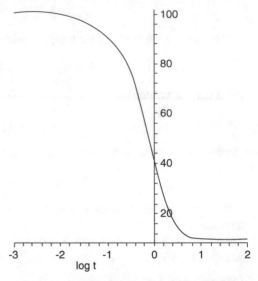

Fig. 10.13 The stress relaxation modulus $E_{\text{rel}}(t)$ for $E_g = 100$, $E_r = 10$, and $\tau = 1$ [1]

value E_g is the short-term elastic modulus, sometimes called the *glassy* modulus, while the lower value E_r is called the *rubbery* modulus reflecting a transition in the strain resistance from bond distortion to conformal extension [1].

The form of (10.25b) is convenient when the stress needed to generate a given strain is desired. The inverse problem of finding the strain generated by a given stress

$$\bar{\varepsilon}(s) = \bar{\sigma}(s)/\mathcal{E} \tag{10.29}$$

is complicated by the form of the equation, with $\mathcal{E}$, as given by (10.25c) in the denominator, making the equation more difficult to invert [1]. The resulting function

$$C_{\text{crp}}(t) = \mathcal{E}(t)/\sigma_0 \tag{10.30}$$

is the creep compliance which is written as [1].

$$C_{\text{crp}}(t) = C_g + \left(C_r - C_g\right)[1 - \exp(-t/\tau_c)] \tag{10.31}$$

and where $C_g = 1/(k_e + k_1)$, $C_r = 1/k_e$, $\tau_c = \tau\,[(k_e + k_1)/k_e]$.

The glassy compliance C_g and the rubbery compliance C_r are the inverse of the respective participating spring stiffnesses, but the characteristic retardation time τ_c is longer than the characteristic relaxation time τ. A representative plot of a creep compliance function is shown in Fig. 10.14.

10.4.4 Dynamic Response

In the case of dynamic loading developed in Sect. 10.2.5, the stress and strain are both in the form of (10.7a) and (10.7b), proportional to $e^{i\omega t}$. All derivatives with respect to time then contain $[i\omega]\,e^{i\omega t}$, and (10.13b) becomes

Fig. 10.14 A creep
compliance function [1]

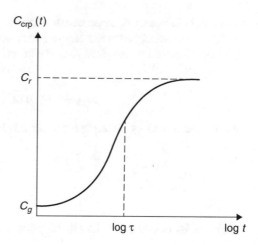

$$k[i\omega]\varepsilon_0^* e^{i\omega t} = (1/\tau + i\omega)\sigma_0^* e^{i\omega t} \tag{10.32}$$

whereupon the complex elastic modulus is

$$E^* = \sigma_0^*/\varepsilon_0^* = \frac{k[i\omega]}{(1/\tau + i\omega)} = \frac{k[i\omega\tau]}{(1 + i\omega\tau)} \tag{10.33}$$

Multiplying through by the complex conjugate of the denominator $(1 - i\omega\tau)$ gives

$$E^* = \frac{k\omega^2\tau^2}{1 + \omega^2\tau^2} + i\frac{k\omega\tau}{1 + \omega^2\tau^2} \tag{10.34}$$

which is in the explicitly separated form of (10.9).

10.4.5 Correspondence Principle

For the solution of elasticity problems for elastic materials, the tractions and displacements generally depend on position, while for viscoelastic materials, the solution carries the added variable of time introduced through the constitutive laws derived in Sect. 10.3. Rather than restarting from basics, it is often possible to use the viscoelastic correspondence principle to adapt an available elasticity solution to viscoelastic materials [4].

When a mechanics problem that is subject to Laplace transform is not altered spatially, the solution for an elastic body will apply to the viscoelastic body as well in the so-called Laplace plane. This is due to the separable form of functions such as

$$\check{T}(x,t) = f(x)g(t) \rightarrow \overline{\check{T}}(x,t) = \bar{f}(x)\bar{g}(s)$$

that do not change spatially on transformation. The major result of this recognition in our present context is that the elastic constants from the elasticity equations can be replaced directly by complex viscoelastic analogs. For example, we consider the Lamé equation (5.7)

$$\mu u_{j,ii} + (\lambda + \mu)u_{i,ij} = f_j = 0$$

and replace $\mu = G$ by the complex shear modulus as defined in (10.10),

$$\mu \rightarrow G^*(i\omega) = G'(\omega) + iG''(i\omega) \tag{10.35}$$

10.5   Viscoelasticity in Biological Tissue

10.5.1   General

The mechanical characteristics of biological tissue are very important to the future of biological science and medicine. This information is used in such wide-ranging applications as tissue engineering, medical simulation, and diagnostic medicine. Biomaterials are very complex in nature as most of them are composite materials composed of several distinct components arranged in a number of ways making mechanical characterization more difficult. Biological tissues are typically aniso-tropic as well. Experimentation on these materials is challenging due to their small size, large deformations, active contraction, damage during dissection, and the lack of an unloaded state in vivo. Most tissues show highly nonlinear stress–strain relationships and pronounced stress relaxation and creep, leading to the consider-ation of soft tissues as viscoelastic in nature [5].

10.5.2   History Dependence

Y.C. Fung, considered the father of biomechanics, developed a history-dependent model of a cylindrical specimen under a tensile load [5]. The stress is a function of both the time t and the stretch ratio λ, the ratio of the final length to the initial length as discussed in Sect. 3.8, 3.66.

The history of the stress response is called the relaxation function introduced in Sect. 10.4.2 and takes the form

$$K(\lambda, t) = G(t)T^{(e)}(\lambda), \quad G(0) = 1. \tag{10.36}$$

The reduced relaxation function is denoted $G(t)$ and is a function of time, while the elastic response is denoted $T^{(e)}(\lambda)$ and is a function of stretch ratio λ. Both of these terms can be determined experimentally. The specimen is in a state of stretch at time τ.

Using superposition, Fung derived this equation for tensile stress as a function of instantaneous stress and stress history:

$$T(t) = T^{(e)}[\lambda(t)] + \int_0^t T^{(e)}[\lambda(t - \tau)]\frac{\partial G(\tau)}{\partial \tau} d\tau. \tag{10.37}$$

The $T^{(e)}$ $[\lambda(t)]$ term represents the instantaneous stress, and the integral term represents the stress history. The stress history term is generally negative since the term $\frac{\partial G(\tau)}{\partial \tau}$ is negative (See Fig. 10.12). Thus, the stress history term tends to decrease the tensile stress. This equation operates under the assumption of linear viscoelasticity as it states that the tensile stress is the sum of all past stresses. Most biological materials do not show linear behavior, but this approach is instructive for small strains.

Equation (10.37) is only a one-dimensional model and needs to be extended to generate a general stress–strain history law for specific biomechanic applications. This may be a difficult process due to the many factors that can cause this relationship to vary between materials, such as material properties, chemical factors (pH, temperature, etc.), species, location in the body, exercise/rest state, and aging [5]. Currently, there is an increasing body of knowledge in characterizing the material properties of biological materials.

10.5.3 Mechanical Properties

The equations governing wave propagation, introduced in Sect. 9.2, are widely used to approximate the mechanical properties of biological tissue using fitted measured displacement data. Because many of these materials exhibit viscoelastic as opposed to purely elastic behavior, the appropriate wave propagation equations are readily found from the classical equations by invoking the correspondence principle introduced in Sect. 10.4.5.

The Lamé equations for vibrations in an unbounded elastic medium are given as (9.1)

$$\mu u_{j,ii} + (\lambda + \mu)u_{i,ij} = \rho \ddot{u}_j \quad (i, j = 1, 2, 3). \tag{9.1}$$

The second term on the l.h.s. is associated with volume change and is known as the dilatational component of motion as identified in (9.14). Shear wave excitation is often used to measure material response in soft tissues which are nearly incompressible [5]. Correspondingly, the dilatational component of the Lamé equation is neglected to give the simpler (9.12):

$$\mu u_{j,ii} = \rho \ddot{u}_j \quad (i,j = 1,2,3) \tag{9.12}$$

For a harmonic steady-state mechanical excitation, the displacements are taken in the form [7].

$$u_k(x,y,z;t) = \left(U'_k + iU''_k\right)e^{i\omega t} + \left(U'_k - iU''_k\right)e^{-i\omega t} \quad (k = 1,2,3) \tag{10.38}$$

where $U'_k(x,y,z;t)$ and $U''_k(x,y,z;t)$ are the real and imaginary components of the complex exponential. Using the correspondence principle (10.35) to replace μ in (9.12) by $G^*(i\omega)$ leads to a coupled set of equations in terms of the harmonic coefficients and the complex shear modulus:

$$G'U'_{j,ii} - G''U''_{j,ii} = -\rho\omega^2 U'_j \quad (i,j = 1,2,3) \tag{10.39a}$$

$$G'U''_{j,ii} + G''U'_{j,ii} = -\rho\omega^2 U''_j \quad (i,j = 1,2,3) \tag{10.39b}$$

The displacements in the three orthogonal directions are uncoupled so that the material property estimates can be made for the acquisition of any single displacement component in the case of material isotropy [7].

The wave equations are widely used for material characterizations due to the relative ease of applying harmonic excitation to specimens and measuring the response. For purely elastic materials, experimental estimates of the displacement field are fitted to (9.12), while for viscoelastic materials, they would be based on (10.39a) and (10.39b).

Now moving to a transverse isotropic material model that might be suitable for a biological application such as that presented in Sect. 4.4, it is first assumed that the local fiber directions are known from detailed physical examination or imaging techniques. Then 3-D displacements estimates are fitted to the local coordinate frame and the second derivatives required for (10.39a) and (10.39b) are computed from smoothed finite difference methods or by differentiating fitted polynomial functions. The technique of fitting polynomials or other shape functions to a defined domain is well known in modern numerical analysis as discussed in Sects. 11.7.1 and 11.7.2.

For incompressible anisotropic materials, we refer to (9.16) with the dilatational component $\nabla \cdot \mathbf{u}$ dropped, as discussed following (3.41):

$$\mu_\perp \nabla^2 \mathbf{u} + \tau_\mu \begin{pmatrix} u_{x,zz} + u_{z,xz} \\ u_{y,zz} + u_{z,yz} \\ \nabla^2 u_z \end{pmatrix} = \rho\ddot{\mathbf{u}} - \zeta\nabla^2\dot{\mathbf{u}} \tag{10.40}$$

The two elastic mechanical parameters, the transverse shear modulus $\mu_\perp$ and the anisotropy parameter $\tau = \mu_\parallel - \mu_\perp$, are estimated by minimizing the error in the approximate solution for shear wave propagation [6]. The viscoelastic response is introduced through the shear viscosity ζ in the last term of (10.40).

Sophisticated noninvasive techniques that introduce propagating shear waves into the brains of mice and measure displacements in order to obtain frequency-dependent storage and loss modulus estimates are discussed in Clayton et al. [7] and Okamoto et al. [8]. In the first study, magnetic resonance elastography (MRE) at comparably higher frequencies is used to compare the viscoelastic shear modulus G^* defined in (10.10) to values obtained from mechanical tests at lower frequencies. In the second study, a novel dynamic shear test (DST) is employed to measure G^* using a viscoelastic form of the one-dimensional wave equation (9.13a) for the purposes of direct comparison to MRE. Close agreement between MRF and DST results at overlapping frequencies indicate that G^* can be locally estimated with MRE over a wide frequency range.

10.6 Exercises

1. Consider the Wiechert model shown in Fig. 10.15. The additional Maxwell arms allow a distribution of relaxation times to be considered.

   (a) Compute the viscoelastic modulus $\mathcal{E}$.
   (b) Compute the relaxation modulus $E_{\text{rel}}(t)$.

2. Expand the exponential forms for the dynamic stress and strain $\left(\sigma(t) = \sigma_0^* e^{\iota \omega t}, \varepsilon(t) = \varepsilon_0^* e^{\iota \omega t} \right)$, and show that

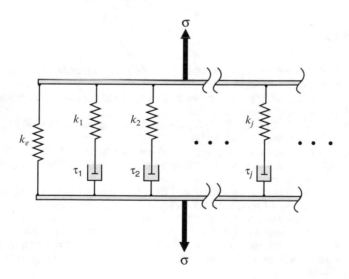

Fig. 10.15 Wiechert model

$$E^* = \frac{\sigma(t)}{\varepsilon(t)} = \frac{\sigma_0 \cos \delta}{\varepsilon_0} + i \frac{\sigma_0 \sin \delta}{\varepsilon_0},$$

where δ is the phase angle between the stress and strain.

3. Using the relation $\bar{\sigma} = \mathcal{E}\bar{\varepsilon}$ for the case of dynamic loading $\varepsilon(t) = \varepsilon_0 \cos \omega t$ and S.L.S. material response $\left(\mathcal{E} = k_e + \dfrac{k_1 s}{\left(s + \frac{1}{\tau}\right)} \right)$, solve for the time-dependent stress $\sigma(t)$. Use this solution to identify the steady-state components of the complex modulus $E^* = E' + iE''$ and the transient component as well. Answer:

$$E^* = \frac{k_1}{1 + \omega^2 \tau^2} e^{-t/\tau} + \frac{k_e + k_1 \omega^2 \tau^2 + k_e \omega^2 \tau^2}{1 + \omega^2 \tau^2} \cos \omega t - \frac{k_1 \omega \tau}{1 + \omega^2 \tau^2} \sin \omega t$$

4. Show that the viscoelastic law for the Voigt form of the standard linear solid (a spring of stiffness $k_v = 1/C_v$ in parallel with a dashpot of viscosity η, and this combination in series with another spring of stiffness $k_g = 1/C_g$) can be written as

$$\varepsilon(t) = C\sigma_0, \quad \text{with} \quad C = \left[C_g + \frac{C_v}{\tau\left(s + \dfrac{1}{\tau}\right)} \right] \quad \text{where } \tau = \eta/k_v.$$

References

1. Roylance D (1999) Engineering viscoelasticity. Module for Engineering/3–11. Massachusetts Institute of Technology, Cambridge, MA
2. Torvik PJ, Bagley R (1984) On the appearance of the fractional derivative in the behavior of real materials. ASME J Appl Mech 51:741–748
3. Singh MP, Chang T-S, Nandan H (2011) Algorithms for seismic analysis of MDOF systems with fractional derivatives. Eng Struct 33(8):2371–2381
4. Flügge W (1967) Viscoelasticity. Blaisdell Publishing Co, Waltham
5. Fung YC (1997) Selected works on biomechanics and aeroelasticity. Part A. World Scientific Publishing Co., Singapore
6. Sinkus R et al (2005) Imaging anisotropic and viscous properties of breast tissue by magnetic resonance-elastography. Magn Reson Med 53:372–387. Wiley Interscience
7. Clayton EH, Garbow JR, Bayly PV (2011) Frequency dependent viscoelastic parameters of mouse brain tissue estimated by MR elastography. Phys Med Biol 56:2391–2406
8. Okamoto RJ, Clayton EH, Bayly PV (2011) Viscoelastic properties of soft gels: comparison of magnetic resonance elastography and dynamic shear testing in the shear wave regime. Phys Med Biol 56:6379–6400

Chapter 11
Energy Principles

Abstract The theory of elasticity may also be developed from energy considerations, leading to the field equations in the form of differential equations. This approach does not promise any obvious computational advantage from the standpoint of *analytical* solutions. However energy methods are extensively developed for the pursuit of *numerical* solutions and are the basis of powerful contemporary programs for solving complex problems in solid mechanics. This development is rather recent and was brought about by the availability of the digital computer. The theoretical underpinnings for energy-based numerical methods are derived, but actual solution techniques are not addressed in any detail since that is a vast field in itself.

11.1 Introduction

The theory of elasticity may be developed from energy considerations, leading to the field equations in the form of differential equations. This approach does not promise any computational advantage from the standpoint of *analytical* solutions, since the same equations found from the classical formulations are produced. However, for the pursuit of *numerical* solutions, energy methods are extensively developed and are the basis of powerful contemporary programs for solving complex problems in solid mechanics facilitated by the emergence of the digital computer. This probably explains the relatively few applications of energy methods to *elasticity* problems found in the classical texts. However, even in the eighteenth century, the mathematician Leonhard Euler contrasted the direct formulation of the governing differential equations, known then as the method of *effective causes*, with the energy approach, known then as the method of *final causes*, in an argument that skillfully blended science and theology (this may have been diplomatic considering the well-known plight of perhaps the first "elastician," Galileo, a century earlier). Euler wrote [1]:

> Since the fabric of the universe is most perfect, and is the work of a most wise Creator, nothing whatsoever takes place in the universe in which some relation of maximum and minimum does not appear. Wherefore there is absolutely no doubt that every effect in the

© Springer International Publishing AG, part of Springer Nature 2018
P. L. Gould, Y. Feng, *Introduction to Linear Elasticity*,
https://doi.org/10.1007/978-3-319-73885-7_11

universe can be explained as satisfactorily from final causes, by the aid of the method of maxima and minima, as it can from the effective causes themselves. . . .

In this chapter, we derive the theoretical underpinnings for energy-based numerical methods, but we do not discuss actual solution techniques since that is a vast field in itself.

11.2   Conservation of Energy

Energy methods are developed from the law of conservation of energy. Assuming that a set of loads is applied slowly and that we have an isothermal deformation process,

$$W_{\mathrm{E}} = W_{\mathrm{V}}, \tag{11.1}$$

where W_{E} is the work done by the external loading and W_{V} is the change in internal energy. With an elastic material, the work W_{E} will be recovered if the loads are removed so that W_{V} may be regarded as stored energy called *strain energy*. We previously introduced this concept in Sect. 4.3 and will develop it further in the next section.

11.3   Strain Energy

11.3.1   Strain Energy Density

We represent the strain energy in the form

$$W_{\mathrm{V}} = \int_{V} W \ dV, \tag{11.2}$$

where the general form of the *strain energy density* W has been stated in (4.8). We consider (4.2a) to write

$$W = \int_{\varepsilon_{ij}} \sigma_{ij} d\varepsilon_{ij}.$$

For the linear elastic isotropic case using (4.18),

$$
\begin{aligned}
W &= \int_{\varepsilon_{ij}} (2\mu\varepsilon_{ij} + \lambda\delta_{ij}\varepsilon_{kk}) \ d\varepsilon_{ij} \\
&= \mu\varepsilon_{ij}\varepsilon_{ij} + \frac{\lambda}{2}(\varepsilon_{kk})^2.
\end{aligned}
\tag{11.3}
$$

It is also convenient to define a dual quantity to W in the manner of (4.2a),

$$\varepsilon_{(ii)} = \frac{\partial W^*}{\partial \sigma_{(ii)}}$$

$$\varepsilon_{ij} = \frac{1}{2}\frac{\partial W^*}{\partial \sigma_{ij}} \qquad i \neq j \tag{11.4}$$

where W^* is called the *complementary* strain energy density. Following (11.3), we have

$$W^* = \int_{\sigma_{ij}} \varepsilon_{ij} d\sigma_{ij}. \tag{11.5}$$

For the isotropic linear elastic case, substituting (4.19) gives

$$W^* = \frac{1}{4\mu}\sigma_{ij}\sigma_{ij} - \frac{\lambda}{4\mu(2\mu + 3\lambda)}(\sigma_{kk})^2 = W. \tag{11.6}$$

The same equation may be obtained from (11.3) using (4.19).

It is of interest to graphically interpret the strain energy and complementary strain energy densities for a one-dimensional elastic, but not necessarily linear, material as shown in Fig. 11.1. For the *linear* case, the densities W and W^* are obviously equal, while for the nonlinear case, they are not.

To illustrate the evaluation of W for a basic problem in elasticity, consider the prismatic bar under axial load previously solved in Sect. 6.2. We first rewrite (11.6) for the principal stress state ($\sigma_{ij} = 0, i \neq j$)

Fig. 11.1 Stress–strain curve for nonlinear elastic material

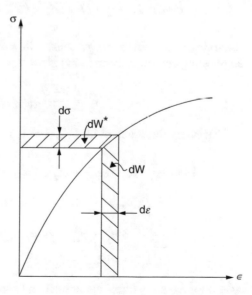

$$W^* = W = \frac{1}{4\mu}\left[1 - \frac{\lambda}{(2\mu + 3\lambda)}\right](\sigma_{kk})^2. \tag{11.7a}$$

In terms of the engineering constants, recognizing (4.24c) gives

$$W = \frac{1}{2E}(\sigma_{kk})^2. \tag{11.7b}$$

When σ_{zz} is the only nonzero stress, (11.7b) reduces to

$$W = \frac{1}{2E}(\sigma_{zz})^2, \tag{11.7c}$$

and, for the bar with $\sigma_{zz} = \gamma Z$ and area $A = BD$,

$$W = \frac{\gamma^2}{2E}z^2. \tag{11.7d}$$

Then,

$$W_V = \frac{\gamma^2}{6}\frac{AL^3}{E} \tag{11.8}$$

from (11.2).

11.3.2   Strain Energy Density of Distortion

Returning to the strain energy density in terms of stress (11.6) and converting to engineering constants using (4.24) gives

$$W = \frac{1+\nu}{2E}\sigma_{ij}\sigma_{ij} - \frac{\nu}{2E}(\sigma_{kk})^2. \tag{11.9}$$

Expanding the equation for 1, 2, 3 $= x, y, z$,

$$
\begin{aligned}
W &= \frac{1+\nu}{2E}(\sigma_{xx}^2 + \sigma_{yy}^2 + \sigma_{zz}^2 + 2\sigma_{xy}^2 + 2\sigma_{yz}^2 + 2\sigma_{zx}^2) \\
&\quad - \frac{\nu}{2E}(\sigma_{xx}^2 + \sigma_{yy}^2 + \sigma_{zz}^2 + 2\sigma_{xx}\sigma_{yy} + 2\sigma_{yy}\sigma_{zz} + 2\sigma_{xx}\sigma_{zz}) \\
&= \frac{1}{2E}[\sigma_{xx}^2 + \sigma_{yy}^2 + \sigma_{zz}^2 - 2\nu(\sigma_{xx}\sigma_{yy} + \sigma_{yy}\sigma_{zz} + \sigma_{xx}\sigma_{zz}) \\
&\quad + 2(1+\nu)(\sigma_{xy}^2 + \sigma_{yz}^2 + \sigma_{zx}^2)]
\end{aligned} \tag{11.10}
$$

where we have separated the normal and shearing contributions. Equation (11.10) may also be written in terms of the stress invariants, (2.59), as

$$W = \frac{1}{2E}[Q_1^2 - 2(1 + \nu)Q_2] \tag{11.11}$$

which emphasizes the nondirectionality of strain energy [2].

Next we consider principal stress components

$$\sigma_{xx} = \sigma^{(1)}, \qquad \sigma_{yy} = \sigma^{(2)}, \qquad \sigma_{zz} = \sigma^{(3)}, \qquad \sigma_{xy} = \sigma_{yz} = \sigma_{zx} = 0 \tag{11.12}$$

and rewrite (11.10) in the form [2]

$$W = \frac{(1 + \nu)}{6E}\left\{ \left[\sigma^{(1)} - \sigma^{(2)}\right]^2 + \left[\sigma^{(2)} - \sigma^{(3)}\right]^2 + \left[\sigma^{(3)} - \sigma^{(1)}\right]^2 \right\}$$
$$+ \frac{1 - 2\nu}{6E}\left[\sigma^{(1)} + \sigma^{(2)} + \sigma^{(3)}\right]^2, \tag{11.13}$$

which may be verified by expansion. The second part of (11.13) represents the mean stress, $1/3[\sigma^{(1)} + \sigma^{(2)} + \sigma^{(3)}]$, which contributes only to volume change as discussed in Sect. 3.5. Therefore the first part represents the strain energy density associated with change in shape or *distortion* and is so defined. Using (4.24a) and (4.24e),

$$W_0 = \frac{1}{12G}\left\{ \left[\sigma^{(1)} - \sigma^{(2)}\right]^2 + \left[\sigma^{(2)} - \sigma^{(3)}\right]^2 + \left[\sigma^{(3)} - \sigma^{(1)}\right]^2 \right\}. \tag{11.14}$$

This quantity is associated with the failure of elastic materials as discussed in Sect. 12.2.2.

The term in brackets is recognized as being proportional to the octahedral shearing stress, (2.86b), so that the strain energy density of distortion may be written as

$$W_0 = \frac{3}{4G}\left(\sigma_{ns}^{oct}\right)^2. \tag{11.15}$$

Therefore the resolution of the strain energy density into the sum of a *volume change* component and a *distortional* component corresponds to the state of stress referred to the octahedral planes.

11.4 Work of External Loading

The elastic body is assumed to be loaded by a system of $\tilde{Q}$ external forces $\mathbf{F}_Q$, where Q indicates the point of application of the force. It is sufficient for our introductory purposes to assume that the forces are concentrated or *discrete*, as opposed to being distributed. Corresponding to each force is a displacement $\mathbf{u}_Q$. By *correspondence*, we mean that $\mathbf{u}_Q$ is at the same point and in the same direction as $\mathbf{F}_Q$. The external work produced is then

$$W_E = \sum_{Q=1}^{\bar{Q}} \frac{1}{2} \mathbf{F}_Q \cdot \mathbf{u}_Q. \tag{11.16}$$

It turns out that the direct application of the principal of conservation of energy is rather limited so that more useful forms are now sought.

11.5 Principle of Virtual Work

11.5.1 Definitions

Many problems in structural mechanics may be solved efficiently by the application of the principle of virtual work (PVW). In this context, the word *virtual* means *not real*, but not necessarily small. As discussed later, the name "work" is somewhat of a misnomer, resulting from the appearance of force–times–distance products having the units of work in the resulting equations.

We consider a body *in equilibrium* under a set of $\tilde{Q}$ applied forces $\mathbf{F}_Q$ as shown in Fig. 11.2a. At any point Q in the body, the state of stress is given by the components of the stress tensor σ_{ij} that satisfy the equilibrium Eq. (2.45). Also indicated are the corresponding displacements $\mathbf{u}_Q$ as well as those at unloaded points that will be of interest later. Now the body is subjected to a *second* system of $\tilde{Q}$ forces *in equilibrium*, $\delta\mathbf{F}_Q$, as shown in Fig. 11.2b. This produces a displaced configuration $\delta\mathbf{u}_Q$ that does not violate the prescribed boundary conditions. (This constraint is not strictly necessary but is sufficient for our purposes.)

Note that the set $\tilde{Q}$ includes *all* referenced points in either system. For unloaded points in Fig. 11.2a, $\mathbf{F}_Q = 0$; and in Fig. 11.2b, $\delta\mathbf{F}_Q = 0$. In the *second* system, the stresses, strains, and displacement components at Q are $\delta\sigma_{ij}$, $\delta\varepsilon_{ij}$, and δu_i,

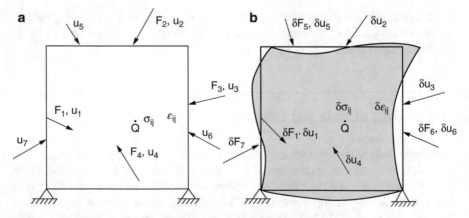

Fig. 11.2 (a) Elastic body in equilibrium under applied forces; (b) elastic body in equilibrium under virtual forces

respectively. The $\delta \mathbf{F}_Q$ are termed *virtual forces*, the $\delta \mathbf{u}_Q$ are called *virtual displacements*, and

$$\delta \varepsilon_{ij} = \frac{1}{2} \left(\delta u_{i,j} + \delta u_{j,i} \right) \tag{11.17}$$

are *virtual strains*. Furthermore, the product of *either* $\mathbf{F}_Q \cdot \delta \mathbf{u}_Q$ or $\delta \mathbf{F}_Q \cdot \mathbf{u}_Q$ is known as *external virtual work*.

Clearly the product is *not* conventional work since each component is due to a different source; that is, $\delta \mathbf{u}_Q$ is *not* due to or caused by $\mathbf{F}_Q$. Since we have stipulated that the virtual forces and virtual distortions (Fig. 11.2b) are applied *after* the actual or *real* loading system and displacements (Fig. 11.2a), we may envisage virtual work as either the product of a virtual force being "dragged through" a real displacement or a virtual displacement "acted through" by a real force. The *actual* work performed by the virtual system,

$$\sum_{Q=1}^{\bar{Q}} \frac{1}{2} \delta \mathbf{F}_Q \cdot \delta \mathbf{u}_Q$$

is of no apparent interest here.

11.5.2   Principle of Virtual Displacements

We begin with the virtual work due to virtual displacements $\delta \mathbf{u}_Q$ applied to a system in equilibrium under a set of surface tractions and body forces

$$\delta W_{\mathrm{E}} = \mathbf{F}_Q \cdot \delta \mathbf{u}_Q = \int_A T_i \delta u_i \ dA + \int_V f_i \delta u_i \ dV. \tag{11.18}$$

Using (2.11) and the divergence theorem in component form (2.32), the surface integral is transformed into

$$\int_A T_i \delta u_i \ dA = \int_A \sigma_{ij} n_j \delta u_i \ dA$$
$$= \int_V \left(\sigma_{ij} \delta u_i \right)_{,j} \ dV. \tag{11.19}$$

Substituting (11.19) into (11.18) gives

$$\delta W_{\mathrm{E}} = \int_V [(\sigma_{ij,j} + f_i) \delta u_i + \sigma_{ij} \delta u_{i,j}] dV$$
$$= \int_V \sigma_{ij} \delta u_{i,j} \ dV, \tag{11.20}$$

Fig. 11.3 Elastic
continuum

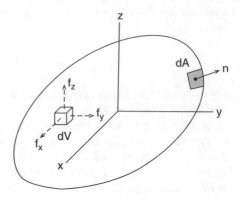

since the first term is the (satisfied) equilibrium Eq. (2.45). Next we introduce
(11.17) into (11.20), recognizing that both i and j become repeated indices. This
produces

$$\delta W_E = \int_V \sigma_{ij} \delta u_{i,j} \; dV = \int_V \sigma_{ij} \delta \varepsilon_{ij} \; dV \tag{11.21}$$

which may be restated as

$$\delta W_E = \delta W_V \tag{11.22a}$$

where

$$\delta W_V = \int_V \sigma_{ij} \delta \varepsilon_{ij} \; dV \tag{11.22b}$$
$$= \text{internal virtual work.}$$

Equations (11.22a) and (11.22b) are the *principle of virtual displacements* (PVD)
derived for a body in equilibrium that is subjected to virtual displacements produced
by a virtual force system also in equilibrium. The important result of this derivation
is the converse: if (11.22a) is satisfied, then the system is in *equilibrium*.

To illustrate an application of the PVD, we re-derive the equations of equilibrium
for an elastic continuum, Fig. 11.3, in terms of displacements. These were previously
found in Sect. 5.2 and are given as (5.7). First we treat the internal virtual work. From
(11.2) and (11.3), the strain energy is

$$W_V = \int_V \left[\mu \varepsilon_{ij} \varepsilon_{ij} + \frac{\lambda}{2} (\varepsilon_{kk})^2 \right] dV. \tag{11.23}$$

Using explicit notation initially, following the presentation of Forray [3] and
selecting $x_1, x_2, x_3 = x, y, z$ as shown on Fig. 11.3, (11.23) becomes

$$W_V = \int_V \left[\left(\frac{\lambda}{2} + \mu \right) \left(\varepsilon_{xx}^2 + \varepsilon_{yy}^2 + \varepsilon_{zz}^2 \right) + 2\mu \left(\varepsilon_{xy}^2 + \varepsilon_{yz}^2 + \varepsilon_{zx}^2 \right) + \lambda \left(\varepsilon_{xx}\varepsilon_{yy} + \varepsilon_{yy}\varepsilon_{zz} + \varepsilon_{zz}\varepsilon_{xx} \right) \right] dV.$$

$$(11.24)$$

Computing the internal virtual work from (11.22b) using (4.2a) for σ_{ij},

$$
\begin{aligned}
\delta W_V &= \frac{\partial W_V(\varepsilon_{ij})}{\partial \varepsilon_{ij}} \delta \varepsilon_{ij} \\
&= \int_V \left[\left(\frac{\lambda}{2} + \mu \right) \left(2\varepsilon_{xx}\delta \varepsilon_{xx} + 2\varepsilon_{yy}\delta \varepsilon_{yy} + 2\varepsilon_{zz}\delta \varepsilon_{zz} \right) \right. \\
&\quad + 2\mu \left(2\varepsilon_{xy}\delta \varepsilon_{xy} + 2\varepsilon_{yz}\delta \varepsilon_{yz} + 2\varepsilon_{zx}\delta \varepsilon_{zx} \right) \\
&\quad + \lambda \left(\varepsilon_{xx}\delta \varepsilon_{yy} + \varepsilon_{yy}\delta \varepsilon_{xx} + \varepsilon_{yy}\delta \varepsilon_{zz} \right. \\
&\quad \left. \left. + \varepsilon_{zz}\delta \varepsilon_{yy} + \varepsilon_{zz}\delta \varepsilon_{xx} + \varepsilon_{xx}\delta \varepsilon_{zz} \right) \right] dV.
\end{aligned}
$$

$$(11.25)$$

Next we recall the strain–displacement relations from (3.14),

$$
\begin{aligned}
\varepsilon_{xx} &= u_{x,x}; & \varepsilon_{yy} &= u_{y,y}; & \varepsilon_{zz} &= u_{z,z} \\
\varepsilon_{xy} &= \frac{1}{2}(u_{x,y} + u_{y,x}); & \varepsilon_{yz} &= \frac{1}{2}(u_{y,z} + u_{z,y}); & \varepsilon_{zx} &= \frac{1}{2}(u_{x,z} + u_{z,x}),
\end{aligned}
$$

$$(11.26)$$

and apply a virtual displacement consisting of only *one* component, say δu_x. The corresponding virtual strain–displacement equations are

$$
\begin{aligned}
\delta \varepsilon_{xx} &= (\delta u_x)_{,x}; & \delta \varepsilon_{yy} &= 0; & \delta \varepsilon_{zz} &= 0 \\
\delta \varepsilon_{xy} &= \frac{1}{2}(\delta u_x)_{,y}; & \delta \varepsilon_{yz} &= 0; & \delta \varepsilon_{zx} &= \frac{1}{2}(\delta u_x)_{,z}.
\end{aligned}
$$

$$(11.27)$$

Substituting (11.26) and (11.27) into (11.25) yields

$$\delta W_V = \int_V \left[\left(\frac{\lambda}{2} + \mu \right) 2u_{x,x}(\delta u_x)_{,x} + \mu \left[(u_{x,y} + u_{y,x})(\delta u_x)_{,y} + (u_{x,z} + u_{z,x})(\delta u_x)_{,z} \right] \right.$$
$$\left. + \lambda \left[(u_{y,y} + u_{z,z})(\delta u_x)_{,x} \right] \right] dV.$$

$$(11.28)$$

Equation (11.28) may be integrated by parts using the following integration formula adapted from [4]:

$$\int_V \left(\mathbf{u}_{,x} \cdot \mathbf{v}_{,x} + \mathbf{u}_{,y} \cdot \mathbf{v}_{,y} + \mathbf{u}_{,z} \cdot \mathbf{v}_{,z} \right) dV = \int_A (\mathbf{v} \cdot \mathbf{u}_{,n}) dA - \int_V \mathbf{v} \cdot \left(\mathbf{u}_{,xx} + \mathbf{u}_{,yy} + \mathbf{u}_{,zz} \right) dV$$

$$(11.29)$$

where $\mathbf{u} = u_x$, u_y, and u_z in turn and $\mathbf{v} = \delta u_x$, so that $\mathbf{u}_{,n}$ becomes $u_{x,n}u_{y,n}u_{z,n}$, respectively. Therefore, (11.28) becomes

$$\delta W_{\mathrm{V}} = -\int_V \left[\left(\frac{\lambda}{2} + \mu \right) 2 u_{x,xx} (\delta u_x) \right.$$

$$+ \mu \left(u_{x,yy} + u_{y,xy} + u_{x,zz} + u_{z,xz} \right) \delta u_x + \lambda \left(u_{y,yx} + u_{z,zx} \right) \delta u_x \right] dV$$

$$\left. + \int_A \left[\left(\frac{\lambda}{2} + \mu \right) 2 u_{x,n} \delta u_x \right. \right.$$

$$\left. + \mu \left(2 u_{x,n} + u_{y,n} + u_{z,n} \right) \delta u_x + \lambda \left(u_{y,n} + u_{z,n} \right) \delta u_x \right] dA, \tag{11.30a}$$

where n is directed along the outer normal to the surface (Fig. 11.3). Since the surface terms enter only into the boundary conditions, we concentrate on the first integral and rearrange (11.30a) as

$$\delta W_{\mathrm{V}} = -\int_V \left[(\lambda + 2\mu) u_{x,xx} + \mu \left(u_{x,yy} + u_{y,xy} + u_{x,zz} + u_{z,xz} \right) \right.$$

$$\left. + \lambda \left(u_{y,yx} + u_{z,zx} \right) \right] \delta u_x dV + [\text{Surface terms}]$$

$$= -\int_V \left[\mu \left(u_{x,xx} + u_{x,yy} + u_{x,zz} \right) + (\lambda + \mu) \left(u_{x,x} + u_{y,y} + u_{z,z} \right)_{,x} \right] \delta u_x \ dV$$

$$+ [\text{Surface terms}]$$

$$= -\int_V \left[\mu \nabla^2 u_x + (\lambda + \mu) (\Delta u)_{,x} \right] \delta u_x \ dV + [\text{Surface terms}], \tag{11.30b}$$

where

$$\nabla^2 (\quad) = \nabla (\quad) \cdot \nabla (\quad) = (\quad)_{,ii} \tag{11.30c}$$

$\nabla^2 (\ )$ was defined in (1.10b), and

$$\Delta u = \operatorname{div} \mathbf{u} = u_{i,i} \tag{11.30d}$$

as defined in (1.11).

We take the external virtual work as given by (11.18) specialized for $u_i = u_x$ as

$$\delta W_{\mathrm{E}} = \int_A T_x \delta u_x dA + \int_V f_x \delta u_x dV$$

$$= \int_V f_x \delta u_x dV + [\text{Surface Terms}] \tag{11.31}$$

and substitute (11.30b) and (11.31) into (11.21). Since the virtual displacement δu_x is arbitrary, we may equate the coefficients of the volume terms in the integrands to get

$$\mu \nabla^2 u_x + (\lambda + \mu)(\Delta u)_{,x} + f_x = 0 \tag{11.32}$$

which is (5.7) with $j = x$. The remaining equations are found by permutation, with a virtual displacement of δu_y and then δu_z.

For completeness, we carry out the same derivation using indicial notation beginning from (11.23)

$$\delta W_V = \int_V \left(2\mu\varepsilon_{ij}\delta\varepsilon_{ij} + \lambda\varepsilon_{kk}\delta\varepsilon_{kk}\right)dV, \tag{11.33a}$$

where

$$\varepsilon_{ij} = \frac{1}{2}\left(u_{i,j} + u_{j,i}\right). \tag{11.33b}$$

With a virtual displacement δu_l,

$$\delta u_l = \delta u_i \delta_{il} = \delta u_j \delta_{jl} \tag{11.33c}$$

and

$$\begin{aligned}
\delta\varepsilon_{ij} &= \frac{1}{2}\left[(\delta u_i)_{,j}\delta_{il} + (\delta u_j)_{,i}\delta_{jl}\right] \\
&= \frac{1}{2}\left[(\delta u_l)_{,j} + (\delta u_l)_{,i}\right],
\end{aligned} \tag{11.33d}$$

(11.33a) becomes

$$\begin{aligned}
\delta W_V &= \int_V \left[2\mu\left[\frac{1}{2}(u_{i,j} + u_{j,i})\right]\frac{1}{2}[(\delta u_l)_{,j} + (\delta u_l)_{,i}] + \lambda(u_{k,k})(\delta u_l)_{,l}\right]dV \\
&= \int_V \left[\frac{\mu}{2}[(u_{l,j} + u_{j,l})(\delta u_l)_{,j} + (u_{i,l} + u_{l,i})(\delta u_l)_{,i}] + \lambda(u_{k,k})(\delta u_l)_{,l}\right]dV.
\end{aligned} \tag{11.33e}$$

On the second line of (11.33e), note that in the term multiplied by $(\delta u_l)_{,j}$, the subscript i in $(u_{i,j} + u_{j,i})$ has been changed to l and in the term multiplied by $(\delta u_l)_{,i}$, the subscript j is changed to l, to match the designated virtual displacement δu_l.

Generalizing the integration formula (11.29), integrating by parts produces typically

$$\int_V u_{i,j}\delta u_{l,j} = u_{i,n}\delta u_l \bigg]_{\text{surface}} - \int_V u_{i,jj}\delta u_l. \tag{11.33f}$$

Applying this formula to the first terms of (11.33e) [which are multiplied by $(\delta u_l)_{,j}$] and integrating with respect to j, then to the second terms [which are multiplied by $(\delta u_l)_{,i}$] and integrating with respect to i, and finally to the last term [which is multiplied by $(\delta u_l)_{,l}$] and integrating with respect to l gives

$$\begin{aligned}
\delta W_V &= -\int_V \left[\frac{\mu}{2}(u_{l,jj} + u_{j,lj} + u_{i,li} + u_{l,ii}) + \lambda(u_{k,k})_{,l}\right]\delta u_l dV \\
&\quad + [\text{Surface Terms}]
\end{aligned} \tag{11.34a}$$

$$\delta W_{\mathrm{V}} = -\int_V \left[\mu(u_l)_{,ii} + (\lambda + \mu)(u_{i,i})_{,l} \right] \delta u_l \ dV \tag{11.34b}$$
$$+ \ [\text{Surface Terms}].$$

In going from (11.34a) to (11.34b), all dummy indices are taken as i. First, the first and fourth terms with factor $\frac{\mu}{2}$ are combined with new factor μ, then the second and third terms with factor $\frac{\mu}{2}$, and the last term with factor λ which is also combined with new factor $(\lambda + \mu)$. δW_{E} is given by (11.31) with $x = l$. Now, setting $\delta W_{\mathrm{V}} = \delta W_{\mathrm{E}}$, equating the volume terms, and changing l to j, we have

$$\mu(u_j)_{,ii} + (\lambda + \mu)(u_{i,i})_{,j} + f_j = 0 \tag{11.35}$$

which is the same as (5.7). Thus, we see that the PVD is an artifice for establishing *equilibrium* relationships.

We return to the discrete loading system shown in Fig. 11.1a and write

$$\delta W_{\mathrm{E}} = \sum_{Q=1}^{\bar{Q}} \mathbf{F}_Q \cdot \delta \mathbf{u}_Q. \tag{11.36}$$

Of interest in (11.36) in the context of work is the absence of the familiar factor of ½ [see (11.16)]. This is explained by insisting that the real loading *precedes* the virtual distortion so that the real forces and displacements are at *full* value before the corresponding virtual effects act. The main use of this form of the PVW is to compute the set of equilibrating forces, one at a time, for a system in which the actual deformations and the corresponding stresses are known. Since the choice of the virtual distortion is open, it is conveniently selected so that the point of application of the single force to be evaluated, say $\mathbf{F}_3$, is given a virtual displacement $\delta \mathbf{u}_3$, while no virtual displacements are permitted at other load points. Equation (11.22a) then reduces to

$$\mathbf{F}_3 \cdot \delta \mathbf{u}_3 = \int_V \sigma_{ij} \delta \varepsilon_{ij} dV \tag{11.37}$$

or, for simplicity, choosing $\delta \mathbf{u}_3$ in the direction of $\mathbf{F}_3$,

$$F_3 = \int_V \sigma_{ij} \delta \varepsilon_{ij} dV. \tag{11.38}$$

A judicious choice of the virtual deformation allows $\mathbf{F}_3$ to be computed. The procedure may be repeated to calculate the entire set of $\tilde{Q}$ loads, provided that the real stresses σ_{ij} and the virtual strains $\delta \varepsilon_{ij}$, corresponding to the particular unit $\delta \mathbf{u}_Q$, can be evaluated. This principle is particularly useful in rigid plastic analysis where the virtual deformation may be discretized at a "plastic hinge."

A contemporary application of the virtual displacement principle is presented by Gladilin et al. [5], where the validity of a *linear elastic* model for soft tissue for predicting the outcome of computer-assisted craniofacial surgery is investigated. The problem is recognized at the outset as nonlinear so that this linear exercise is only an initial phase. Both a differential equation formulation in terms of strains based on (5.19) and an energy formulation are presented. The later approach entails a variational problem of energy minimization (see Sect. 11.6) based on transforming the equilibrium equation in terms of stress σ_{ij}, (5.1), to a form in terms of strains ε_{ij}. The strains are in turn computed from displacements acquired using tomographic data from a CT scan. Then, test displacements u_j^* are selected as weighting functions, and the resulting equilibrium equation is solved based on a piecewise application of the Rayleigh–Ritz method as discussed in Sects. 11.7.2 and 11.7.4, the so-called finite element method with an adaptive grid. The resulting equations match the PVD where the u_j^* are chosen as interpolation functions between the nodes of the mesh at which the errors are evaluated; the u_j^* terms represent essentially virtual displacements.

As described above, the transformation of the weighted equilibrium equation from stresses to strains is carried out in (11.39) starting with the integrated product of the real and the virtual system. The first term $\sigma_{ij,i}u_j^*$ is rewritten as a simple product derivative in the volume domain. Then, the product term that is created, $(\sigma_{ij}u_j^*)_{,i}$, is transformed into a surface integral by Green's theorem (1.14) with the divergence given in indicial form by (1.17b). This becomes essentially the external VW and is zero in the absence of stresses on the boundary. The second term $\sigma_{ij}u_{j,i}^*$ is the key and is transformed into $\sigma_{ij}\varepsilon_{ij}^*$ from the strain–displacement relation (5.2) and the summation convention. The body forces $f_j = 0$.

$$\int_V (\sigma_{ij,i} + f_j)u_j^*\ dV = \int_V \left[(\sigma_{ij}u_j^*)_{,i} - \sigma_{ij}u_{j,i}^* + f_j u_j^*\right]\ dV$$

$$= \int_A \sigma_{ij}n_i u_j^*\ dA - \int_V \sigma_{ij}\varepsilon_{ij}^*\ dV + \int_V f_j u_j^*\ dV = 0.$$

$$(11.39)$$

After using the constitutive relationship (4.18) to replace σ_{ij} by ε_{ij}, the equilibrium condition in the volume domain is the internal VW that is used to monitor the linearization error

$$\int_V \left(2\mu\varepsilon_{ij}\varepsilon_{ij}^* + \lambda\delta_{ij}\varepsilon_{kk}\varepsilon_{kk}^*\right)\ dV = 0. \qquad (11.40)$$

In the cited study, this linear elastic approach apparently produced a useful first approximation to the patient's postoperative appearance but led to substantial errors as the deformations became larger.

11.5.3   Principle of Virtual Forces

First we consider the virtual work produced by virtual surface tractions δT_i and virtual body forces δf_i:

$$
\begin{aligned}
\delta W_E^* &= \int_A \delta T_i u_i dA + \int_V \delta f_i u_i dV \\
&= \int_V \delta\sigma_{ij}\varepsilon_{ij} dV \\
&= \delta W_V^*
\end{aligned}
\tag{11.41}
$$

following the same steps as in the previous Eqs. (11.19) through (11.22a) and (11.22b). This is the so-called principle of virtual forces (PVF) which describes *compatibility* between internal strains and external displacements.

We now consider the external virtual work produced by virtual forces $\delta\mathbf{F}_Q$:

$$
\delta W_E^* = \sum_{Q=1}^{\bar{Q}} \delta\mathbf{F}_Q \cdot \mathbf{u}_Q.
\tag{11.42}
$$

In this form, the principle is useful for computing displacements, one at a time, for a system in which the actual stresses and the corresponding strains are known. If the displacement $\mathbf{u}_3$, is sought, we choose $\delta\mathbf{F}_3 = 1$ with all other $\delta\mathbf{F}_Q = 0$, and (11.41) becomes

$$
u_3 = \int_V \delta\sigma_{ij}\varepsilon_{ij} dV.
\tag{11.43}
$$

Equation (11.43) may be solved, provided that the real strains ε_{ij} corresponding to the particular $\delta\mathbf{F}_Q$ (in this case $\delta\mathbf{F}_3 = 1$) can be evaluated.

A somewhat trivial example is selected to illustrate this technique. The bar shown in Fig. 6.1 is again considered where it is desired to compute the axial deformation along the centerline $u_z(0, 0, z)$. A unit virtual force produces a stress $\delta\sigma_{ij} = \delta\sigma_{zz} = \frac{1}{A}$, and the corresponding strain, given by (6.4), is $\varepsilon_{zz} = \frac{1}{E}\gamma z$. Then, (11.43) becomes

$$
\begin{aligned}
u_z(0,0,z) &= \int_L^z \int_A \frac{1}{A}\frac{1}{E}\gamma z \, dA \, dz \\
&= \left[\frac{\gamma z^2}{2E}\right]_L^z \\
&= \frac{\gamma}{2E}(z^2 - L^2).
\end{aligned}
\tag{11.44}
$$

which checks with (6.19c).

11.5.4 Reciprocal Theorems

We again refer to the elastic body in Fig. 11.2a subjected, in turn, to two sets of forces $\mathbf{F}_Q$ and $\mathbf{F}_R^*$ which produce displacements $\mathbf{u}_Q$ and $\mathbf{u}_R^*$, respectively. The corresponding displacement components, stresses, and strains are $(u_i, \sigma_{ij}, \varepsilon_{ij})$ and $\left(u_i^*, \sigma_{ij}^*, \varepsilon_{ij}^*\right)$, respectively. Two sequences are considered:

1. $\mathbf{F}_Q$ applied first and then $\mathbf{F}_R^*$
2. $\mathbf{F}_R^*$ applied first and then $\mathbf{F}_Q$

For the first case, $\mathbf{F}_Q$ may be considered as the real system and $\mathbf{F}_R^*$ as the virtual system; then (11.36), (11.22a), and (11.22b) give

$$\sum_{Q=1}^{\bar{Q}} \mathbf{F}_Q \cdot \mathbf{u}_Q^* = \int_V \sigma_{ij} \varepsilon_{ij}^* dV. \tag{11.45a}$$

For the second sequence,

$$\sum_{R=1}^{\bar{R}} \mathbf{F}_R^* \cdot \mathbf{u}_R = \int_V \sigma_{ij}^* \varepsilon_{ij} dV. \tag{11.45b}$$

In each case, the linear elastic stress strain law (4.3) is applicable. Therefore,

$$\sigma_{ij} = E_{ijkl} \varepsilon_{kl}, \tag{11.46a}$$

$$\sigma_{ij}^* = E_{ijkl} \varepsilon_{kl}^*. \tag{11.46b}$$

After substitution of (11.46a) and (11.46b) into the r.h.s. of (11.45a) and (11.45b) respectively, recognition that the dummy indices may be interchanged, and realization that $E_{ijkl} = E_{klij}$, it is evident that the r.h.s. of the later equations are identical. Then, equating the l.h.s. gives

$$\sum_{Q=1}^{\bar{Q}} \mathbf{F}_Q \cdot \mathbf{u}_Q^* = \sum_{R=1}^{\bar{R}} \mathbf{F}_R^* \cdot \mathbf{u}_R, \tag{11.47a}$$

which is known as Betti's law, after Enrico Betti. That is, if a linearly elastic body is subjected to two separate loading systems, the sum of the products of the *forces* on the *first* system and their corresponding *displacements* on the *second* system is equal to the sum of the products of the *forces* on the *second* system and their corresponding *displacements* on the *first* system. As a special case, let each loading system be a single unit force, applied at Q and R, respectively. Of course, additional forces are required to maintain equilibrium, but these are assumed to act at points of zero displacement. Then (11.47a) reduces to

$$\mathbf{u}_Q^* = \mathbf{u}_R, \tag{11.47b}$$

which is known as Maxwell's law of reciprocal displacements after the nineteenth-century physicist James Clerk Maxwell. That is, on a linear elastic body in equilibrium with two points Q and R designated, the displacement at Q due to a unit force at R is equal to the displacement at R due to a unit force at Q—a remarkable result. Moreover, the unit "forces" do not have to even be of the same type, that is, $\mathbf{F}_Q$ may be an actual force and $\mathbf{F}_R^*$ a unit moment, whereupon $\mathbf{u}_Q^*$ is a displacement, but $\mathbf{u}_R$ is a rotation. Still, they are equal. Maxwell's law is widely used in classical structural mechanics.

11.6   Variational Principles

11.6.1   Definitions

The PVW may be recast in the form of variational theorems to remove the necessity of dealing with the virtual system per se. We concisely define the *variation* operation $\delta(\ )$ as an arbitrary linear *increment* of a *functional* (a generalized function) that satisfies all constraints on the body, including maintenance of equilibrium. As a simple example, a virtual displacement δu_i may be interpreted as a variation on the displacement field u_i. Corresponding to δu_i, we have the variation of the strain field, $\delta \varepsilon_{ij}$.

The variation of a function is computationally the same as the differential, so that

$$\delta(u_i u_i) = \delta(u_i)^2 = 2u_i \delta u_i = 2u_i \delta_{ij} \delta u_j$$

where δ_{ij} is the Kroencker δ and

$$\delta(\varepsilon_{kk} \varepsilon_{kk}) = \delta(\varepsilon_{kk})^2 = 2\varepsilon_{kk} \delta \varepsilon_{kk} = 2\varepsilon_{kk} \delta_{ij} \delta \varepsilon_{ij}, \quad \text{etc.}$$

11.6.2   Principle of Minimum Total Potential Energy

We now proceed to compute the resulting variation of the internal strain energy (11.2) with the strain energy density in the form of (11.3)

$$\delta W_V = \delta \int_V \left[\mu \varepsilon_{ij} \varepsilon_{ij} + \frac{\lambda}{2} (\varepsilon_{kk})^2 \right] dV$$

$$= \int_V \left(\left[\mu 2 \varepsilon_{ij} \delta \varepsilon_{ij} + \frac{\lambda}{2} 2 \varepsilon_{kk} \delta_{ij} \delta \varepsilon_{ij} \right] dV, \right.$$

or, using (4.18),

$$\delta W_V = \int_V \sigma_{ij} \delta \varepsilon_{ij} \, dV. \tag{11.48}$$

In a similar manner, the external virtual work δW_E may be equated to the work done by the surface tractions T_i and body forces f_i during the variation using (11.18). It is convenient to represent these contributions in terms of potential functions [6]

$$T_i = -\frac{\partial \Im}{\partial u_i} \quad \text{and} \quad f_i = -\frac{\partial f}{\partial u_i} \tag{11.49}$$

When such functions exist, (11.18) may be written as

$$\begin{aligned}
\delta W_E &= \int_A T_i \delta u_i dA + \int_V f_i \delta u_i dV \\
&= -\int_A \frac{\partial \Im}{\partial u_i} \delta u_i dA - \int_V \frac{\partial f}{\partial u_i} \delta u_i dV \\
&= -\delta \left(\int_A \Im dA - \int_V f \, dV \right) \\
&= -\delta V_E,
\end{aligned} \tag{11.50}$$

where V_E is defined as the potential of the external forces, given by

$$V_E = \int_A \Im dA + \int_V f \, dV \tag{11.51}$$

Further, it is assumed that the surface and body forces are vector point functions of *position* only; that is, they are *not* dependent on the deformation of the body, either in magnitude or direction. This type of force field is said to be *conservative* and encompasses a large majority of problems in solid mechanics. With $\Im = -\Im_i u_i$ and $f = -f_i u_i$, (11.51) becomes

$$V_E = -\int_A \Im_i u_i dA - \int_V f_i u_i dV. \tag{11.52}$$

Then the principle of minimum total potential energy (PMPE) is properly written as

$$\delta \Pi = \delta(W_V + V_E) = 0, \tag{11.53}$$

in which Π is called the potential energy *functional*. The form of (11.53) emphasizes that both terms of Π are subjected to the same variation and that the variation is performed *after* W_V and V_E are evaluated and combined. This is a so-called *extremum* statement since only the first variation is expressed. To show the actual *minimum* character, the second variation would have to be considered, but this is beyond our scope.

This principle implies that the *correct displacement state* of all those states that may satisfy the boundary conditions is that which makes the total potential energy a minimum. The great utility of the PMPE is that it is relatively straightforward *to approximate* displacement fields by *shape functions* that satisfy the kinematic boundary conditions and to scale these fields by minimizing the resulting total potential energy. We illustrate this point in a later section.

11.6.3 Principle of Minimum Complementary Energy

Recalling that two versions of the virtual work principle were derived depending on which form of external virtual work was manipulated, it follows that a dual to the PMPE can be created by focusing on the stress field σ_{ij} and the variation $\delta\sigma_{ij}$.

With the strain energy density in the form of (11.6) and the system subjected to an arbitrary $\delta\sigma_{ij}$, we have

$$\delta W_V^* = \delta \int_V \left[\frac{1}{4\mu}\sigma_{ij}\sigma_{ij} - \frac{\lambda}{4\mu(2\mu + 3\lambda)}(\sigma_{kk})^2 \right] dV$$

$$= \int_V \left[\frac{1}{2\mu}\sigma_{ij}\delta\sigma_{ij} - \frac{\lambda}{2\mu(2\mu + 3\lambda)}\sigma_{kk}\delta_{ij}\delta\sigma_{ij} \right] dV,$$

or, using (4.19),

$$\delta W_V^* = \int_V \varepsilon_{ij}\delta\sigma_{ij} \ dV \tag{11.54}$$

where δW_V^* is the internal *virtual* work, as derived in (11.39).

Analogous to (11.50), the external virtual work is represented in potential form by

$$\delta W_E^* = \delta\left(\int_A \Im_i u_i dA + \int_v f_i u_i dV \right) = -\delta V_E^*, \tag{11.55}$$

where $V_E^* = V_E$ as given by (11.52), for conservative fields.

Finally, the principle of minimum complementary energy (PMCE) is properly written as

$$\delta\Pi^* = \delta(W_V^* + V_E^*) = 0, \tag{11.56}$$

where Π^* is called the complementary energy functional. Again, it is apparent that the variations are performed on Π^* rather than on the individual terms. This principle implies that the *correct stress state* of all those that satisfy equilibrium is that which makes the total complementary energy a minimum. In contrast to the attractiveness of selecting displacement functions that satisfy compatibility, it may be more difficult to find stress distributions that satisfy equilibrium.

For a linear elastic body, $W_V^* = W_V$ and $V_E^* = V_E$. However, the *variations* of these quantities that appear in the foregoing principles are not equal since in the PMPE, the *displacements* are varied, while in the PMCE, the *stresses* are varied.

11.7 Direct Variational Methods

11.7.1 Motivation

As we mentioned in the introductory comments to this chapter, the accelerated prominence of energy methods for the solution of elasticity problems is a fairly recent development. The underlying basis of most of the computer-based techniques, which have come to be known as *finite element* methods, is the previously derived variational principles together with the appropriate numerical analysis algorithms. Using the PMPE as an example, functions that satisfy the kinematic constraints are selected to represent the dependent variables which are most frequently generalized displacements. These so-called shape or comparison functions are scaled in accordance with the PMPE to provide the best approximation to equilibrium. Recall how a sine wave may closely match the deformation of a simply supported beam, both under static loading and in free vibration.

In the course of introducing shape functions to represent the generalized displacements, the extremum problem of the calculus of variations is transformed into the maximum–minimum problem of the classical calculus. Such techniques are termed *direct* variational methods. The treatment of the vibration of elastic bodies by Lord Rayleigh (John W. Strutt) in *The Theory of Sound* [7] is probably the origin of this approach, and it is fitting that the *Rayleigh–Ritz method* be recognized as the most prominent direct method.

11.7.2 Rayleigh–Ritz Method

To illustrate the Rayleigh–Ritz idea, we focus on (11.53) where both W_V and V_E are regarded as functions of the displacement field u_i. We select kinematically admissible polynomials, i.e., that satisfy the displacement boundary conditions, for *each* u_i consisting of n_i terms

$$
\begin{aligned}
u_i \cong \bar{u}_i = {} & C_{i0} + C_{i1}x + C_{i2}y + C_{i3}z + C_{i4}x^2 + \cdots \\
& + C_{ij}x^m y^p z^q + \cdots + C_{in_i}x^{\tilde{m}}y^{\tilde{p}}z^{\tilde{q}} \quad (i = 1, 2, 3),
\end{aligned}
\tag{11.57}
$$

where the C_{ij} are as yet undetermined coefficients; u_i represents u_x, u_y, and u_z, in turn; and m, p, and q are the exponents of x, y, and z up to $\tilde{m}$, $\tilde{p}$, and $\tilde{q}$, in each polynomial.

After the introduction of the three (11.57) into (11.2), expressed in terms of strains and then displacements using (11.3) and (3.14), and then into (11.52) and (11.53), the subsequent integration and minimization lead to

$$\delta\Pi(\bar{u}_i) = \sum_{j=1}^{n_i} \frac{\partial \Pi}{\partial C_{ij}} \delta C_{ij}, \tag{11.58}$$

where the usual summation convention applies on i. Since the variations δC_{ij} are arbitrary, Eq. (11.58) may be satisfied only if

$$\frac{\partial \Pi}{\partial C_{ij}} = 0 \qquad i = 1,\ 2,\ 3; \ \ j = 1, 2, \ldots, n_i, \tag{11.59}$$

which produce a set of three $(n_1 + n_2 + n_3)$ simultaneous *algebraic* equations for the unknown coefficients. With the coefficients C_{ij} evaluated, (11.57) represents the displacements over the domain; the strains and stresses may be computed accordingly from the laws of the theory of elasticity. A similar technique may be applied to minimize the functional Π^* given by (11.56).

11.7.3 Torsion of Rectangular Cross Section

We return to the St. Venant torsion solution for an illustration of the Rayleigh–Ritz method. Here, we examine a rectangular cross section as shown in Fig. 11.4, an elusive problem by analytical means as we noted in Sect. 6.4.3.

First, we have the strain energy W_V defined by (11.2) with the density W given by (11.4). Based on the developments in Sect. 6.4.2, the formulation is in terms of the stress function ϕ. From (11.6) and (6.65),

Fig. 11.4 Rectangular cross section

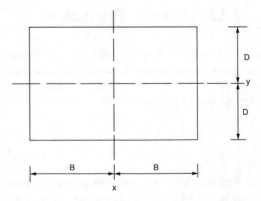

$$W = \frac{1}{4\mu}\sigma_{ij}\sigma_{ij}$$

$$= \frac{2}{4\mu}[(\sigma_{xz})^2 + (\sigma_{yz})^2] \qquad (11.60)$$

$$= \frac{1}{2G}[(\phi_{,y})^2 + (\phi_{,x})^2],$$

and the strain energy *per unit length* is given by

$$W_V = \frac{1}{2G} \int\int \left[(\phi_{,y})^2 + (\phi_{,x})^2\right] dx\, dy. \qquad (11.61)$$

Next, the corresponding potential of the applied loads is given in the form of (11.52) using (6.84)

$$V_E = -M\alpha = -2\alpha \int_A \phi \; dA. \qquad (11.62)$$

Forming the functional as defined in (11.53), we have

$$\Pi = \int\int \left\{ \frac{1}{2G}\left[(\phi_{,y})^2 + (\phi_{,x})^2\right] - 2\alpha\phi \right\} dx\, dy. \qquad (11.63)$$

At this point, we could address (11.63) directly, seeking a function ϕ that is a solution to $\delta\Pi = 0$ which is a traditional problem in the calculus of variations. However, we follow the Rayleigh–Ritz approach whereby we assume a solution in the form [8]

$$\psi \cong \phi - (x^2 - D^2)(y^2 - B^2) \sum_m \sum_n C_{mn} x^m y^n. \qquad (11.64)$$

This approximation satisfies the condition that ϕ must vanish on the boundaries $x = \pm D$, $y = \pm B$. The solution proceeds by substituting (11.64) into (11.63) and integrating over the area to get the approximate functional $\overline{\Pi}$. Thus, the variational problem is transformed into a calculus problem for which the equations

$$\frac{\partial \overline{\Pi}}{\partial C_{mn}} = 0 \qquad (11.65)$$

are used to obtain a set of simultaneous equations for the C_{mn} coefficients.

We do not carry out a detailed general solution for this problem; however, we may illustrate with a coarse approximation, a single term of (11.64), and a square cross section $B = D$

$$\overline{\phi} = C_0(x^2 - B^2)(y^2 - B^2). \qquad (11.66)$$

After evaluating

$$\overline{\phi}_{,y} = 2C_0\left(x^2 - B^2\right)y$$
$$\overline{\phi}_{,x} = 2C_0\left(y^2 - B^2\right)x, \tag{11.67}$$

we are ready to substitute (11.66) and (11.67) into (11.63). However, since we may anticipate integrating to find $\overline{\Pi}$ and then taking $\dfrac{\partial \overline{\Pi}}{\partial C_0}$, it is expedient to commute these operations and write the minimum condition as

$$\frac{\partial \overline{\Pi}}{\partial C_0} = \int_{-B}^{B} \int_{-B}^{B} \left\{ \frac{1}{2G}\left[2\overline{\phi}_{,y}\frac{\partial\left(\overline{\phi}_{,y}\right)}{\partial C_0} + 2\overline{\phi}_{,x}\frac{\partial\left(\overline{\phi}_{,x}\right)}{\partial C_0}\right] - 2\alpha\frac{\partial\overline{\phi}}{\partial C_0}\right\}dx\,dy = 0 \tag{11.68}$$

$$\int_{-B}^{B} \int_{-B}^{B} \left\{ \frac{1}{2G}\left[4C_0\left(x^2 - B^2\right)^2 y^2 + 4C_0\left(y^2 - B^2\right)^2 x^2\right] - 2\alpha\left(x^2 - B^2\right)\left(y^2 - B^2\right)\right\}\,dx\,dy = 0. \tag{11.69}$$

After integration and simplification, this gives

$$C_0 = \frac{5}{8}\frac{G\alpha}{B^2}, \tag{11.70}$$

and from (6.84),

$$M_z = 2\int_A \phi\,dA = 2\int_{-B}^{B} \int_{-B}^{B} \frac{5}{8}\frac{G\alpha}{B^2}\left(x^2 - B^2\right)\left(y^2 - B^2\right)\,dx\,dy$$
$$= \frac{20}{9}G\alpha B^4, \tag{11.71}$$

which differs by just over 1% from the correct solution [8].

Timoshenko and Goodier [8] have also used a three-term approximation,

$$\overline{\phi} = \left(x^2 - B^2\right)\left(y^2 - B^2\right)\left[C_0 + C_1\left(x^2 + y^2\right)\right], \tag{11.72}$$

where C_1 is a combined coefficient due to symmetry. The resulting M_z is but 0.15% from the correct value. However, the corresponding error in the maximum shear stress is about 4%. The distribution of the shearing stresses is quite similar to the elliptical cross section, as previously illustrated in Sect. 6.4.3, with maxima at the center of the sides decreasing to zero at the corners.

Considering that M_z was approximated almost exactly, the relatively large error in the maximum shear stress may be surprising; however, we recall that the stresses are obtained by *differentiation* of ϕ (6.65), so that they may indeed be less accurate. This demonstrates that extreme care should be exercised in measuring convergence for approximate solutions; specifically the most highly differentiated quantities should be considered. In this case, more terms in the approximation will improve convergence.

11.7.4 Commentary

Although a considerable mathematical simplification is accomplished in principle with the Rayleigh–Ritz technique, two rather formidable obstacles are evident once more complex problems are investigated. First, it may be difficult to find kinematically admissible representations for displacement functions, i.e., functions that satisfy the boundary conditions, over the *entire* domain. Second, if appropriate functions are found, the resulting set of simultaneous algebraic equations may be rather large if suitable accuracy is to be achieved.

These obstacles have been overcome by two modern developments. The first is the notion that perhaps it is sufficient for the approximate displacement functions to span over only a portion or *element* of the continuum and to be constrained to satisfy interelement continuity. This was apparently recognized early on by Courant and Hilbert [9], but was not fully exploited, perhaps since it exacerbates the second obstacle by producing even more simultaneous equations. The ultimate solution is achieved by the use of a high-speed digital computer with its ability for "number-crunching." Properly ordered, tens of thousands of simultaneous equations may be solved routinely.

The prophecy of Euler, quoted in the introductory section of this chapter, has been richly fulfilled by the development of the finite element method. Yet, we should not dismiss the rigorous analytical solutions demanded by the theory of elasticity as archaic, for the finite element solutions are still approximations. They need to be calibrated with selected analytical solutions, both numerically and judgmentally, to establish confidence and to quantify errors. In the opinion of the authors, *total* reliance on finite element methods or other numerical techniques that are implemented in "black box" computer programs is surely *unwise*, *probably uneconomical*, and possibly *unsafe*.

11.8 Computational Example with MATLAB

We illustrate the use of MuPAD to compute the theoretical formulations of displacement, stress, and strain based on the energy principal. Consider the right half of the thin plate shown as Problem 7.8. The boundary conditions, which represent a diaphragm, are given by

$$u_x = 0 \text{ along } y = 0 \text{ and } y = b$$
$$u_y = 0 \text{ along } x = 0 \text{ and } x = a$$

For a constant body force $f_y = p$, consider the PVD and the Rayleigh–Ritz method with displacement functions in the form

$$u_x = \sum_{n=1}^{\infty} \left\{ C_{0n} + \sum_{m=1}^{\infty} C_{mn} \cos\left(\frac{m\pi x}{a}\right) \right\} \sin\left(\frac{n\pi y}{b}\right)$$

$$u_y = \sum_{m=1}^{\infty} \left\{ D_{0m} + \sum_{n=1}^{\infty} D_{mn} \cos\left(\frac{n\pi y}{b}\right) \right\} \sin\left(\frac{m\pi x}{a}\right)$$

If we take $n = 1$ and $m = 2$, then we can compute the strain, the internal work, and the stress as follows:

- First, define variables.

```
mu  := Symbol::mu
```
μ
```
lambda := Symbol::lambda
```
λ

- Then, we input the u_x and u_y with $n = 1$ and $m = 2$.

```
ux := (C01 + C11*cos(PI*x/a)) * sin(PI*y/b)
```
$$\sin\left(\frac{\pi y}{b}\right)\left(C01 + C11 \cos\left(\frac{\pi x}{a}\right)\right)$$
```
uy := (D01 + D11*cos(PI*y/b)) * sin(PI*x/a)
```
$$\sin\left(\frac{\pi x}{a}\right)\left(D01 + D11 \cos\left(\frac{\pi y}{b}\right)\right)$$

- The strains ε_{xx}, ε_{yy}, and ε_{xy} can be calculated by differentiation.

```
exx := diff(ux, x)
```
$$-\frac{\pi\, C11 \sin\left(\frac{\pi x}{a}\right) \sin\left(\frac{\pi y}{b}\right)}{a}$$
```
eyy := diff(uy, y)
```
$$-\frac{\pi\, D11 \sin\left(\frac{\pi x}{a}\right) \sin\left(\frac{\pi y}{b}\right)}{b}$$
```
exy := (1/2) * (diff(ux, y) + diff(uy, x))
```
$$\frac{\pi \cos\left(\frac{\pi y}{b}\right)\left(C01 + C11 \cos\left(\frac{\pi x}{a}\right)\right)}{2b} + \frac{\pi \cos\left(\frac{\pi x}{a}\right)\left(D01 + D11 \cos\left(\frac{\pi y}{b}\right)\right)}{2a}$$

- Internal virtual work W_v.

$$
\begin{aligned}
&\texttt{w := mu*(exx\^{}2 + eyy\^{}2 + 2*exy\^{}2) + lambda/2*(exx + eyy)\^{}2}\\[4pt]
&\lambda\left(\frac{\pi\,C11\,\sigma_2\,\sigma_1}{a}+\frac{\pi\,D11\,\sigma_2\,\sigma_1}{b}\right)^{2}\Big/2
+\mu\left(2\left(\frac{\pi\,\sigma_3\left(C01+C11\,\sigma_4\right)}{2\,b}+\frac{\pi\,\sigma_4\left(D01+D11\,\sigma_3\right)}{2\,a}\right)^{2}
+\frac{C11^{2}\,\pi^{2}\,\sigma_2^{2}\,\sigma_1^{2}}{a^{2}}+\frac{D11^{2}\,\pi^{2}\,\sigma_2^{2}\,\sigma_1^{2}}{b^{2}}\right)
\end{aligned}
$$

where

$$\sigma_1 = \sin\left(\frac{\pi\,y}{b}\right)$$

$$\sigma_2 = \sin\left(\frac{\pi\,x}{a}\right)$$

$$\sigma_3 = \cos\left(\frac{\pi\,y}{b}\right)$$

$$\sigma_4 = \cos\left(\frac{\pi\,x}{a}\right)$$

- Total internal strain energy of the plate domain.

$$
\begin{aligned}
&\texttt{W\_plate := int(int(w,x=0..a), y=0..b)}\\[4pt]
&\frac{\pi^{2}\,(2\,\lambda\,C11\,D11\,a+2\,\mu\,C11\,D11\,a)}{8\,a}
+\frac{\pi^{2}\,b\,\left(\lambda\,C11^{2}+2\,\mu\,C11^{2}+2\,\mu\,D01^{2}+\mu\,D11^{2}\right)}{8\,a}\\[10pt]
&\qquad\qquad +\frac{\pi^{2}\,\left(2\,\mu\,C01^{2}\,a^{2}+\mu\,C11^{2}\,a^{2}+\lambda\,D11^{2}\,a^{2}+2\,\mu\,D11^{2}\,a^{2}\right)}{8\,a\,b}
\end{aligned}
$$

- Potential of external forces V_E.

$$
\begin{aligned}
&\texttt{VE := -int(int(p*uy, x=0..a), y=0..b)}\\[4pt]
&-\frac{2\,D01\,a\,b\,p}{\pi}
\end{aligned}
$$

- Total potential energy for Rayleigh–Ritz method.

$$
\begin{aligned}
&\texttt{W\_total := W\_plate + VE}\\[4pt]
&\frac{\pi^{2}\,(2\,\lambda\,C11\,D11\,a+2\,\mu\,C11\,D11\,a)}{8\,a}
+\frac{\pi^{2}\,b\,\left(\lambda\,C11^{2}+2\,\mu\,C11^{2}+2\,\mu\,D01^{2}+\mu\,D11^{2}\right)}{8\,a}\\[10pt]
&\qquad +\frac{\pi^{2}\,\left(2\,\mu\,C01^{2}\,a^{2}+\mu\,C11^{2}\,a^{2}+\lambda\,D11^{2}\,a^{2}+2\,\mu\,D11^{2}\,a^{2}\right)}{8\,a\,b}-\frac{2\,D01\,a\,b\,p}{\pi}
\end{aligned}
$$

- Calculate C01, C11, D01, and D11 based on potential energy minimization.

```
eq1 := diff(W_total, C01)
```
$$\frac{\mu\,C01\,\pi^2\,a}{2\,b}$$

```
eq2 := diff(W_total, C11)
```
$$\frac{\pi^2\,(2\,\lambda\,D11\,a+2\,\mu\,D11\,a)}{8\,a}+\frac{\pi^2\,b\,(2\,\lambda\,C11+4\,\mu\,C11)}{8\,a}+\frac{\mu\,C11\,\pi^2\,a}{4\,b}$$

```
eq3 := diff(W_total, D01)
```
$$\frac{\mu\,D01\,\pi^2\,b}{2\,a}-\frac{2\,a\,b\,p}{\pi}$$

```
eq4 := diff(W_total, D11)
```
$$\frac{\pi^2\,(2\,\lambda\,C11\,a+2\,\mu\,C11\,a)}{8\,a}+\frac{\pi^2\,(2\,\lambda\,D11\,a^2+4\,\mu\,D11\,a^2)}{8\,a\,b}+\frac{\mu\,D11\,\pi^2\,b}{4\,a}$$

```
sol := linsolve({eq1=0, eq2=0, eq3=0, eq4=0}, {C01, C11, D01, D11})
```
$$\left[C01=0,\ C11=0,\ D01=\frac{4\,a^2\,p}{\mu\,\pi^3},\ D11=0\right]$$

- Evaluate u, ε, and σ.

```
subs(ux, sol)
```
$$0$$

```
subs(uy, sol)
```
$$\frac{4\,a^2\,p\,\sin\left(\frac{\pi x}{a}\right)}{\mu\,\pi^3}$$

```
subs(exx, sol)
```
$$0$$

```
subs(eyy, sol)
```
$$0$$

```
subs(exy, sol)
```
$$\frac{2\,a\,p\,\cos\left(\frac{\pi x}{a}\right)}{\mu\,\pi^2}$$

```
sigxy := 2*mu*subs(exy, sol)
```
$$\frac{4\,a\,p\,\cos\left(\frac{\pi x}{a}\right)}{\pi^2}$$

11.9 Exercises

11.1 Consider the right half of the thin plate shown as Problem 7.8. The boundary conditions, which represent a diaphragm, are given by

$$u_x = 0 \text{ along } y = 0 \text{ and } y = b$$
$$u_y = 0 \text{ along } x = 0 \text{ and } x = a$$

For a constant body force $f_y = p$, consider the PVD and the Rayleigh–Ritz method with displacement functions in the form

$$u_x = \sum_{n=1}^{\infty} \left\{ C_{on} + \sum_{m=1}^{\infty} C_{mn} \cos\left(\frac{m\pi x}{a}\right) \right\} \sin\left(\frac{n\pi y}{b}\right)$$

$$u_y = \sum_{m=1}^{\infty} \left\{ D_{om} + \sum_{n=1}^{\infty} D_{mn} \cos\left(\frac{n\pi y}{b}\right) \right\} \sin\left(\frac{m\pi x}{a}\right)$$

Assuming $m = n = 1$, compute the displacements, strains, and stresses for a square plate with $a = b$.

11.2 Use the PVF to verify the solution (6.62) for the rate of twist in the circular rod subject to a constant torque shown in Fig. 6.5.

11.3 Resolve the St. Venant torsion of a square cross section discussed in Sect. 11.7.3 using the improved approximation (11.72), and evaluate the shear stresses. Compare these results to the values obtained from the one-term approximation.

11.4 Show that (11.6) can be written in terms of the engineering material constants as [10]

$$W = \frac{1}{2E}(\sigma_{11}^2 + \sigma_{22}^2 + \sigma_{33}^2) - \frac{\nu}{E}(\sigma_{11}\sigma_{22} + \sigma_{11}\sigma_{33} + \sigma_{22}\sigma_{33})$$

for a triaxial stress state, $\sigma_{ij} = 0 \ (i \neq j)$.

11.5 A state of plane strain relative to the (x, y) plane has a strain energy density function given by

$$W = \frac{1}{2}b_{11}\varepsilon_{xx}^2 + b_{22}\varepsilon_{yy}^2 + b_{33}\varepsilon_{xy}^2 + 2b_{12}\varepsilon_{xx}\varepsilon_{yy} + 2b_{13}\varepsilon_{xx}\varepsilon_{xy} + 2b_{23}\varepsilon_{yy}\varepsilon_{xy},$$

where the b_{ij} are elastic coefficients. Derive the equations of equilibrium in terms of the displacements $u_x(x, y)$ and $u_y(x, y)$ including the effects of body forces [11].

References

1. Timoshenko S (1953) History of strength of materials. McGraw-Hill Book Company Inc., New York
2. Boresi AP et al (1978) Advanced methanics of materials, 3rd edn. Wiley, New York
3. Forray MJ (1968) Variational calculus in science and engineering. McGraw-Hill Book Company Inc., New York
4. Pierce BO, Foster RM (1950) A short table of integrals, 4th edn. Ginn and Company, Boston. Formula 969
5. Gladilin E, Zachow S, Deuflhard P, Hege H-C (2001) Validation of a linear elastic model for soft tissue simulation in craniofacial surgery, In: Proceedings of SPIE, vol. 4319. http://spiedl.org. Berlin
6. Tauchert TR (1974) Energy principles in structural mechanics. McGraw-Hill Book Company Inc., New York
7. Rayleigh L (1945) The theory of sound. Dover, New York
8. Timoshenko S, Goodier JW (1951) Theory of elasticity, 2nd edn. McGraw-Hill Book Company Inc., New York
9. Courant R, Hilbert D (1966) Methods of mathematical physics, vol 1. Wiley-Interscience Publications, New York
10. Gopu VKA (1987) Validity of distortion-energy-based strength criterion for timber members. J Struct Eng ASCE 113(12):2475–2487
11. Boresi AP, Chong FP (1987) Elasticity in engineering mechanics. Elsevier, New York

Chapter 12
Strength and Failure Criteria

Abstract The application of the rigorous methods of analysis embodied in the theory of elasticity is naturally of interest to engineers. Within this largely theoretical subject, it is of interest to introduce the fundamental basis of such applications, namely, the comparison of the analytical *results* obtained from an elasticity solution to the expected *capacity* of the resisting material.

Traditionally, the capacity of a specific material has been stated in terms of a *failure criterion or strength*, whereby exceeding a critical value of a controlling parameter marks the limit of the functional range. It is recognized that other failure mechanisms, such as serviceability exceedence or fatigue, may govern. Here, only pointwise criteria in consort with elasticity theory are presented and discussed.

12.1 Introduction

The application of the rigorous methods of analysis is embodied in the theory of elasticity. While a broad treatment of engineering applications is beyond our scope, it is of interest to introduce the fundamental basis of such applications, namely, the comparison of the analytical *results* obtained from an elasticity solution to the expected *capacity* of the resisting material.

Historically, the capacity of a specific material has been stated in terms of a *failure criterion* whereby exceeding a critical value of a controlling parameter marks the limit of the functional range. More recently, it has been recognized that factors other than an explicit extreme failure criterion may govern the limit, such as excessive flexibility or fatigue. Therefore we choose to add the term *strength criterion*, which implies that there may be additional considerations in assessing the utility of the material. Furthermore, we restrict the discussion to criteria that are correlated to the pointwise focus of the theory and thereby exclude geometric instability considerations which are strongly controlled by topological parameters.

© Springer International Publishing AG, part of Springer Nature 2018
P. L. Gould, Y. Feng, *Introduction to Linear Elasticity*,
https://doi.org/10.1007/978-3-319-73885-7_12

12.2   Isotropic Materials

12.2.1   Classical Tests

For isotropic materials, it is helpful to discuss the measures of capacity with respect to tests that are easily performed on small specimens. Most common are the uniaxial tensile test and the torsion test, which imparts a state of pure shear. The state of stress in the $x-y$-plane from each of these tests is shown in Fig. 12.1. Within each element, the state of stress for a rotation of $45°$ is shown, since this orientation produces the maximum shear stress in the tension test and the maximum extensional stress in the shear test.

For calculation purposes, we assume a circular cross section of radius B for each test with an area $A = \pi B^2$ and polar moment of inertia, (6.60b), of $J_c = \frac{\pi B^4}{2}$. It is of interest in what follows to carry out a rather complete analysis and comparison of key parameters in the two classical tests. The parameters of interest are measures of *stress*, *strain*, and *energy* and are designated by the superscript P for tensile tests and the superscript T for torsion tests.

We first consider stress quantities. In the tensile test, the load produces an axial stress

$$\sigma_{xx}^{P} = \frac{P}{A} = \frac{P}{\pi B^2},\tag{12.1}$$

while in the torsion test, the torque produces a shearing stress, (6.61),

$$\sigma_{xy}^{T} = \frac{TB}{J_c} = \frac{2T}{\pi B^2}.\tag{12.2}$$

With a $45°$ counterclockwise rotation to the $x'y'$-axes, the respective stresses are

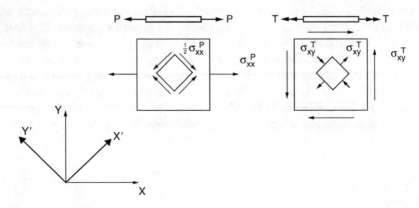

Fig. 12.1 Tension and torsion tests

$$\sigma^P_{x'y'} = -\frac{1}{2}\sigma^P_{xx} \tag{12.3}$$

from (2.13) with $\alpha_{x'x} = \frac{\sqrt{2}}{2}$ and $\alpha_{y'x} = -\frac{\sqrt{2}}{2}$ for the tensile test and

$$\sigma^T_{x'x'} = \sigma^T_{xy} \tag{12.4}$$

for the torsion test since the stress state is pure shear.

We also evaluate the octahedral shearing stresses for each case using (2.86b)

$$\sigma^{\text{oct } P}_{ns} = \frac{1}{3}\left[(\sigma^P_{xx})^2 + (-\sigma^P_{xx})^2\right]^{1/2}$$
$$= \frac{\sqrt{2}}{3}\sigma^P_{xx} \tag{12.5}$$

and referring to Fig. 12.1 with $\sigma^{(1)} = \sigma^T_{xy}, \sigma^{(2)} = -\sigma^T_{xy}$

$$\sigma^{\text{oct } T}_{ns} = \frac{1}{3}\left[(\sigma^T_{xy} + \sigma^T_{xy})^2 + (-\sigma^T_{xy})^2 + (-\sigma^T_{xy})^2\right]^{1/2}$$
$$= \frac{\sqrt{6}}{3}\sigma^T_{xy}. \tag{12.6}$$

Turning to the strains, for the tensile test the maximum normal strain is

$$\varepsilon^P_{xx} = \frac{\sigma^P_{xx}}{E}, \tag{12.7}$$

while the corresponding value in the torsion test may be evaluated from $\varepsilon_{11} = \varepsilon_{xx}$ (4.29b) referred to the rotated axes,

$$\varepsilon^T_{xx} = \frac{1}{E}[\sigma^T_{xy} - \nu(-\sigma^T_{xy})] = \frac{1+\nu}{E}\sigma^T_{xy}. \tag{12.8}$$

Now we consider energy measures. The strain energy density for a uniaxial load is given by (11.7b), which is for the tensile test

$$W = \frac{1}{2E}\left(\sigma^P_{xx}\right)^2 \tag{12.9}$$

and for the torsion test

$$W = \frac{1}{2E}\left[(\sigma^T_{xy})^2 + (-\sigma^T_{xy})^2\right]$$
$$= \frac{1}{E}(\sigma^T_{xy})^2. \tag{12.10}$$

Finally, we have the strain energy density of distortion as derived in (11.14)

$$W_0 = \frac{1}{12G}\left[(\sigma_{xx}^P)^2 + (-\sigma_{xx}^P)^2\right]$$
$$= \frac{1}{6G}(\sigma_{xx}^P)^2$$

(12.11)

for the tensile test and

$$W_0 = \frac{1}{12G}\left[(2\sigma_{xy}^T)^2 + (-\sigma_{xy}^T)^2 + (-\sigma_{xy}^T)^2\right]$$
$$= \frac{1}{2G}(\sigma_{xy}^T)^2$$

(12.12)

for the torsion test. Recall from (11.15) that the octahedral shearing stress and W_0 are related.

12.2.2 Failure Theories

Failure loads for tension and torsion tests and the corresponding stresses are P and T, respectively, σ_{xx}^P, etc. Next, strength theories are postulated whereby a *single* parameter, such as a state of stress, strain, and/or energy at a point in an elastic body, is correlated with the like state in either the tensile or torsion test situation. Six such theories, referred to a tension test, are summarized in Table 12.1 [1].

It is obvious that a parallel table could be produced with the torsion test limits replacing the tension test values. An interesting format for comparing the dual statements is presented in Table 12.2 after Boresi et al. [1]. Experimental evidence for metals suggests that the shear yield stress σ_{xy}^T is in the range of 0.5–0.58 of the tensile yield stress σ_{xx}^P, giving credence to the maximum shearing stress and octahedral shearing stress criteria. For more complex materials, it is advantageous to work with nondirectional or invariant quantities such as energy.

12.2.3 Invariants

A direct relationship between the invariants of the stress tensor, which are stated in terms of the principal stresses in (2.61), and the strain energy of distortion or octahedral shearing stress criterion may be established.

Initially, the hydrostatic stresses are discounted as a contributor to the plastic deformation of metals in accordance with experimental observations [2]. As a result, the mean normal stress as defined in (3.48) is not considered initially. Since this

Table 12.1 Theories of failure referred to a tensile test

Theory/criterion	Statement	Applicability
Maximum principal stress (Rankine criterion)	Inelastic action begins when the maximum principal stress at any point, $\sigma^{(1)}$, reaches the tensile yield stress of the material, σ_{xx}^P	Normal and shear stresses acting on other planes have no influence. Limited to materials that fail by brittle fracture since, for ductile materials, σ^T is much less than σ_{xx}^P
Maximum shearing stress (Tresca or Coulomb criterion)	Inelastic action begins when the maximum shearing stress at any point in a member, σ_{ns}^{max}, reaches the value which occurs in a tensile specimen at the onset of yielding, $\frac{1}{2}\sigma_{xx}^P$	Justified for ductile materials in which fairly high shears are developed
Maximum strain (St. Venant criterion)	Inelastic action begins when the maximum (normal) strain of any point in a member, $\varepsilon^{(1)}$, reaches the value which occurs in a tensile specimen at the onset of yielding, $\frac{1}{E}\sigma_{xx}^P$	Improvement over maximum principal stress criterion since it accounts for biaxial and triaxial stress states. Does not reliably predict failures in ductile materials but is widely used for brittle materials such as concrete
Strain energy density (Beltrami and Haigh criterion)	Inelastic action begins when the strain energy density at any point, W, is equal to the value which occurs in a tensile specimen at the onset of yielding, $\frac{1}{2E}\left(\sigma_{xx}^P\right)^2$	Not applicable for biaxial and triaxial stress states
Strain energy density of distortion (Huber, von Mises, Hencky criterion)	Inelastic action begins when the strain energy density of distortion at any point, W_0, is equal to the value which occurs in a tensile specimen at the onset of yielding, $\frac{1}{6G}\left(\sigma_{xx}^P\right)^2$	Accounts for the delay of inelastic action under large hydrostatic stresses. Associates failure with energy absorbed in changing shape, since hydrostatic stresses are associated with volume change only
Octahedral shearing stress	Inelastic action begins when the octahedral shearing stress at any point, σ_{ns}^{oct}, is equal to the value which occurs in a tensile specimen at the onset of yielding, $\frac{\sqrt{2}}{3}\sigma_{xx}^P$	Same as strain energy of distortion

quantity is proportional to the first invariant, we concentrate on the second and third invariants defined in (2.61) or (2.63) in indicial form

$$J_2 = \frac{1}{2}\left[\sigma^{(1)^2} + \sigma^{(2)^2} + \sigma^{(3)^2}\right] \tag{12.13}$$

Table 12.2 Comparison of tension and torsion limits

Failure criterion	Value from tensile test	Value from torsion test	Relationship between σ_{xx}^P and σ_{xy}^T if criterion is correct for both states $(2) = (3)$
(1)	(2)	(3)	(4)
Maximum principal stress	$\sigma^{(1)} = \sigma_{xx}^P$	$\sigma^{(1)} = \sigma_{xy}^T$	$\sigma_{xy}^T = \sigma_{xx}^P$
Maximum shearing stress	$\sigma_{ns}^{\max} = \dfrac{1}{2}\sigma_{xx}^P$	$\sigma_{ns}^{\max} = \sigma_{xy}^T$	$\sigma_{xy}^T = 0.5\,\sigma_{xx}^P$
Maximum strain	$\varepsilon^{(1)} = \dfrac{1}{E}\sigma_{xx}^P$	$\varepsilon^{(1)} = \dfrac{1+v}{E}\sigma_{xy}^T$	$\sigma_{xy}^T = \dfrac{1}{1+v}\sigma_{xx}^P$
Strain energy density	$W = \dfrac{1}{2E}\left(\sigma_{xx}^P\right)^2$	$W = \dfrac{1}{E}\left(\sigma_{xy}^T\right)^2$	$\sigma_{xy}^T = 0.707\sigma_{xx}^P$
Strain energy density of distortion	$W_0 = \dfrac{1}{6G}\left(\sigma_{xx}^P\right)^2$	$W_0 = \dfrac{1}{2G}\left(\sigma_{xy}^T\right)^2$	$\sigma_{xy}^T = 0.577\sigma_{xx}^P$
Octahedral shearing stress	$\sigma_{ns}^{\text{oct}} = \dfrac{\sqrt{2}}{3}\sigma_{xx}^P$	$\sigma_{ns}^{\text{oct}} = \dfrac{\sqrt{6}}{3}\sigma_{xy}^T$	$\sigma_{xy}^T = 0.577\sigma_{xx}^P$

and

$$J_3 = \frac{1}{3}\left[\sigma^{(1)^3} + \sigma^{(2)^3} + \sigma^{(3)^3}\right]. \tag{12.14}$$

In Sect. 3.5, the stress tensor was represented as the sum of a mean normal stress component $\boldsymbol{\sigma}_M$ and a stress deviator component $\boldsymbol{\sigma}_D$, (3.47)–(3.50). Following a construction suggested by Prager and Hodge [3], the normal stress on the octahedral surface is entirely due to $\boldsymbol{\sigma}_M$, while $\boldsymbol{\sigma}_D$ produces only the octahedral shearing stress. Hence the Cartesian components of the octahedral shearing stress can be computed from (2.21a) as

$$\sigma_{ns(k)}^{\text{oct}} = T_k \quad k = 1,2,3 \tag{12.15}$$

since $\sigma_{nn(k)}^{\text{oct}} = 0$. T_k in each case is found as

$$T_k = \sigma_D^{(k)} n_k \tag{12.16}$$

following (2.11) written for principal stresses and (3.53) with $\sigma_D^{(k)} = T_k$ and $\sigma_{ij} = 0$ $(i \neq j)$. With

$$n_k = \frac{1}{\sqrt{3}} \quad k = 1,2,3 \tag{12.17}$$

from (2.85), the components of T_k are

$$T_1 = \sigma_D^{(1)} \frac{1}{\sqrt{3}}, \quad T_2 = \sigma_D^{(2)} \frac{1}{\sqrt{3}}, \quad T_3 = \sigma_D^{(3)} \frac{1}{\sqrt{3}} \qquad (12.18)$$

and from (2.86b),

$$\sigma_{ns}^{oct} = \frac{1}{3} \left[\left(\sigma_D^{(1)}\right)^2 + \left(\sigma_D^{(2)}\right)^2 + \left(\sigma_D^{(3)}\right)^2 \right]^{1/2} = \left[\frac{2}{3} J_{2D} \right]^{1/2} \qquad (12.19)$$

in view of (12.13). The octahedral shearing stress is thus shown to be proportional to the second invariant of the stress deviator. Likewise, the strain energy density of distortion is related to that quantity.

12.3 Yield Surfaces

12.3.1 General

Since the state of stress at a point is specified by the value of six independent stress components σ_{ij} referred to an arbitrary set of orthogonal coordinate axes x, y, and z, it might seem that a general stress criterion would be a function in six-dimensional space, $f(\sigma_{xx}, \sigma_{yy}, \sigma_{zz}, \sigma_{xy}, \sigma_{yz}, \sigma_{zx}) = 0$. Choosing the principal axes as the basis eliminates the shear stresses σ_{ij} ($i \neq j$) and does not reduce generality for isotropic systems. This produces yield criterion in terms of a three-dimensional principal stress space $f(\sigma^{(1)}, \sigma^{(2)}, \sigma^{(3)}) = 0$. The states of stress referred to both sets of axes are shown in Fig. 12.2a.

It is informative to demonstrate these criterions using yield surfaces, which are an essential component of the theory of plasticity [4]. The yield surface is produced by all points that satisfy a particular strength criterion. In the three-dimensional principal stress space, the surface may be visualized as a prism with the axis along the *space diagonal* $\sigma^{(1)} = \sigma^{(2)} = \sigma^{(3)}$, which is the ray OC oriented as $\left(\frac{1}{\sqrt{3}} \mathbf{e}_x + \frac{1}{\sqrt{3}} \mathbf{e}_y + \frac{1}{\sqrt{3}} \mathbf{e}_z \right)$ on the bottom diagram in Fig. 12.2a. Since the stress state $\boldsymbol{\sigma}$ may be resolved into a mean normal component $\boldsymbol{\sigma}_M$ and a stress deviator component $\boldsymbol{\sigma}_D$ as discussed in Sect. 3.5, (3.47), the cross section of the prism essentially represents $\boldsymbol{\sigma}_D$. Any point on the yield surface may be represented by the sum of the mean and deviatoric components. From (3.47) to (3.52), we have

$$\left(\sigma^{(1)}, \sigma^{(2)}, \sigma^{(3)} \right) = (\sigma_m, \sigma_m, \sigma_m) + \left(\sigma_D^{(1)}, \sigma_D^{(2)}, \sigma_D^{(3)} \right). \qquad (12.20)$$

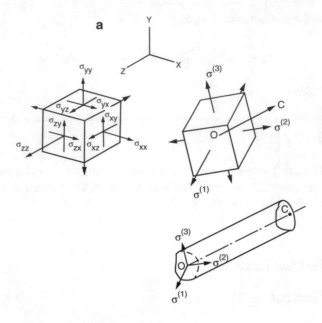

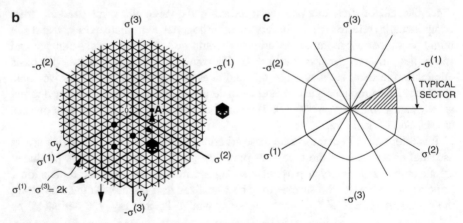

Fig. 12.2 (**a**) Three-dimensional state of stress and space diagonal. (**b**) Plan view of the π-plane [4]. (**c**) C-curve in the π-plane and axes of symmetry [4]

For the principal stresses determined in Sect. 2.6, $\sigma^{(1)} = -1168$, $\sigma^{(2)} = 1380$, and $\sigma^{(3)} = 988$:

$$\sigma_m = \frac{1}{3}(-1,168 + 1,380 + 988) = 400$$

$$\sigma_D^{(1)} = -1,168 - 400 = -1,568$$

$$\sigma_D^{(2)} = 1,380 - 400 = 980$$

$$\sigma_D^{(3)} = 988 - 400 = 588.$$

Check $\sigma_D^{(1)} + \sigma_D^{(2)} + \sigma_D^{(3)} = -1,568 + 980 + 588 = 0$ after (3.54).

The cross section of the prism may be plotted on any plane perpendicular to the space diagonal. Such planes have the equation $\sigma^{(1)} + \sigma^{(2)} + \sigma^{(3)} =$ constant [4]. As suggested by Calladine [4], the plane passing through the origin is conveniently selected and called the π-plane, Fig. 12.2b, and the corresponding cross-sectional boundary is the so-called C-curve, Fig. 12.2c. Also, a triangular grid may be inscribed on the π-plane, which represents a view of a cubic lattice in the first octant as shown by the shaded unit cube and on the inset of Fig. 12.2b. Point A represents a principal stress state $\sigma^{(i)} = (1, 4, 2)$. Sliding along the space diagonal gives parallel cross sections corresponding to other values of the hydrostatic stress σ_M. Various yield conditions can be represented on the π-plane.

12.3.2 Tresca Yield Condition

The maximum shearing stress criteria in Table 12.1 may be stated in terms of the principal stresses by (2.89) which is rewritten for our purposes here as

$$\sigma^{(i)} - \sigma^{(j)} = 2\sigma_{ns}^{\max} = 2k, \tag{12.21}$$

where k is the limiting value of the shear stress. Taking for the moment

$$\sigma^{(1)} > \sigma^{(2)} > \sigma^{(3)}, \tag{12.22}$$

the projection of the stress point, (12.21) with $i = 1$, $j = 3$, lies between the projections of the $(+) \sigma^{(1)}$-axis and the $(-) \sigma^{(3)}$-axis on the π-plane, and

$$\sigma^{(1)} - \sigma^{(3)} = 2k \tag{12.23}$$

is a straight line perpendicular to the bisector of the boundaries of the region.[1] Each of the other five permutations of (12.22) will produce similar lines, and the composite C-hexagon is shown in Fig. 12.3.

Within the theory of elasticity, we focus on the states of stress within or impinging on the yield surface. Beyond that point, plastic deformation will ensue and a broader theory is required. This is briefly elaborated on in Sect. 12.5.

---

[1]To plot a point within the grid defined in (12.23) on Fig. 12.2b, we find $\sigma^{(1)}$ on the $+\sigma^{(1)}$-axis and move down the grid parallel to the $-\sigma^{(3)}$-axis. Then we find $\sigma^{(3)}$ on the $-\sigma^{(3)}$-axis and move along the grid parallel to the $+\sigma^{(1)}$-axis to locate the intersection. Note that the arbitrary $\sigma^{(1)}$ and $\sigma^{(3)}$ points represented by the large dots fall within the yield surface, so do not indicate failure. Values of $\sigma^{(1)}$ and $-\sigma^{(3)}$ corresponding to the heavier lines on the grid would place the point on the yield surface indicating failure.

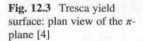

Fig. 12.3 Tresca yield surface: plan view of the π-plane [4]

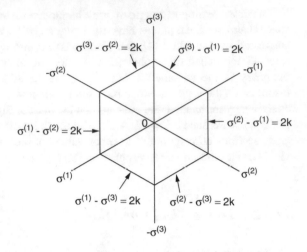

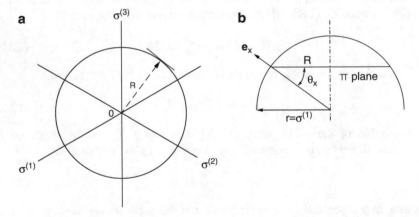

Fig. 12.4 (**a**) von Mises yield surface (plan view of π-plane) [4]. (**b**) Section normal to π-plane

12.3.3   von Mises Yield Condition

The yield condition attributed to R. von Mises is represented by a circle replacing the hexagon on the π-plane, as shown in Fig. 12.4a. The circle is the intersection of a sphere of radius r

$$\left(\sigma^{(1)}\right)^2 + \left(\sigma^{(2)}\right)^2 + \left(\sigma^{(3)}\right)^2 = r^2 \tag{12.24}$$

in the stress space and the plane

$$\sigma^{(1)} + \sigma^{(2)} + \sigma^{(3)} = 0. \tag{12.25}$$

Since (12.25) is identically satisfied by the components of $\boldsymbol{\sigma}_D$, the required equation of the sphere is

$$\left(\sigma_D^{(1)}\right)^2 + \left(\sigma_D^{(2)}\right)^2 + \left(\sigma_D^{(3)}\right)^2 = R^2. \tag{12.26}$$

The radius of the circle R is evaluated to make the circle coincide with the Tresca hexagon in the π-plane for states of pure tension and compression. With the yield stress in pure tension equal to σ_Y, we compute the radius as the projection of the radius vector $\sigma^{(1)}\mathbf{e}_x = \sigma_Y\mathbf{e}_x$ onto the π-plane as shown in Fig. 12.4b. Denoting the angle of inclination between the vector and the π-plane as θ_x, we have

$$\mathbf{e}_x \cdot \left(\frac{1}{\sqrt{3}}\mathbf{e}_x + \frac{1}{\sqrt{3}}\mathbf{e}_y + \frac{1}{\sqrt{3}}\mathbf{e}_z\right) = \cos\left(\frac{\pi}{2} - \theta_x\right) = \frac{1}{\sqrt{3}} \tag{12.27}$$

since the axis of the prism is along the space diagonal which is perpendicular to the π-plane. Thus $\left(\frac{\pi}{2} - \theta_x\right)$ is the angle between the vector and the space diagonal. Equation (12.27) simplifies to

$$\sin\theta_x = \frac{1}{\sqrt{3}} \tag{12.28a}$$

or

$$\theta_x = \cos^{-1}\sqrt{\frac{2}{3}} \tag{12.28b}$$

from which

$$R = r\cos\theta_x = \sigma_Y\sqrt{\frac{2}{3}} \tag{12.29}$$

as shown in Fig. 12.4b. Equation (12.26) becomes

$$\left(\sigma_D^{(1)}\right)^2 + \left(\sigma_D^{(2)}\right)^2 + \left(\sigma_D^{(3)}\right)^2 = \frac{2}{3}\sigma_Y^2, \tag{12.30a}$$

which may also be written as [4]

$$\left[\sigma^{(1)} - \sigma^{(2)}\right]^2 + \left[\sigma^{(2)} - \sigma^{(3)}\right]^2 + \left[\sigma^{(3)} - \sigma^{(1)}\right]^2 = 2\sigma_Y^2. \tag{12.30b}$$

In terms of the stress deviator invariant given by (12.13) with $\sigma^{(i)} = \sigma_D^{(i)}$, (12.30a) becomes

$$\sigma_Y = \sqrt{3J_{2D}}. \tag{12.31}$$

The Tresca criterion (12.21), written in terms of the yield stress $\sigma_Y = 2\,k$, is

$$\text{max of } \left| \sigma^{(1)} - \sigma^{(2)} \right|, \quad \left| \sigma^{(2)} - \sigma^{(3)} \right|, \quad \left| \sigma^{(3)} - \sigma^{(1)} \right| = \sigma_Y. \tag{12.32}$$

Comparing (12.30b) to (12.32), Calladine [4] observed that the Mises condition is related to the r.m.s. of the principal stress differences, while the Tresca condition considers only the largest absolute value. He also calculates that the differences between the two criteria are at most 15%.

12.3.4 General Criterion for Isotropic Media

The Tresca and von Mises yield conditions are based on the stress deviator $\boldsymbol{\sigma}_D$ and thus are essentially independent of the mean or hydrostatic stress $\boldsymbol{\sigma}_M$. This is appropriate for ductile materials but is not sufficiently broad to describe all isotropic materials, specifically those where the behavior is dependent on the hydrostatic pressure and the third stress invariant, (12.14). Examples of such materials cited by Podgorski [5] include plain concrete and sand.

The general failure criterion is expressed in terms of the octahedral shearing stress as

$$A_0 + A_1 \sigma_{ns}^{\text{oct}} + A_2 \left(\sigma_{ns}^{\text{oct}} \right)^2 = 0, \tag{12.33a}$$

where A_0 is a function of hydrostatic pressure only and A_1 and A_2 depend on the second and/or third stress invariants. According to Podgorski [5], the von Mises yield condition derived in the previous section fits into this form with

$$A_0 = -\sqrt{\frac{2}{3} J_{2D}}, \tag{12.33b}$$

$$J_{2D} = \frac{1}{2} \sigma_{D_{ij}} \sigma_{D_{ij}}$$
$$= \frac{1}{2} \left[\left(\sigma_D^{(1)} 2 \right) + \left(\sigma_D^{(2)} 2 \right) + \left(\sigma_D^{(3)} 2 \right) \right] \tag{12.33c}$$

and $A_1 = 1$.

Checking based on (12.19), we have

$$A_0 = -\left\{ \frac{2}{3} \times \frac{1}{2} \left[\left(\sigma_D^{(1)} 2 \right) + \left(\sigma_D^{(2)} 2 \right) + \left(\sigma_D^{(3)} 2 \right) \right] \right\}^{1/2} = -\sigma_{ns}^{\text{oct}}. \tag{12.33d}$$

He also gives the requisite constants for the Tresca conditions. While both the von Mises and Tresca conditions represent yield surfaces with constant cross sections along the main stress axis, OC, in Fig. 12.2a, yield conditions that reflect dependence on the hydrostatic stress would be non-prismatic. Conical, pyramidal, and paraboloidal surfaces are represented by (12.33) [5].

12.4 Anisotropic Materials

12.4.1 Objectives

For anisotropic materials, it is desirable to have a criterion that is invariant with respect to coordinate transformation; treats interaction terms (such as between normal and shear stresses) as independent components; accounts for the difference in strengths due to positive and negative stresses; and can be specialized to account for different material symmetries, multidimensional space, and multiaxial stresses [6]. On the other hand, the sixth-order tensor which would emerge from linking all of the stress components is foreboding, and a simpler form is preferred.

12.4.2 Failure Surface

As proposed by Tsai and Wu [6], the equation of a failure surface can be written as the sum of a second-order tensor and a fourth-order tensor:

$$f(\sigma_k) = F_i\sigma_i + F_{ij}\sigma_i\sigma_j = 1 \tag{12.34}$$

in which $i, j, k = 1, 2, \ldots, 6$; F_i and F_{ij} are the components of the strength tensors; and the single subscripted stress σ_i represents the following components of stress:

$$\sigma_1 = \sigma_{11}, \quad \sigma_2 = \sigma_{22}, \quad \sigma_3 = \sigma_{33}$$
$$\sigma_4 = \sigma_{23}, \quad \sigma_5 = \sigma_{13}, \quad \sigma_6 = \sigma_{12}.$$

The ordering was apparently chosen to allow planar problems to be considered in the 2–3 planes so that plane stress can be extracted as the second, third, and fourth rows and columns without reordering.

The linear terms σ_i are useful to describe the difference between positive and negative (tension and compression) stress-induced failures and also load reversal phenomena, such as the Bauschinger effect which is a reduction of the yield stress upon loading in the direction opposite from the previous direction [5]. The quadratic terms $\sigma_i\sigma_j$ define an ellipsoid in the stress space similar to the curve shown in Fig. 3.3a, which may be considered as a generalization of the sphere described in Sect. 12.3.3, Fig. 12.4(b).

Following the symmetry argument presented for the generalized Hooke's law in Sect. 4.3, $\mathbf{F}_{ij}$ is taken as symmetric, and (12.34) can be written in matrix form as

$$\boldsymbol{\sigma}^T\mathbf{F}_i + \boldsymbol{\sigma}^T\mathbf{F}_{ij}\boldsymbol{\sigma} = 1 \tag{12.35}$$

in which

$$\mathbf{F}_i = \{F_1 \ F_2 \ F_3 \ F_4 \ F_5 \ F_6\}, \tag{12.36a}$$

$$\mathbf{F}_{ij} = \begin{bmatrix} F_{11} & F_{12} & F_{13} & F_{14} & F_{15} & F_{16} \\ F_{12} & F_{22} & F_{23} & F_{24} & F_{25} & F_{26} \\ F_{13} & F_{23} & F_{33} & F_{34} & F_{35} & F_{36} \\ F_{14} & F_{24} & F_{34} & F_{44} & F_{45} & F_{46} \\ F_{15} & F_{25} & F_{35} & F_{45} & F_{55} & F_{56} \\ F_{16} & F_{26} & F_{36} & F_{46} & F_{56} & F_{66} \end{bmatrix}, \tag{12.36b}$$

and

$$\boldsymbol{\sigma} = \{ \sigma_1 \quad \sigma_2 \quad \sigma_3 \quad \sigma_4 \quad \sigma_5 \quad \sigma_6 \}. \tag{12.36c}$$

It is necessary to constrain the terms of $\mathbf{F}_{ij}$ such that the diagonal terms are positive and dominate the interaction terms. The latter is expressed in terms of a stability condition:

$$F_{(ii)}F_{(jj)} - F_{ij}^2 \geq 0. \tag{12.37}$$

12.4.3 Specializations

Again, guided by the reductions for material symmetry used in Sect. 4.3, the matrices $\mathbf{F}_i$ and $\mathbf{F}_{ij}$ may be specialized for some special cases. In the context of composite materials, an orthotropic material is deemed specifically orthotropic if the principal geometric axes coincide with the principal material axes. For a specifically orthotropic material, (4.13), we assume that the sign of the shear stress does not change the failure stress to eliminate F_4, F_5, and F_6 and the shear–normal coupling terms F_{14}, F_{15}, F_{16}, F_{24}, F_{25}, F_{26}, F_{34}, F_{35}, and F_{36}. Further, if shear stresses are all uncoupled, we eliminate F_{45}, F_{46}, and F_{56}. The remaining are three terms in $\mathbf{F}_i$ and nine in $\mathbf{F}_{ij}$, the same terms as in $\mathbf{C}_{ij}$ in (4.13a). The off-diagonal coefficients F_{12}, F_{13}, and F_{23} represent coupling between the normal strengths.

Another common case in composites is transverse or planar isotropy, which refers to a plane such as 2–3 shown in Fig. 12.5. The indices associated with the isotropic plane are identical, i.e.,

$$F_2 = F_3, \quad F_{12} = F_{13}, \quad F_{22} = F_{33}, \quad F_{55} = F_{66}. \tag{12.38}$$

Thus, (12.35) becomes

$$\boldsymbol{\sigma}^T \begin{Bmatrix} F_1 \\ F_2 \\ F_2 \\ 0 \\ 0 \\ 0 \end{Bmatrix} + \boldsymbol{\sigma}^T \begin{bmatrix} F_{11} & F_{12} & F_{12} & 0 & 0 & 0 \\ F_{12} & F_{22} & F_{23} & 0 & 0 & 0 \\ F_{12} & F_{23} & F_{22} & 0 & 0 & 0 \\ & & & F_{44} & & \\ & & & & F_{55} & \\ & & & & & F_{55} \end{bmatrix} \boldsymbol{\sigma} = 1. \tag{12.39}$$

Fig. 12.5 Equivalent states of stress in pure shear and tension–compression

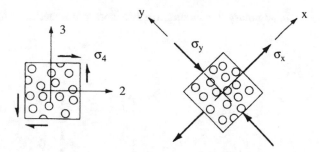

We note that the two stress states shown in Fig. 12.5, pure shear and equal principal tension and compression, are equivalent. Therefore, we may evaluate (12.39), first with $\sigma_4(\sigma_{23}) = \bar{\sigma}$, then with $\sigma_2(\sigma_{22}) = \bar{\sigma}$, and finally with $\sigma_3(\sigma_{33}) = -\bar{\sigma}$, and equate the results. This gives an additional relationship between the components

$$F_{44} = 2(F_{22} - F_{23}) \tag{12.40}$$

so that the transverse isotropic case retains two independent F_i components and five F_{ij} components.

Extending (12.39) and (12.40) to the other two orthogonal planes creates a totally isotropic case with $F_{11} = F_{22}$, $F_{44} = F_{55}$, and $F_{12} = F_{23}$ and the components related by (12.39). This leaves one independent component in F_i and two in F_{ij}. Further, if internal stresses are not considered, the F_i component is dropped to produce the expected two-parameter isotropic equation.

Another specialization proposed by Tsai and Wu [6] is to eliminate the failures due to the hydrostatic stress state $\sigma^{(i)} = \bar{\sigma}$, as defined in Sect. 2.8.5. The rationale for this assumption, which is known as zero volume change or *incompressibility*, has been presented in preceding sections. For the general case, the assumption is implemented by setting $\sigma_1 = \sigma_2 = \sigma_3 = \bar{\sigma}$ in (12.35), factoring the F_{ij} coefficients of $\bar{\sigma}$, and zeroing the factor. This gives

$$F_{11} + F_{22} + F_{33} + 2(F_{12} + F_{23} + F_{13}) = 0. \tag{12.41}$$

For the isotropic case, the direct stress components corresponding to (1, 2, 3) become identical and likewise the shear components (4, 5, 6). Thus, we have

$$\begin{aligned} F_{22} &= F_{33} = F_{11} \\ F_{55} &= F_{66} = F_{44} \\ F_{12} &= F_{23} = F_{13} = -\frac{1}{2}F_{11} \end{aligned} \tag{12.42}$$

and, from (12.40),

$$F_{44} = 2F_{11}\left[1 - \left(-\frac{1}{2}\right)\right] = 3F_{11}. \tag{12.43}$$

In summary, for incompressible isotropic materials with zero initial stress,

$$\mathbf{F}_i = 0 \tag{12.44}$$

and

$$\mathbf{F}_{ij} = F_{11} \begin{bmatrix} 1 & -\dfrac{1}{2} & -\dfrac{1}{2} & 0 & 0 & 0 \\ & 1 & -\dfrac{1}{2} & 0 & 0 & 0 \\ & & 1 & 0 & 0 & 0 \\ & & & 3 & 0 & 0 \\ & & & & 3 & 0 \\ & & & & & 3 \end{bmatrix}. \tag{12.45}$$

Another case of practical interest is that of plane stress. Considering the 1–2 plane and (12.35) and (12.36a, 12.36b, and 12.36c), we retain

$$\mathbf{F}_i = \{ F_1 \quad F_2 \quad F_6 \} \tag{12.46a}$$

and

$$\mathbf{F}_{ij} = \begin{bmatrix} F_{11} & F_{12} & F_{16} \\ & F_{22} & F_{26} \\ & & F_{66} \end{bmatrix}, \tag{12.46b}$$

a total of nine independent components. If the material is specifically orthotropic,

$$F_6 = F_{16} = F_{26} = 0 \tag{12.47}$$

as discussed earlier in this section. Thus, two and four components are retained in $\mathbf{F}_i$ and $\mathbf{F}_{ij}$, respectively.

12.4.4   Evaluation of Components

The format presented in (12.34), or in matrix form in (12.35), requires the evaluation of the components of the strength tensors, $\mathbf{F}_i$ and $\mathbf{F}_{ij}$. Tsai and Wu [6] present a detailed discussion of the evaluation of the key components which is of interest in providing some physical meaning to strength theories. It is helpful to expand (12.34) in explicit form as

$$F_1\sigma_1 + F_2\sigma_2 + F_3\sigma_3 + F_4\sigma_4 + F_5\sigma_5 + F_6\sigma_6$$
$$+ F_{11}\sigma_1^2 + 2F_{12}\sigma_1\sigma_2 + 2F_{13}\sigma_1\sigma_3 + 2F_{14}\sigma_1\sigma_4 + 2F_{15}\sigma_1\sigma_5 + 2F_{16}\sigma_1\sigma_6$$
$$+ F_{22}\sigma_2^2 + 2F_{23}\sigma_2\sigma_3 + 2F_{24}\sigma_2\sigma_4 + 2F_{25}\sigma_2\sigma_5 + 2F_{26}\sigma_2\sigma_6$$
$$+ F_{33}\sigma_3^2 + 2F_{34}\sigma_3\sigma_4 + 2F_{35}\sigma_3\sigma_5 + 2F_{36}\sigma_3\sigma_6$$
$$+ F_{44}\sigma_4^2 + 2F_{45}\sigma_4\sigma_5 + 2F_{46}\sigma_4\sigma_6$$
$$+ F_{55}\sigma_5^2 + 2F_{56}\sigma_5\sigma_6$$
$$+ F_{66}\sigma_6^2 = 1$$

$$(12.48)$$

accounting for the symmetry relationship in (12.36a), (12.36b), and (12.36c).

To start, assume that a uniaxial stress is imposed on a specimen oriented along the 1-axis and the measured tensile and compressive failure stresses are σ_1^T and σ_1^C. Substitution into (12.48) with all other stresses equal to zero gives

$$F_1\sigma_1^T + F_{11}\left(\sigma_1^T\right)^2 = 1$$
$$- F_1\sigma_1^C + F_{11}\left(\sigma_1^C\right)^2 = 1,$$

$$(12.49)$$

which yields

$$F_1 = \frac{1}{\sigma_1^T} - \frac{1}{\sigma_1^C}$$

and

$$F_{11} = \frac{1}{\sigma_1^T\sigma_1^C}.$$

$$(12.50)$$

Uniaxial tension and compression tests along the two and three axes give

$$F_2 = \frac{1}{\sigma_2^T} - \frac{1}{\sigma_2^C}$$
$$F_{22} = \frac{1}{\sigma_2^T\sigma_2^C}$$

$$(12.51)$$

and

$$F_3 = \frac{1}{\sigma_3^T} - \frac{1}{\sigma_3^C}$$
$$F_{32} = \frac{1}{\sigma_3^T\sigma_3^C},$$

$$(12.52)$$

where $\sigma_2^T, \sigma_2^C, \sigma_3^T$, and σ_3^C are the corresponding failure stresses.

Next, pure shear is imposed in the three horizontal planes producing

$$F_4 = \frac{1}{\sigma_{23}^{S+}} - \frac{1}{\sigma_{23}^{S-}}, \qquad F_{44} = \frac{1}{\sigma_{23}^{S+}\sigma_{23}^{S-}}$$

$$F_5 = \frac{1}{\sigma_{13}^{S+}} - \frac{1}{\sigma_{13}^{S-}}, \qquad F_{55} = \frac{1}{\sigma_{13}^{S+}\sigma_{13}^{S-}} \qquad (12.53)$$

$$F_6 = \frac{1}{\sigma_{12}^{S+}} - \frac{1}{\sigma_{12}^{S-}}, \qquad F_{66} = \frac{1}{\sigma_{12}^{S+}\sigma_{12}^{S-}},$$

where $\sigma_{23}^{S+}, \sigma_{23}^{S-}; \sigma_{13}^{S+}, \sigma_{13}^{S-};$ and $\sigma_{12}^{S+}, \sigma_{12}^{S-}$ are the positive and negative pure shear strengths in the 2–3, 1–3, and 1–2 planes, respectively.

The preceding procedures established all of the components of $\mathbf{F}_i$ and the diagonal components of $\mathbf{F}_{ij}$ and are relatively straightforward in principle. The off-diagonal components relate to the interaction of two stress components and require more complicated tests.

A biaxial tension with

$$\sigma_1 = \sigma_2 = \sigma_{12}^T; \quad \sigma_3 = \sigma_4 = \sigma_5 = \sigma_6 = 0 \qquad (12.54)$$

in (11.48) gives

$$\sigma_{12}^T(F_1 + F_2) + \left(\sigma_{12}^T\right)^2(F_{11} + F_{22} + 2F_{12}) = 1 \qquad (12.55)$$

from which

$$F_{12} = \frac{1}{2\left(\sigma_{12}^T\right)^2}\left[1 - \sigma_{12}^T\left(\frac{1}{\sigma_1^T} - \frac{1}{\sigma_1^C} + \frac{1}{\sigma_2^T} - \frac{1}{\sigma_2^C}\right) - \left(\sigma_{12}^T\right)^2\left(\frac{1}{\sigma_1^T\sigma_1^C} + \frac{1}{\sigma_2^T\sigma_2^C}\right)\right].$$
$$(12.56)$$

Also, biaxial compression $\sigma_1 = \sigma_2 = \sigma_{12}^2$ could have been applied. Similarly, F_{23} and F_{31} can be determined using biaxial states $\sigma_2 = \sigma_3 = \sigma_{23}^T$ or σ_{23}^C and $\sigma_1 = \sigma_3 = \sigma_{13}^T$ or σ_{13}^C. There are other tests, specifically 45° specimens, which are preferred by some investigators for determining the interaction terms as discussed by Tsai and Wu [6].

For anisotropic materials, the components F_{16} are determined from an axial–torque combination performed on a tubular specimen with the 1-axis along the tube. This produces the desired tension–shear combination

$$\sigma_1 = \sigma_6 = \sigma_{16}^{TS}; \quad \sigma_2 = \sigma_3 = \sigma_4 = \sigma_5 = 0. \qquad (12.57)$$

Equation (12.48) becomes

$$\sigma_{16}^{TS}(F_1 + F_6) + \left(\sigma_{16}^{TS}\right)^2(F_{11} + F_{66} + 2F_{16}) = 1 \qquad (12.58)$$

from which

$$F_{16} = \frac{1}{2(\sigma_{16}^{TS})^2}\left[1 - \sigma_{16}^{TS}\left(\frac{1}{\sigma_1^T} - \frac{1}{\sigma_1^C} + \frac{1}{\sigma_{12}^{S+}} - \frac{1}{\sigma_{12}^{S-}}\right) - (\sigma_{16}^{TS})^2\left(\frac{1}{\sigma_1^T\sigma_1^C} + \frac{1}{\sigma_{12}^{S+}\sigma_{12}^{S-}}\right)\right].$$

$$(12.59)$$

Also compression–shear $\sigma_1 = \sigma_6 = \sigma_{16}^{CS}$ could have been used. Components F_{26} can be determined by the same test with the 1-axis along the circumference of the tube. If the tube axis coincides with the material symmetry axis, a specifically orthotropic material is produced for which F_{16} and F_{26} are zero.

Finally it was indicated that such tests as the biaxial and the axial torque can be performed using either tension or compression. This leads to redundant measurements of the interaction terms, which may be used to establish the validity and accuracy of the procedure.

As a numerical illustration, Tsai and Wu [6] considered a unidirectional graphite–epoxy composite as a specifically orthotropic material. The 1-axis is oriented along the fibers, and the 2-axis transverses to the fibers.

The strengths are

$$\sigma_1^T = 150\,\text{ksi}, \quad \sigma_1^C = 100\,\text{ksi},$$
$$\sigma_2^T = 6\,\text{ksi}, \quad \sigma_2^C = 17\,\text{ksi},$$
$$\sigma_{12}^S = 10\,\text{ksi}$$

which gives in (12.46a) and (12.46b)

$$\{F_i\} = \{F_1 \ F_2 \ F_6\} = \{-0.003 + 0.108 \ 0\}$$

and

$$[F_{ij}] = \begin{bmatrix} 0.00007 & +0.0008 & 0 \\ & 0.0098 & 0 \\ & & 0.01 \end{bmatrix},$$

where F_{12} has been bounded as $F_{12} = \pm\sqrt{F_{11}F_{22}} = \pm0.0008$ by (12.37). A more precise determination of F_{12} would require a combined test as discussed previously.

It is evident that the proper determination of material constants requires careful and appropriate testing and evaluation.

An up-to-date failure theory, applicable to classical as well as to recently developed materials, is presented by a leading contemporary researcher in the field, R. Christiansen [7].

12.5   Failure of Structures

The preceding sections are directed toward the establishment of strength limitations based on the *initiation* of inelastic action. Barring premature failure due to loss of stability or fatigue, the loads that produce the yield condition at the critical point can be expected to be lower bounds on the actual capacity of the structure. That is, the entire system may possess the capability to resist loads beyond those corresponding to the criterion. Moreover, the concept of yield is associated with ductile materials, while brittle materials are also of considerable practical interest.

As suggested by Bažant and Mazars [8], the two general classical theories for the prediction of complete structural failure are (1) *plasticity*, which describes material failure distributed over a plastic zone and occurring throughout the entire plastic zone simultaneously, and (2) *fracture mechanics*, which describes material failure that is concentrated into a very small fracture process zone that propagates throughout the structure. Basically, plasticity would seem to be appropriate for ductile materials, while fracture mechanics would apply to brittle materials.

For complex materials such as cement-based aggregate composites (concrete, mortar, fiber-reinforced concrete), rocks and soils, and various fibrous and particulate composites, a more elaborate theory may be appropriate. It has been suggested that the material failure begins simultaneously over a larger zone but then localizes into a relatively small zone that propagates throughout the structure. This concept is complicated by size effects but promises to somewhat unify some of the basic theories into a more general concept applicable for the prediction of the total failure of arbitrary structures.

12.6   Stress Analysis Example

Consider a square shaft with a side dimension of 20 mm and yield stress $\sigma_y = 300$ MPa. If the axial load $P = 40$ kN,

(a) Determine the torque T that can be applied to the shaft to initiate yielding.
(b) Determine the torque that can be applied if the shaft is designed for a factor of safety of 2.0 for both P and T. Base the calculations on the octahedral shear stress criterion (1 MPa = 1 N/mm^2).

1. *Axial load stress*

$$\sigma_n = \frac{P}{A} = \frac{40 \text{ kN}}{(20)^2 \text{mm}^2} = 100 \text{ MPa}.$$

2. *Maximum shear stress for torque*

$$\tau_{\max} = \frac{T}{k_1 B H^2}.$$

For square-shaped cross section, $B = H = L$ and $k_1 = 0.208$ (from reference table). With T in Nm,

$$\tau_{\max} = \frac{T \cdot 1000 \ \text{mm/m}}{0.208(8000) \ \text{mm}^2} = \frac{T}{1.664} \, \text{MPa}.$$

To evaluate σ^{oct} using (12.6),

$$\sigma^{\text{oct}} = \frac{1}{3} \sqrt{\sigma_n^2 + \sigma_n^2 + 6\tau_{\max}^2} = \frac{1}{3} \sqrt{100^2 + 100^2 + 6\left(\frac{T}{1.664}\right)^2}.$$

With $\sigma^{\text{oct}} = \sigma_y = 300$ MPa, we have $T = 603.8$ Nm.
If F.S. $= 2$, $\sigma^{\text{oct}} = \sigma_y/2 = 150$ MPa, we have $T = 290.2$ Nm.

12.7   Computational Examples Using MATLAB

Example 1
In Sect. 12.2, several failure criteria were introduced. In this section, we use MuPAD to illustrate the derivation of octahedral stress in terms of stress components and invariants and verify the corresponding criterion in Table 12.2.

- First, define stress tensor symbols:

```
sig := Symbol::sigma

    σ

sig11 := Symbol::subScript(sig, 11):
sig22 := Symbol::subScript(sig, 22):
sig33 := Symbol::subScript(sig, 33):
sig12 := Symbol::subScript(sig, 12):
sig23 := Symbol::subScript(sig, 23):
sig13 := Symbol::subScript(sig, 13):

sigxxP := Symbol::subSuperScript(sig, xx, P):
sigxyT := Symbol::subSuperScript(sig, xy, T):

sig := matrix([[sig11, sig12, sig13], [sig12, sig22, sig23], [sig13, sig23, sig33]])
```

$$\begin{pmatrix} \sigma_{11} & \sigma_{12} & \sigma_{13} \\ \sigma_{12} & \sigma_{22} & \sigma_{23} \\ \sigma_{13} & \sigma_{23} & \sigma_{33} \end{pmatrix}$$

- Based on the definition of octahedral stress (2.86), we calculate the eigenvalues of the stress tensor:

```
sigEign := linalg::eigenvalues(sig)
```

$$\left\{ \frac{\sigma_{11}}{3} + \frac{\sigma_{22}}{3} + \frac{\sigma_{33}}{3} + \frac{\sigma_3}{\sigma_2} + \sigma_2, \; \frac{\sigma_{11}}{3} + \frac{\sigma_{22}}{3} + \frac{\sigma_{33}}{3} - \frac{\sigma_3}{2\sigma_2} - \frac{\sigma_2}{2} - \sigma_1, \; \frac{\sigma_{11}}{3} + \frac{\sigma_{22}}{3} + \frac{\sigma_{33}}{3} - \frac{\sigma_3}{2\sigma_2} - \frac{\sigma_2}{2} + \sigma_1 \right\}$$

where

$$\sigma_1 = \frac{\sqrt{3}\left(\frac{\sigma_3}{\sigma_2} - \sigma_2 \right) i}{2}$$

$$\sigma_2 = \left(\sigma_5 - \sigma_4 + \sqrt{\left(\sigma_4 - \sigma_5 + \sigma_8 + \sigma_7 + \sigma_6 - \sigma_{12}\,\sigma_{13}\,\sigma_{23} - \frac{\sigma_{11}\,\sigma_{22}\,\sigma_{33}}{2} \right)^2 - \sigma_3^{\,3}} - \sigma_8 - \sigma_7 - \sigma_6 + \sigma_{12}\,\sigma_{13}\,\sigma_{23} + \frac{\sigma_{11}\,\sigma_{22}\,\sigma_{33}}{2} \right)^{1/3}$$

$$\sigma_3 = \frac{\left(\sigma_{11} + \sigma_{22} + \sigma_{33} \right)^2}{9} - \frac{\sigma_{11}\,\sigma_{22}}{3} - \frac{\sigma_{11}\,\sigma_{33}}{3} - \frac{\sigma_{22}\,\sigma_{33}}{3} + \frac{\sigma_{12}^{\,2}}{3} + \frac{\sigma_{13}^{\,2}}{3} + \frac{\sigma_{23}^{\,2}}{3}$$

$$\sigma_4 = \frac{\left(\sigma_{11} + \sigma_{22} + \sigma_{33} \right)\left(-\sigma_{12}^{\,2} - \sigma_{13}^{\,2} - \sigma_{23}^{\,2} + \sigma_{11}\,\sigma_{22} + \sigma_{11}\,\sigma_{33} + \sigma_{22}\,\sigma_{33} \right)}{6}$$

$$\sigma_5 = \frac{\left(\sigma_{11} + \sigma_{22} + \sigma_{33} \right)^3}{27}$$

$$\sigma_6 = \frac{\sigma_{12}^{\,2}\,\sigma_{33}}{2}$$

$$\sigma_7 = \frac{\sigma_{13}^{\,2}\,\sigma_{22}}{2}$$

$$\sigma_8 = \frac{\sigma_{11}\,\sigma_{23}^{\,2}}{2}$$

- Calculate the octahedral normal stress `sigOctn` and shear stress `sigOcts`:

```
sigOctn := 1/3*(sigEign[1] + sigEign[2] + sigEign[3])
```
$$\frac{\sigma_{11}}{3} + \frac{\sigma_{22}}{3} + \frac{\sigma_{33}}{3}$$
```
sigOcts := Simplify(1/3*((sigEign[1]-sigEign[2])^2 + (sigEign[2]-sigEign[3])^2 + (sigEign[3]-sigEign[1])^2)^(1/2))
```
$$\frac{\sqrt{2}\,\sqrt{\sigma_{11}^{\,2} - \sigma_{11}\,\sigma_{22} - \sigma_{11}\,\sigma_{33} + 3\,\sigma_{12}^{\,2} + 3\,\sigma_{13}^{\,2} + \sigma_{22}^{\,2} - \sigma_{22}\,\sigma_{33} + 3\,\sigma_{23}^{\,2} + \sigma_{33}^{\,2}}}{3}$$

- Calculate the first and second invariants I_1 and I_2. We verify the octahedral shear stress which can also be written by $\frac{1}{3}\sqrt{2I_1^2 - 6I_2}$:

```
I1 := linalg::tr(sig)
```
$$\sigma_{11} + \sigma_{22} + \sigma_{33}$$
```
I2 := (1/2)*(I1^2 - linalg::tr(sig*sig))
```
$$\frac{\left(\sigma_{11} + \sigma_{22} + \sigma_{33} \right)^2}{2} - \frac{\sigma_{11}^{\,2}}{2} - \sigma_{12}^{\,2} - \sigma_{13}^{\,2} - \frac{\sigma_{22}^{\,2}}{2} - \sigma_{23}^{\,2} - \frac{\sigma_{33}^{\,2}}{2}$$
```
simplify((1/3)*sqrt(2*I1^2 - 6*I2) - sigOcts)
```
`0`

- Finally, use the octahedral stress to derive the tension and torsion limits in Table 12.2. For tensile test, we have $\sigma_{11} = \sigma_{xx}^P$; therefore, the octahedral shear stress is

```
simplify(subs(sigOcts, sig11 = sigxxP, sig22=0, sig33=0, sig12=0, sig13=0, sig23=0))
```
$$\frac{\sqrt{2}\sqrt{\sigma_{xx}^{P\,2}}}{3}$$

- For shear test, we have $\sigma_{12} = \sigma_{xy}^T$; therefore, the octahedral shear stress is

```
simplify(subs(sigOcts, sig11 = 0, sig22=0, sig33=0, sig12=sigxyT, sig13=0, sig23=0))
```
$$\frac{\sqrt{6}\sqrt{\sigma_{xy}^{T\,2}}}{3}$$

Example 2

We now solved the problem in Sect. 12.6 using the computational method shown in Example 1. The bar with a square cross section was under axial and torsion load, which has a stress state of $\sigma_{11} = 100$ MPa, $\sigma_{12} = T/1.664$, and $\sigma_{22} = \sigma_{33} = \sigma_{13} = \sigma_{23} = 0$. With the octahedral stress calculated from Example 1, we can substitute the above stress components and solve T directly.

With $\sigma^{\text{oct}} = \sigma_y = 300$ MPa, we have $T = 603.8$ Nm.

```
solve(subs(sigOcts, sig11 = 100, sig22=0, sig33=0, sig12=T/1.664, sig13=0, sig23=0)-300, T)
{ -603.7974053, 603.7974053}
```

If F.S. $= 2$, $\sigma^{\text{oct}} = \sigma_y/2 = 150$ MPa, we have $T = 290.2$ Nm.

```
solve(subs(sigOcts, sig11 = 100, sig22=0, sig33=0, sig12=T/1.664, sig13=0, sig23=0)-150, T)
{ -290.2078336, 290.2078336}
```

12.8 Exercises

12.1 A closed-ended thin-walled cylinder of titanium alloy ($\sigma_Y = 800$ MPa) has an inside diameter of 38 mm and a wall thickness of 2 mm. The cylinder is subjected to an internal pressure $p = 20.0$ MPa and an axial load $P = 40.0$ kN. Determine the torque T that can be applied to the cylinder if the factor of safety for design is $FS = 2.00$. The design is based on the maximum shearing stress criterion of failure assuming that failure occurs at the initiation of yielding [1]. Hint: the maximum shear stress for torsion of a thin-walled cylinder is $\dfrac{2T}{\pi D_m^2 t}$, where D_m is the mean diameter of the cylinder.

12.2 Resolve Problem 12.1 based on the maximum octahedral shearing stress criterion.

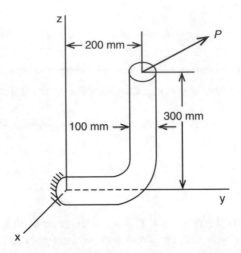

12.3 The 100-mm-diameter bar shown is made of a ductile steel that has a yield stress $\sigma_Y = 420$ MPa. The free end of the bar is subjected to a load P with a direction of $\begin{bmatrix} 1/\sqrt{3} & 1/\sqrt{3} & 1/\sqrt{3} \end{bmatrix}^T$. Using the maximum octahedral shearing stress criterion of failure, determine the magnitude of P that will initiate yielding.

12.4 A thin-walled tube is fabricated from a brittle metal having an ultimate tensile strength $\sigma_{xx}^P = 300$ MPa and an ultimate compressive strength $\sigma_{xx}^P = -700$ MPa. The outer and inner radii are B = 105 mm and A = 100 mm, respectively. Calculate the maximum torque M_z that can be applied without causing failure from the following failure criteria:

(a) Maximum principal stress theory
(b) Maximum absolute shear stress theory
(c) Maximum octahedral shearing stress theory

The maximum torque M_z and maximum shearing stresses τ are related through (6.61):

$$\tau = \frac{M_z r}{J_c}.$$

The torsional constant J_c for the thin-walled tube can be found by subtracting the value calculated for the inner radius from that calculated for the outer radius [9].

12.5 Consider the π-plane shown in Fig. 12.2b. Verify that the lines which bisect the projections of the axes of principal stress correspond to state of pure shear. (Hint: compute the principal deviatoric stresses corresponding to a state of pure shear and plot them on the π-plane, or consider any two vectors corresponding to known states of principal stress in the π-plane, add them to establish the direction of the resultant, and examine the corresponding state of principal stress [4]).

12.6 Verify the equality of (12.30a) and (12.30b) and (12.31).

References

1. Boresi AP et al (1978) Advanced mechanics of materials, 3rd edn. Wiley, New York
2. Meek JL, Lin WJ (1990) Geometric and material nonlinear analysis of thin-walled beam-columns. J Struct Eng ASCE 116(6):1473–1489
3. Prager W, Hodge PG (1951) Theory of perfectly plastic solids. Wiley, New York
4. Calladine CR (1985) Plasticity for engineers. Ellis Howard, Ltd., Chichester
5. Podgórski J (1985) General failure criterion for isotropic media. J Eng Mech, ASCE 111 (2):188–201
6. Tsai SW, Wu EM (1971) A general theory of strength for anisotropic materials. J Compos Mater 5:58–80
7. Christensen RM (2013) The theory of materials failure. Oxford University Press
8. Bažant ZP, Mazars J (1990) France—US workshop on strain localization and size effect due to cracking and damage. J Eng Mech, ASCE 116(6):1412–1424
9. Ugural AC, Fenster SK (1987) Advanced strength and applied elasticity, 2nd edn. Elsevier, New York

Chapter 13
Epilogue: Something New

Abstract Since the theory of elasticity originated primarily in the first half of the nineteenth century, most of the original work supporting the presentation in this introductory text is many decades old. Occasionally, new developments emerge, even in a mature field. A recent contribution, countering the classical notion that a material cannot have a negative Poisson's ratio, is the development of a substance called antirubber that bulges in the middle when the ends are pulled. Also, modern material science has recently explored methods to coax metallic alloys into shapes that were only possible from molded plastics. The materials are known as "Glassimetals" referring to the alloys called metallic glasses and possess properties that promise to be useful as structural components or protective casings. Also in the area of manufacturing new materials, 3-D printing has opened up many possibilities including one that optimizes filament alignment using elasticity concepts.

Throughout this book, we have emphasized the cornerstones of the classical theory of elasticity: *equilibrium*, *compatibility*, and *constitutive laws*. While the first two cornerstones remain constant within the linear range, the third is constantly expanding with the development of new materials and the characterization of biological substances. We have introduced several examples throughout the book with sufficient basic detail to enable the reader to examine the current literature. In this concluding chapter, we offer a few more examples with only a brief discussion.

13.1 Negative Poisson Ratio

A novel recent contribution challenging one of the basic conclusions of elasticity, and notable here because it is attributed to an elementary course in elasticity theory, was developed by Prof. R. Lakes of the University of Iowa. Spurred by the common notion that a real material cannot have a negative Poisson's ratio as noted in Sect. 4.3, he developed a substance called antirubber that bulges in the middle when the ends are pulled, an indication of a negative Poisson's ratio. This was accomplished by collapsing or folding cells of polyurethane into a configuration, which is solidified after heating. Then, as the foam is pulled apart, the cells unfold and expand in the direction being pulled and in lateral directions producing the negative Poisson's ratio.

The process is shown in Fig. 13.1, which may be one of the only fundamental developments in the theory of elasticity reported in a newspaper article [1].

13.2   Metallic Glass

Modern materials science has recently explored methods to coax metallic alloys into shapes that were only possible from molded plastics. The materials are known commercially as "Liquidmetals" or" Glassimetals," referring to the materials called metallic glasses, and possess properties that promise to be useful as structural components or protective casings.

From a more fundamental standpoint in the study of solid matter, solids with long-range atomic order and periodicity are referred to as "crystal," while those without such order (with local order more like a liquid) are termed "amorphous" or "glassy." For many years, it was assumed that solid metals would only exist in a crystalline state [2]. However, it has since been shown that metallic glasses can be produced in a wide variety of systems, from the first metallic glass in 1960, to the first bulk form over a cm thick in 1984, to the commercially used series of Vitreloy 1 through the recent Vitreloy 106A. Their amorphous structure provides a higher elasticity and coefficient of restitution (a property pertaining to rebound speed) than their crystalline counterparts and together with a lack of grain boundaries, increases strength and reduces susceptibility to corrosion [3].

Applications are numerous, including computer cases, artificial joint replacements, and sporting goods, but producing such materials in bulk form remains problematic. Because their critical cooling rates are so high and the overall cooling rates are reduced as material thickness (the linear distance away from the cooling surface) increases, there is a practical limit on possible dimensions for metallic glass parts. The rare set of alloys that have shown good bulk formability are expensive (e.g., palladium- or zirconium-based alloys). For comparison, in 1998, zirconium averaged \$250/kg, and palladium was roughly \$290/ounce, while aluminum was only 65 cents/1b, and iron was about 5.5 cents/kg. Research is required to determine what causes certain compositions to form metallic glasses more easily than others, with the ultimate goal of optimizing and economizing the process and understanding the fundamental governing physics behind their formation. A recent development was led by Prof. W.J. Johnson and graduate student D. Hofmann of Caltech. The process is accomplished using rapid heating by an electric current that produces uniform temperatures throughout the materials. "We took a metallic glass, which is considered a brittle material, and showed that by making a designed composite out of it, we can span the entire space of toughness," Hofmann remarks. "The tougher it is, the harder it is to drive a crack through it. Now we have ductility and toughness," he claims [4].

Antirubber, Roderic Lakes of the
University of Iowa has developed a
material that, when pulled apart on
two sides, expands on all sides. When
pushed together, it contracts on all
sides. It also absorbs energy more
effectively than other materials.

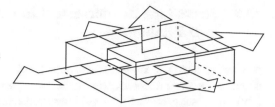

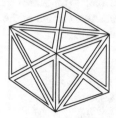

Uncompressed cell. Antirubber is created by
transforming the open-cell structure in normal
polyurethane or metal foam.

Collapsed cell. The foam is compressed by putting
pressure on it from all directions. The cells
collapse or fold in upon themselves, and after
heating, are solidified into the collapsed
configuration.

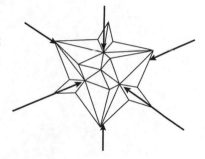

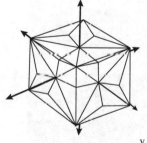

Expanded cell. It's the new, collapsed internal
structure that produces the antirubber
characteristics. As the foam is pulled apart, the
cells unfold and open up in the direction being
pulled as well as in all other directions.

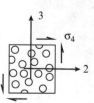

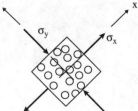

Fig. 13.1 Antirubber

13.3   Stress Line Reinforcement in Additive Manufacturing (3-D Printing)

It is logical that there are various manufactured materials that attempt to optimize the structural resistance to the stresses induced by a particular loading condition. The theory of elasticity suggests that principal stress trajectories (stress lines) within the system provide the most efficient resistance paths, but such patterns are difficult to replicate in a traditional system constructed on a rectilinear or circular platform. An example of the stress lines for a shell with a vertical point load at the apex is shown in Fig. 13.2 .

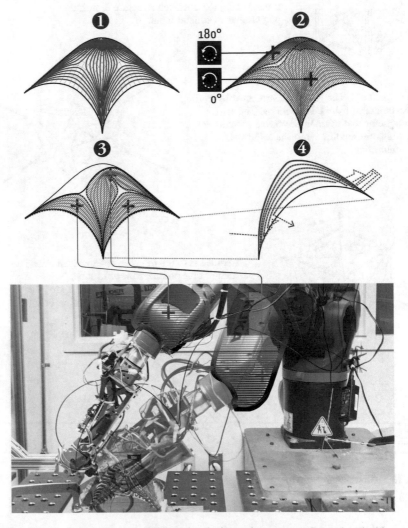

Fig. 13.2 Process to convert stress line-based paths into print programs Reproduced with permission [5]. (1) Stress lines within one layer, (2) global clustering according to print surface orientation, (3) local clustering by geometric and robot movement consideration, (4) sequencing and linking of stress line-based paths

Additive manufacturing (AM), commonly called 3-D printing, has become popular for fabricating objects of increasingly complex geometry. In a traditional AM process known as fused deposition modeling (FDM), material is deposited in layers parallel to the horizontal test bed using a crisscross pattern. This leads to anisotropic products with strength and ductility behavior dependent on filament orientation, which does not necessarily correlate with structural demands [5].

As shown on Fig. 13.2, an integrated software and hardware process augments the basic FDM technique by explicitly adding material along trajectories of principal stress using a robot arm. Plane or curved relatively thin structural types, such as some of the 2-D forms considered in Chap. 7 and the plates and shells described in Chap. 8, would seem to be good candidates for this application. With respect to the continuity of the theory of elasticity, the developers attribute the concept to optimization studies performed by the early twentieth-century engineer AGM Mitchell [6], whose renown brother and collaborator was cited earlier in Chap. 5, and apply the principal stress calculations of the theory of elasticity to convert stress line-based paths into print programs as described in thefigure [5].

References

1. Chicago Sunday Tribune (1989) Chicago, IL
2. Spaepen F, Turnbull D (1984) Metallic glasses. Ann Rev Phys Chem 35:241–263
3. (2012, Febraury 6) Liquidmetal alloys unique material properties: high yield strength and superior elastic limit. Liquidmetal Product Design and Development. Liquidmetal Technologies. Web. http://www.liquidmetal.com/technology/properties-comparison/
4. New York Times (2011, June 5) New York, NY and http://media.caltech.edu/press_releases/13110
5. Tam KM, Mueller CT, Coleman JR, Fine NW (2016) Stress line additive manufacturing (SLAM) for 2.5-D Shells. J IASS Int Assoc Shell Spat Struct 57(4):249–259
6. Michell AGM (1904) The limits of economy of material in frame-structures. Philos Mag 47 (8):589–597

Appendix I
Basic Commands of MATLAB

Let $a = \begin{bmatrix} 1 & 2 & 3 \end{bmatrix}$ and $b = \begin{bmatrix} 4 & 5 & 6 \end{bmatrix}$ be two vectors and $A = \begin{bmatrix} 1 & 2 & 3 \\ 4 & 5 & 6 \\ 7 & 8 & 9 \end{bmatrix}$ and

$B = \begin{bmatrix} 1 & 0 & 0 \\ 0 & 1 & 0 \\ 0 & 0 & 1 \end{bmatrix}$ be two matrices. The variable assignment commands are:

```
>> a = [1, 2, 3], b=[4, 5, 6]
a =
   1    2    3
b =
   4    5    6
>> A = [1, 2, 3; 4, 5, 6; 7, 8, 9], B = eye(3)
A =
   1    2    3
   4    5    6
   7    8    9
B =
   1    0    0
   0    1    0
   0    0    1
```

Note the eye ( ) is a function to generate unit matrix We illustrate the basic MATLAB commands using the above vectors and matrices.

© Springer International Publishing AG, part of Springer Nature 2018
P. L. Gould, Y. Feng, *Introduction to Linear Elasticity*,
https://doi.org/10.1007/978-3-319-73885-7

I.1 Operators

+ Plus, addition operator.

```
>> a+b
ans =
    5   7   9
>> A+B
ans =
    2   2   3
    4   6   6
    7   8   10
```

− Minus, subtraction operator.

```
>> a-b
ans =
-3   -3   -3
>> A-B
ans =
    0   2   3
    4   4   6
    7   8   8
```

* Scalar and matrix multiplication operator.

```
>> A*B % matrix multiplication
ans =
    1   2   3
    4   5   6
    7   8   9
```

.* Array multiplication operator.

```
>> A.*B % multiply each element separately
ans =
    1   0   0
    0   5   0
    0   0   9
```

^ Scalar and matrix exponentiation operator.

```
>> A^2 % multiple matrix A by itself
ans =
    30    36    42
    66    81    96
   102   126   150
```

.^ Array exponentiation operator

```
>> A.^2 % each element of the matrix is squared
ans =
     1    4    9
    16   25   36
    49   64   81
```

\ Left-division operator.

```
>> A\B
ans =
   1.0e+16 *
    0.3153  -0.6305   0.3153
   -0.6305   1.2610  -0.6305
    0.3153  -0.6305   0.3153
```

/ Right-division operator.

```
>> A/B
ans =
     1    2    3
     4    5    6
     7    8    9
```

.\ Array left-division operator.

```
>> A.\B
ans =
   1.0000        0        0
        0   0.2000        0
        0        0   0.1111
```

./ Array right-division operator.

```
>> A./B
ans =
1 Inf Inf
Inf 5 Inf
Inf Inf 9
```

: Colon generates regularly spaced elements and represents an entire row or column.

```
>> A(1,:) % first row of the matrix A
ans =
     1    2    3
>> A(:,1) % first column of the matrix A
ans =
     1
     4
     7
```

' Transpose operator.

```
>> A'
ans =
   1   4   7
   2   5   8
   3   6   9
```

== Relational operator: equal to.

```
>> a==b % compare each element
ans =
   0   0   0
```

~ = Relational operator: not equal to.

```
>> a~=b % compare each element
ans =
   1   1   1
```

< Relational operator: less than.

```
>> a<b
ans =
   1   1   1
```

<= Relational operator: less than or equal to.

```
>> a<=b
ans =
   1   1   1
```

> Relational operator: greater than.

```
>> a>b
ans =
   0   0   0
```

> = Relational operator: greater than or equal to.

```
>> a>=b
ans =
   0   0   0
```

& Logical operator: AND.>> (a-1)&b

```
ans =
0 1 1
```

| Logical operator: OR.

```
>> (a-1)|b
ans =
    1   1   1
```

~ Logical operator: NOT.

```
>> ~((a-1)&b)
ans =
    1   0   0
```

xor Logical operator: EXCLUSIVE OR.

```
>> xor((a-1),b)
ans =
    1   0   0
```

I.2 Input/Output Commands

disp displays contents of an array or string.

```
>> disp(A)   % displace matrix A in the command prompt
    1    2    3
    4    5    6
    7    8    9
```

fprintf performs formatted writes to screen or file.

```
>> fprintf('%1.2f\n', A) % printout elements of A with 2 decimal
points
1.00
4.00
7.00
2.00
5.00
8.00
3.00
6.00
9.00
```

I.3 Array Commands

find finds indices of nonzero elements.

```
>> find(a>1) % find indices of vector a which is larger than 1
ans =
    2   3
```

length computes number of elements.

```
>> length(a)
ans =
    3
```

linspace creates regularly spaced vector.

```
>> linspace(1, 10, 5) % create vector starts with 1 ends with 10 with
5 equally spaced elements
ans =
    1.0000   3.2500   5.5000   7.7500  10.0000
```

max returns the largest element.

```
>> max(a)
ans =
    3
>> max(A) % for matrix, it operates on each column
ans =
    7   8   9
```

min returns the smallest element.

```
>> min(a)
ans =
    1
>> min(A) % for matrix, it operates on each column
ans =
    1   2   3
```

size computes array size.

```
>> size(A) % A is a 3 by 3 matrix
ans =
    3   3
```

sum sums all elements at each column.

```
>> sum(a)
ans =
   6
```

det computes determinant of an array.

```
>> det(A-B)
ans =
   32
```

inv computes inverse of a matrix.

```
>> inv(A-B)
ans =
  -0.5000    0.2500          0
   0.3125   -0.6563    0.3750
   0.1250    0.4375   -0.2500
```

I.4 Plot Command

The mostly used commands for plotting in MATLAB are listed as follows:

axis sets axis limits.
grid displays gridlines.
plot generates x-y plot.
print prints plot or saves plot to a file
title puts text at top of plot.
xlabel adds text label to x-axis.
ylabel adds text label to y-axis.

We illustrate the usage of these commands by plotting the stress of a point directly beneath the load (Eq. 7.93a, with $\theta = 0$): $\sigma_{xx} = -\frac{2P}{\pi H}$.

```
% set P to a unit force
P = 1;
% plot the variation with a depth from 0 to 10
H = 1:0.1:10;
% calcuate the stress
sig_xx = -2*P./(pi*H);

% first open a newfigure
fh = figure;
% plot figure
plot(H, sig_xx);
% set x and y axes labels
xlabel('H');
ylabel('\sigma_{xx}');
% set x-axis limit to [1 10], y-axis limit to [-1 0]
axis([1 10 -1 0]);
% open the plot grid
grid on;
% set the figure title
title('Distribution of \sigma_{xx}')
% print figure into a .tif file
print(fh, '-dtiff', 'sigxx.tif')
```

The plotted figures are as follows:

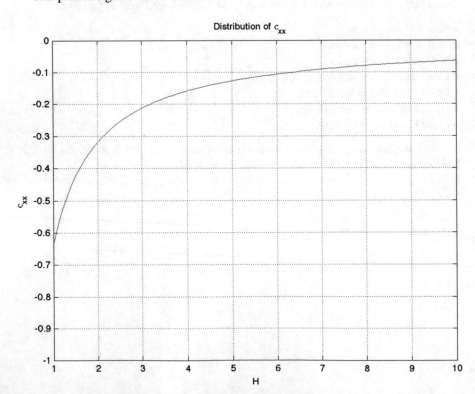

Appendix II
Basic Commands of MuPAD

Computing with Numbers

The algorithmic operators in MuPAD are the same as in MATLAB; however, MuPAD does not have the element-by-element operation (dot operation in MATLAB). We used the same variables in Appendix I to illustrate the basics. Let $a = \begin{bmatrix} 1 & 2 & 3 \end{bmatrix}$ and $b = \begin{bmatrix} 4 & 5 & 6 \end{bmatrix}$ be two vectors and $A = \begin{bmatrix} 1 & 2 & 3 \\ 4 & 5 & 6 \\ 7 & 8 & 9 \end{bmatrix}$ and $B = \begin{bmatrix} 1 & 0 & 0 \\ 0 & 1 & 0 \\ 0 & 0 & 1 \end{bmatrix}$ be two matrices. The variable assignment commands are:

```
a := matrix([1, 2, 3]); b := matrix([[4, 5, 6]])
  ⎛1⎞
  ⎜2⎟
  ⎝3⎠

(4 5 6)
a := transpose(a)
(1 2 3)
```

```
A := matrix([[1, 2, 3], [4, 5, 6] , [7, 8, 9]]); B := matrix([[1, 0, 0], [0, 1, 0], [0, 0, 1]])
  ⎛1 2 3⎞
  ⎜4 5 6⎟
  ⎝7 8 9⎠

  ⎛1 0 0⎞
  ⎜0 1 0⎟
  ⎝0 0 1⎠
```

Using the assignment operator :=, we have assigned the expression to an identifier. Note that in MuPAD, the vectors and matrices were generated by function matrix() and the column vector is generated by using "[]," while row vector is generated by applying the bracket twice. The matrix is generated by assigning each row element and brackets them together.

© Springer International Publishing AG, part of Springer Nature 2018
P. L. Gould, Y. Feng, *Introduction to Linear Elasticity*,
https://doi.org/10.1007/978-3-319-73885-7

Algorithmic operations are based on the matrix operations. Unlike MATLAB, MuPAD does not have element-by-element dot operations.

$$a + b$$
$$(5\ 7\ 9)$$

$$\text{transpose}(a) * b$$
$$\begin{pmatrix} 4 & 5 & 6 \\ 8 & 10 & 12 \\ 12 & 15 & 18 \end{pmatrix}$$

$$A + B$$
$$\begin{pmatrix} 2 & 2 & 3 \\ 4 & 6 & 6 \\ 7 & 8 & 10 \end{pmatrix}$$

$$A * B$$
$$\begin{pmatrix} 1 & 2 & 3 \\ 4 & 5 & 6 \\ 7 & 8 & 9 \end{pmatrix}$$

We may notice the advantage of using MuPAD is not to do matrix computations but symbolic manipulations.

Symbolic Computation

MuPAD has powerful symbolic calculation functions. In this introduction, we illustrate the main functions that are often used in the elasticity context, such as differentiation, integration, solving equations, etc.

Let $u_x = 3yx^2 + 6$; we use this expression to illustrate the symbolic computation: First assign u_x:

$$\text{ux} := 3*x^2*y + 6$$
$$3\,y\,x^2 + 6$$

Calculate $u_{x,\,x}$:

$$\text{ux\_x} := \text{diff(ux, x)}$$
$$6\,x\,y$$

MuPAD offers the system function diff for differentiating expressions.

Calculate $u_{x,\,xx}$:

```
ux_xx := diff(ux_x, x)
  6 y
```

We can also use

```
ux_xx :=diff(ux, x, x)
  6 y
```

Calculate $u_{x,\,xy}$:

```
diff(ux, x, y)
  6 x
```

You can compute integrals by using int for both definite and indefinite integrals.
Calculate $\int u_x dx$:

```
int(ux, x)
  y x³ + 6 x
```

Calculate $\int u_x dy$:

```
int(ux, y)
  3 y (y x² + 4)
  ─────────────
        2
```

Calculate $\displaystyle\int_0^1 u_x dx$:

```
int(ux, x=0..1)
  y + 6
```

Calculate $\displaystyle\int_0^1 u_x dy$:

```
int(ux, y=0..1)
  3 x²
  ──── + 6
   2
```

If you try to compute the indefinite integral of an expression and it cannot be
represented by elementary functions, then int returns the call symbolically:

```
int(1/(exp(x^2) + 1), x = 0..1)
```

$$\int_0^1 \frac{1}{e^{x^2}+1}\, dx$$

We can obtain a floating-point approximation by applying float:

```
float(%)
```
0.41946648

where the symbol % is an abbreviation for the previously computed expression.

Visualization of the functions is as convenient as in MATLAB. The relevant MuPAD functions for generating graphics are the plot command and the routines from the graphics library plot.

Plot $\int_0^1 u_x dy$ for $x = [-1\ 1]$:

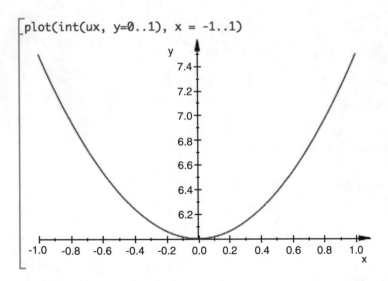

```
plot(int(ux, y=0..1), x = -1..1)
```

Plot u_x for $x = [0, \pi]$, $y = [0, \pi]$:

```
plot(ux, x = 0..PI, y = 0..PI, #3D)
```

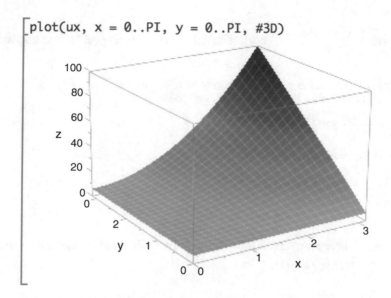

We can also call the plot library for the 3-D plot:

```
plot(plot::Function3d(ux, x = 0..PI, y = 0..PI))
```

Solving equations are common practice in elasticity. In MuPAD, we use the function "solve" for equation and equation systems.

Solve $\sin\left(\frac{wH}{c_{p1}}\right) = 0$ for w:

```
solve(sin(w/cp1*H)=0, w, IgnoreSpecialCases)
```
$$\left\{\frac{\pi\, c_{p1}\, k}{H} \;\middle|\; k \in \mathbb{Z}\right\}$$

Note that ignorespecialcases is to ignore special case such as $H = 0$. If we do not specify this, we will get results for all the special cases:

```
solve(sin(w/cp1*H)=0, w)
```
$$\begin{cases} \mathbb{C} & \text{if } H = 0 \\ \left\{\frac{\pi\, c_{p1}\, k}{H} \;\middle|\; k \in \mathbb{Z}\right\} & \text{if } H \neq 0 \end{cases}$$

Solve $\begin{cases} x+y=a \\ x-ay=b \end{cases}$ for $\{x, y\}$, with both special cases specified and unspecified:

```
equations := {x + y = a, x - a*y = b}: unknowns := {x, y}:
solve(equations, unknowns, IgnoreSpecialCases)
```
$$\left\{\left[x = \frac{a^2+b}{a+1}, y = \frac{a-b}{a+1}\right]\right\}$$

```
solve(equations, unknowns)
```
$$\begin{cases} \left\{\left[x = \frac{a^2+b}{a+1}, y = \frac{a-b}{a+1}\right]\right\} & \text{if } a \neq -1 \\ \{[x = -z-1, y = z]\} & \text{if } a = -1 \wedge b = -1 \\ \varnothing & \text{if } a = -1 \wedge b \neq -1 \end{cases}$$

Another important application is to solve differential equations, such as $y'' = -\left(\frac{w^2}{c^2}\right)y$, for $y = y(x)$:

```
diffEq := ode(y''(x) = -(w^2/c^2)*y(x), y(x))
```
$$\text{ode}\left(y''(x) + \frac{w^2\, y(x)}{c^2}, y(x)\right)$$

```
ux := Simplify(rewrite(solve(diffEq), cos))
```
$$\left\{C5\left(\cos\left(\frac{w\,x}{c}\right) + \sin\left(\frac{w\,x}{c}\right)i\right) + C6\left(\cos\left(\frac{w\,x}{c}\right) - \sin\left(\frac{w\,x}{c}\right)i\right)\right\}$$

where Simplify is to simplify the expression and rewrite is to rewrite the expression in terms of cosine functions.

Index

© Springer International Publishing AG, part of Springer Nature 2018
P. L. Gould, Y. Feng, *Introduction to Linear Elasticity*,
https://doi.org/10.1007/978-3-319-73885-7